KV-396-508

Biology of the Nitrogen Cycle

Biology of the Nitrogen Cycle

Edited by:

Hermann Bothe
University of Cologne, Cologne, Germany

Stuart J. Ferguson
University of Oxford, Oxford, United Kingdom

William E. Newton
The Virginia Polytechnic Institute and State University, Blacksburg, USA

ELSEVIER

Amsterdam • Boston • Heidelberg • London • New York • Oxford
Paris • San Diego • San Francisco • Singapore • Sydney • Tokyo

Elsevier
Radarweg 29, PO Box 211, 1000 AE Amsterdam, The Netherlands
The Boulevard, Langford Lane, Kidlington, Oxford OX5 1GB, UK

First edition 2007

Library of Congress Cataloging-in-Publication Data
A catalog record for this book is available from the Library of Congress

British Library Cataloguing in Publication Data
A catalogue record for this book is available from the British Library

ISBN-13: 978-0-444-52857-5
ISBN-10: 0-444-52857-1

Transferred to digital print 2007
Printed and bound by CPI Antony Rowe, Eastbourne

Contents

List of contributors

Ambus, Per, Risø National Laboratory, Biosystems Department, Frederiksborgvej 399, DK-4000 Roskilde, Denmark

Bedmar, Eulogio J., Departamento de Microbiología del Suelo y Sistemas Simbióticos, Estación Experimental del Zaidín, CSIC. P.O. Box 419, 18080-Granada, Spain

Bergman, Birgitta, Department of Botany, Stockholm University, SE-10691 Stockholm, Sweden

Bakken, Lars, Department of Plant and Environmental Sciences, Norwegian University of Life Sciences, P.O. Box 5003, N-1432 Aas, Norway

Boeckx, Pascal, Faculty of Bioscience Engineering, Ghent University, Coupure 653, B-9000 Gent, Belgium

Bonete, Maria José, Division of Biochemistry and Molecular Biology, University of Alicante, 03080 Alicante, Spain

Bothe, Hermann, Botanical Institute, The University of Cologne, Gyrhofstr. 15, D-50923 Cologne, Germany

Braker, Gesche, Max Planck Institute for Terrestrial Microbiology, Karl-von-Frisch-Strasse, 35043 Marburg, Germany

Casella, Sergio, Dipartimento di Biotecnologie Agrarie, University of Padova, Agripolis, Viale dell'Università 16, 35020 Legnaro, Padova, Italy

Cole, Jeff A., School of Biosciences, University of Birmingham, Birmingham B15 2TT, UK

Cutruzzolà, Francesca, Department of Biochemical Sciences "A.Rossi Fanelli", University of Rome La Sapienza, P.le A. Moro 5 00185 Rome, Italy

Delgado, María J., Departamento de Microbiología del Suelo y Sistemas Simbióticos, Estación Experimental del Zaidín, CSIC P.O. Box 419, 18080-Granada, Spain

De Vries, Simon, Laboratory of Biotechnology, Delft University of Technology, Delft 2628BC, The Netherlands

Dörsch, Peter, Department of Plant and Environmental Sciences, Norwegian University of Life Sciences, P.O. Box 5003, N-1432 Aas, Norway

Drake, Harold, Department of Ecological Microbiology, University of Bayreuth, Dr. Hans-Frisch-Strasse 1-3, D-95440 Bayreuth, Germany

Engel, Marion, GSF – National Research Center for Environment and Health, Institute of Soil Ecology, Ingolstaedter Landstrasse 1, 85764 Neuherberg, Germany

Ferguson, Stuart J., Department of Biochemistry, University of Oxford, South Parks Road, Oxford OX1 3QU, UK

Flores, Enrique, Instituto de Bioquímica Vegetal y Fotosíntesis, C.S.I.C. – Universidad de Sevilla, Avda. Américo Vespucio 49, E-41092 Sevilla, Spain

Forchhammer, Karl, Institut für Mikrobiologie und Molekularbiologie, Justus-Liebig-Universität Giessen, D-35392 Giessen, Germany

Hallin, Sara, Department of Microbiology, Swedish University of Agricultural Sciences, P.O. Box 7025, 750 07 Uppsala, Sweden

Jetten, Mike S.M., Department of Microbiology, IWWR, Faculty of Science, Radboud University Nijmegen, The Netherlands

Kaiser, Werner, Julius v. Sachs Institut, Universität Würzburg, Julius v. Sachs Platz, 78654 Würzburg, Germany

Körner, Heinz, Institute of Applied Biosciences, Division of Molecular Microbiology, University Karlsruhe, D-76128 Karlsruhe, Germany

Lindgren, Per-Eric, Department of Molecular and Clinical Medicine, Division of Medical Microbiology, Linköping University, SE-581 85 Linköping, Sweden

Mahne, Ivan, Department of Food Science and Technology, University of Ljubljana, Biotechnical Faculty, Jamnikarjeva 101, 1000 Ljubljana, Slovenia

Martínez-Espinosa, Rosa Maria, Division of Biochemistry and Molecular Biology, University of Alicante, 03080 Alicante, Spain

Masepohl, Bernd, Lehrstuhl für Biologie der Mikroorganismen, Ruhr-Universität Bochum, D-44780 Bochum, Germany
Mohan, Sudesh B., School of Biosciences, University of Birmingham, Birmingham B15 2TT, UK
Montoya, Joseph, P., School of Biology, Georgia Institute of Technology, Atlanta, GA 30332, USA
Moreno-Vivián, Conrado, Departamento de Bioquímica y Biología Molecular, Edificio Severo Ochoa, 1[a] planta, Campus de Rabanales, Universidad de Córdoba, E-14071 Córdoba, Spain
Mrkonjic Fuka, Mirna, GSF – National Research Center for Environment and Health, Institute of Soil Ecology, Ingolstaedter Landstrasse 1, 85764 Neuherberg, Germany
Munch, Jean Charles, GSF – National Research Center for Environment and Health, Institute of Soil Ecology, Ingolstaedter Landstrasse 1, 85764 Neuherberg, Germany

Newton, William E., Department of Biochemistry, The Virginia Polytechnic Institute and State University, Blacksburg, VA 24061, USA

Op den Camp, Huub J.M., Department of Microbiology, IWWR, Faculty of Science, Radboud University Nijmegen, The Netherlands

Pawlowski, Katharina, Department of Botany, Stockholm University, SE- 10691 Stockholm, Sweden
Philippot, Laurent, INRA – University of Bourgundy, Environmental and Soil Microbiology, CMSE, 17, rue Sully, B.V. 86510, 21065 Dijon Cedex, France
Pouvreau, Laurice A.M., Laboratory of Biotechnology, Delft University of Technology, Delft 2628BC, The Netherlands
Prosser, Jim I., School of Biological Sciences, University of Aberdeen, Cruickshank Building, St. Machar Drive, Aberdeen AB24 3UU, UK

Richardson, David J., Centre for Metalloprotein Spectroscopy and Biology, School of Biological Sciences, University of East Anglia, Norwich Research Park, Norwich NR4 7TJ, UK
Rinaldo, Serena, Department of Biochemical Sciences "A.Rossi Fanelli", University of Rome La Sapienza, P.le A. Moro 5, 00185 Rome, Italy

Schloter, Michael, GSF – National Research Center for Environment and Health, Institute of Soil Ecology, Ingolstaedter Landstrasse 1, 85764 Neuherberg, Germany
Suharti, Laboratory of Biotechnology, Delft University of Technology, Delft 2628BC, The Netherlands

Stacey, Gary, National Center for Soybean Biotechnology, Divisions of Plant Science and Biochemistry, Department of Molecular Microbiology and Immunology, Christopher S. Bond Life Sciences Center, University of Missouri, Columbia, MO 65211, USA

Stres Blaž, University of Ljubljana, Biotechnical Faculty, Department of Food Science and Technology, Jamnikarjeva 101, 1000 Ljubljana, Slovenia

Strous, Marc, Department of Microbiology, IWWR, Faculty of Science, Radboud University Nijmegen, The Netherlands

Tischner, Rudolf, Albrecht von Haller Institut für Pflanzenwissenschaften, University of Göttingen, Untere Karspüle 2, 37073 Göttingen, Germany

Tonderski, Karin, Department of Physics, Chemistry and Biology, Division of Biology, Linköping University, SE-581 83 Linköping, Sweden

Van Cleemput, Oswald, Faculty of Bioscience Engineering, Ghent University, Coupure 653, B-9000 Gent, Belgium

Vanderleyden, Jos, Centre for Microbial and Plant Genetics, Department of Microbial and Molecular Systems, Faculty of Bioscience Engineering, Kasteelpark Arenberg 20, B-3001 Leuven, Belgium

Van Dommelen, Anne, Centre for Microbial and Plant Genetics, Department of Microbial and Molecular Systems, Faculty of Bioscience Engineering, Kasteelpark Arenberg 20, B-3001 Leuven, Belgium

Van Spanning, Rob J.M., Department of Molecular Cell Physiology, Faculty of Earth and Life Sciences, Free University, De Boelelaan 1087, 1081 HV Amsterdam, The Netherlands

Velthof, Gerard L., Alterra, Wageningen University and Research Centre, P.O. Box 47, 6700 AA Wageningen, The Netherlands

Zechmeister-Boltenstern, Sophie, Federal Research and Training Centre for Forest, Natural Hazards and Landscape (BFW), Seckendorff-Gudent-Weg 8, A-1131 Vienna, Austria

Zehr, Jonathan P., Department of Ocean Sciences, University of California, Santa Cruz, CA 95064, USA

Zumft, Walter G., Institute of Applied Biosciences, Division of Molecular Microbiology, University Karlsruhe, D-76128 Karlsruhe, Germany

Preface

Nitrogen (N) is a major element of all organisms and it accounts for 6.25% of their dry mass on average. In biology, N undergoes a variety of oxidations and reductions that produce compounds with oxidation states ranging from +5 (as in nitrate, NO_3^-) to −3 (as in ammonia, NH_3). These nitrogen cycle, redox reactions are performed in different ways by different organisms, and the reactions in total make up the biological N-cycle (depicted on the front cover and in Figure 0-1). All of these reactions are performed by bacteria, archaea and some specialized fungi. There is only one example of a higher life form performing such a reaction; the remarkable exception is *assimilatory NO_3^- reduction*, which also occurs in plants. In assimilatory NO_3^- reduction, NO_3^- is reduced via nitrite (NO_2^-) to the NH_4^+ ion. In general, NH_4^+ is then utilized for the synthesis of glutamine as the first organic N-containing molecule formed. Glutamine is the N-donor for the synthesis of other amino acids and heterocyclic N-compounds.

In addition to NO_3^- acting as the source of N-atoms incorporated into cells through NO_3^- assimilation, NO_3^- can also serve as an electron acceptor to eliminate excess reductant through *dissimilatory NO_3^- reduction*. This pathway uses NO_3^- rather than oxygen (O_2) as the respiratory electron acceptor under anaerobic conditions. In the first step, NO_3^- is reduced to NO_2^- in a reaction catalyzed by several different types of reductases, as discussed in several chapters of this book. The subsequent reduction of NO_2^- to nitric oxide (NO), nitrous oxide (N_2O) and finally to dinitrogen (N_2) involves the action of a sequence of specific enzymes also extensively described in the following chapters. Because this respiratory process results in the N-atom being finally excreted as gas by organisms, the terms *dissimilatory NO_3^- reduction, NO_3^- respiration or denitrification* tend to be used equivalently in the literature, although strictly speaking dissimilatory refers to non-assimilatory reactions which are not directly coupled to generation of

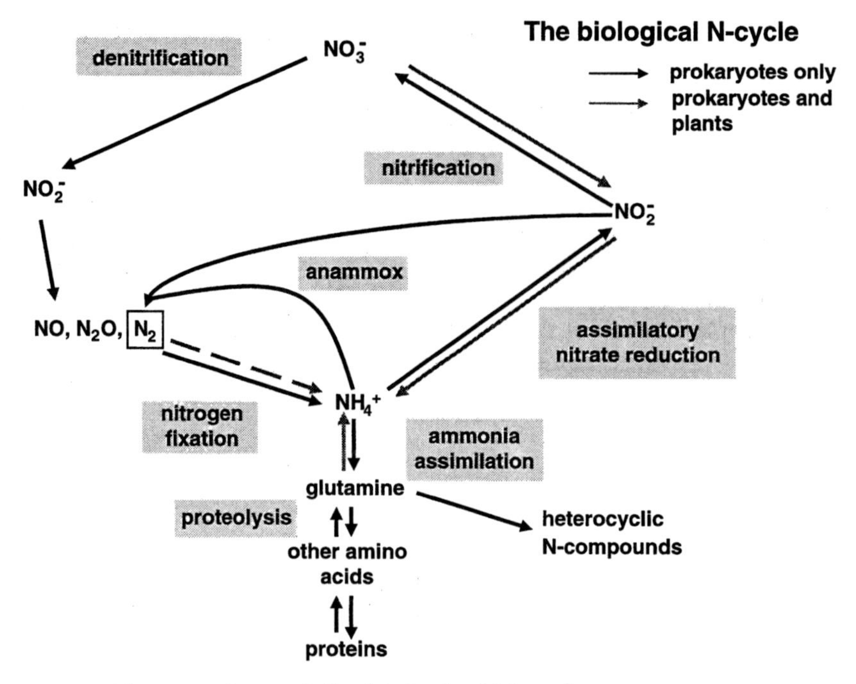

Figure 0-1 The reactions of the biological N-cycle.

protonmotive force (see Moreno-Vivian and Ferguson (1998) Mol. Microbiol. 29: 664). An alternative fate for nitrite, typically occurring in the Enterobacteriaceae, is reduction to NH_4^+, which is then excreted; this process is called *NO_3^-/NO_2^- ammonification*. The names NO_3^- and NO_2^- reductases are used in respect to several distinct types of enzymes, assimilatory and dissimilatory, which, in the case of NO_2^- reductases, may even have different products, NO or NH_4^+; the reader needs to keep this in mind.

All of the reactions mentioned above involve reduction. However, specialized organisms can oxidize either NH_4^+ or NO_2^- to meet their demands for energy and reducing equivalents by using a pathway called *nitrification*. Nitrification was originally believed to be restricted to autotrophic organisms, some of which oxidize NH_4^+ and others NO_2^- with O_2. Recent evidence suggests, however, that a whole range of heterotrophic, as yet mainly uncultured, nitrifying organisms thrives in soils. Moreover, members of the planctomycetes have recently been found that both oxidize NH_4^+ and utilize NO_2^- as the respiratory electron acceptor in a process called *anammox*. Because the end product of anammox is N_2, such organisms offer exciting perspectives in wastewater treatment. A poorly studied subject is *proteolysis*, through which organisms degrade proteins and other N-containing compounds to meet both their N-demand and presumably also their energy requirement. The final connection in the N-cycle is *(di)nitrogen fixation*, which allows certain bacteria and archaea to reduce N_2 to NH_4^+ to provide their N-requirement. This reductive reaction is catalyzed by the enzyme complex called nitrogenase. N_2 fixation proceeds not only in free-living prokaryotes but also in bacteria in symbiosis with plants and is, therefore, of particular interest also in plant biology.

All of the reactions mentioned above as well as the environmental issues outlined below are extensively discussed in this book, with all chapters written by internationally recognized experts in the field.

Components of the N-cycle affect life in various ways. For example, both nitrous oxide (N_2O) and nitric oxide (NO), produced mainly in denitrification

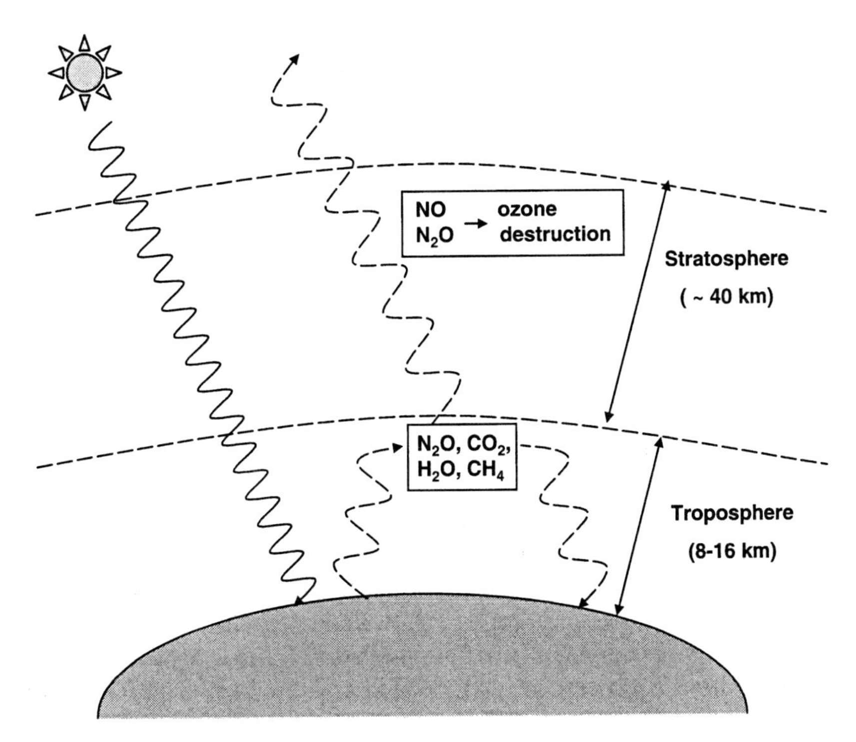

Figure 0-2 The impact of N_2O, NO and other gases on the greenhouse effect and the destruction of the ozone layer. The sun's rays on reaching the Earth's surface are absorbed. The Earth emits rays, at longer wavelengths, to the upper troposphere. There the rays are absorbed by the greenhouse gases (N_2O, CO_2, CH_4 and H_2O) and then reemitted to travel back to the Earth or onto the stratosphere and beyond. In the upper stratosphere N_2O is oxidized to NO by the action of UV light. NO destroys the ozone layer which protects living things against the sun's UV-radiation. We thank U. Sehy and J.-C. Munch, Neuherberg, who originally supplied the figure and D. Stopar, Ljubljana, who modified it for us.

or nitrification, have a severe impact on our atmosphere (Figure 0-2). N_2O is approximately 320-fold more efficient than CO_2 in contributing to global warming (the greenhouse effect). This gas concentrates in the troposphere and acts to reflect the sun's rays back to the earth. In the stratosphere, N_2O and NO destroy the ozone layer that protects us against UV light. Further, the NO produced can be chemically oxidized to nitrogen dioxide (NO_2), which is hydrated to HNO_2 and HNO_3 which fall to earth as constituents of acid rain. Moreover, when entering soil as fertilizer and if not utilized by plants or microorganisms, NO_3^- and NO_2^- are readily leached from soils and enter the groundwater to become a major concern in water purity. In particular, NO_2^- causes toxicity by interacting with hemoglobin and other human molecules, and NO (formed from NO_2^-) is an important signal molecule in many animal and plant reactions. Sewage plants also make use of this biochemistry by using a two-stage process for N-disposal in which NH_4^+ is first oxidized to NO_3^-, followed by reduction to the inert and innocuous N_2.

The impetus for this book comes from the members of Action 856 of ESF COST of the European Community under the title "Denitrification in Agriculture, Air and Water Pollution" (http://www.cost856.de). This group has met three times a year since 2002 and supports short-term scientific missions (STSMs) for younger scientists to learn new techniques in partner laboratories. COST 856 is one of about 200 presently constituent Actions. COST is the acronym for European **CO**operation in the field of **S**cientific and **T**echnical Research and is an intergovernmental European network for international cooperation in research. Established by the Ministerial Conference in November 1971, COST is presently used by the scientific communities of 35 European countries to cooperate in common research projects supported by national funds. The funds provided by COST – less than 1% of the total value of the projects – support the COST cooperation networks (COST Actions) through which, with only around €20 million per year, more than 30,000 European scientists are involved in research having a total value which exceeds €2 billion per year, indicating the financial worth of the European added value which COST achieves.

A "bottom up approach" (the initiative of launching a COST Action comes from the European scientists themselves), "à la carte participation" (only countries interested in the Action participate), "equality of access" (participation is open also to the scientific communities of countries not belonging to the European Union) and "flexible structure" (easy implementation and light management of the research initiatives) are the main characteristics of COST. As precursor of advanced multidisciplinary research, COST has a very important role for the realisation of the European Research Area (ERA); anticipating and complementing the activities of the Framework Programmes; constituting a "bridge" toward the scientific communities of emerging countries; increasing the mobility of researchers across Europe and; fostering the establishment of "Networks of Excellence" in many key scientific domains, such as physics, chemistry, telecommunications and information science, nanotechnologies, meteorology, environment, medicine and health, forests, agriculture and social sciences. It covers basic and more applied research and also addresses issues of pre-normative nature or of societal importance.

The editors gratefully recognize COST and its Action 856. The editors would also like to thank Elsevier, in particular Betty Daniels, Anne Russum and Joyce Vlietstra in Amsterdam, for publishing the book so expertly within the short period of half a year.

There is no recent book on the N-cycle currently available on the market and, owing to the rapid developments in this field through the use of molecular techniques and the broad implications of the N-cycle reactions on ecology, agriculture and health, the production of this volume is timely. This book is aimed at addressing the needs of students at the advanced undergraduate and graduate levels and serving as a handbook for researchers in the field. We hope that the readers will like our organization of the subject matter and find the book to be a valuable source of new information on the many and varied facets of the N-cycle.

Hermann Bothe, Köln, Germany
Stuart J. Ferguson, Oxford, England
William E. Newton, Blacksburg, VA, USA

Part I

Denitrification

Chapter 1

Introduction to the Biochemistry and Molecular Biology of Denitrification

Rob J.M. van Spanning, David J. Richardson and Stuart J. Ferguson

1.1 Introduction

The denitrification part of the N-cycle transforms nitrate (NO_3^-) into N_2 gas. This is a reductive process and thus is a form of respiration; it occurs in four stages, NO_3^- to nitrite (NO_2^-), NO_2^- to nitric oxide (NO), NO to nitrous oxide (N_2O) and N_2O to N_2. All steps within this metabolic pathway are catalysed by complex multisite metalloenzymes with characteristic spectroscopic and structural features [1]. In recent years, high-resolution crystal structures have become available for these enzymes with the exception of the structure for NO-reductase [2]. Further it should be noted that there may be more than one kind of reductase for each step. In general, the proteins required for denitrification are only produced under (close to) anaerobic conditions, and if anaerobically grown cells are exposed to O_2 then the activities of the proteins are inhibited. Thus, for

Biology of the Nitrogen Cycle
Edited by H. Bothe, S.J. Ferguson and W.E. Newton

denitrifying organisms, respiration of O_2 usually occurs in preference to the use of N-oxides or oxyanions. The principal aim of this chapter is to give an overview of the biochemistry and genetics of denitrification in such organisms. At the end of each section we shall briefly consider aspects of denitrification that occur in the archaea and certain fungi. Denitrification has been mostly studied in *Paracoccus denitrificans* and *Pseudomonas stutzeri* and so we will describe denitrification for each of these organisms in turn before considering to what extent general principles can be discerned.

1.2 Proteins of denitrification

1.2.1 Paracoccus denitrificans

In all bacteria the enzymes of denitrification receive e^- from the respiratory chain system that is part of the cytoplasmic membrane. In other words, denitrification is a form of respiration and shares respiratory chain components with the e^- transport system that delivers e^- to O_2 via terminal oxidases [3, 4]. As far as denitrification-specific components are concerned, we need to start at the ubiquinol/ubiquinone component of the chains. Reduction of ubiquinone to ubiquinol occurs using e^- originating from reductants such as NADH, fatty acids, succinate, etc. In denitrification, ubiquinol can be directly oxidized by a membrane-bound respiratory NO_3^--reductase (colour Figure A). There is a crystal structure for the corresponding enzyme, usually known as Nar, from *Escherichia coli* and thus we know in some detail how the enzyme functions. In brief, the ubiquinol is oxidized towards the periplasmic surface of the membrane, with the release of H^+ to the periplasm but transfer of e^- across the membrane to the active site, which is located on a globular domain that protrudes into the cytoplasm. More detail of this enzyme is given in Chapter 2, but the key point to note here is that transfer of e^- through Nar, together with H^+ release and uptake at the two sides of the membrane, generates a H^+-motive force across the membrane. The location of the site of NO_3^- reduction on the cytoplasmic side of the membrane requires a transport system for NO_3^- (Figure A, see Color Plate Section). This is believed to be provided by NarK proteins [5]. In *P. denitrificans* there are two such proteins fused together. Current evidence indicates that one of these proteins catalyses NO_3^- symport with one or more H^+. This would allow entry of NO_3^- into the cell to initiate respiration. In the steady state the NO_3^- import would be in exchange for NO_2^- export to the periplasm, a process that would be e^--neutral and thus not affected by, nor dissipating, the H^+-motive force across the membrane, which, as in all bacteria, has a membrane potential with polarity negative inside the cell.

Export of NO_2^- to the periplasm is needed because that is the location of the NO_2^--reductase of the denitrification system. In *P. denitrificans* this is a cytochrome cd_1-type NO_2^--reductase (see Chapter 3).

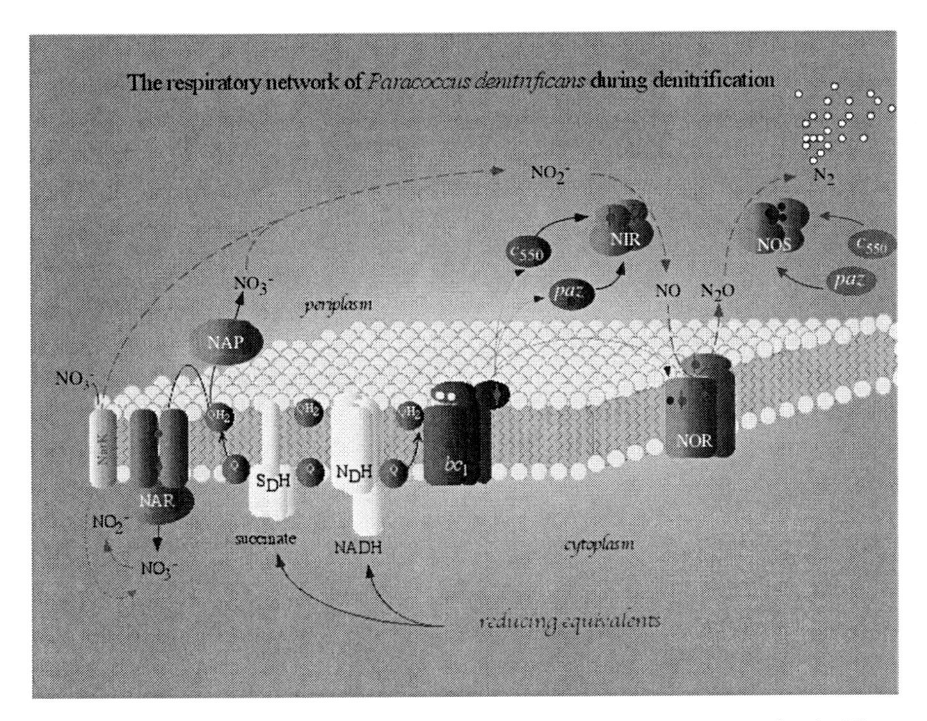

Figure A Scheme of a full denitrification process in *Paracoccus denitrificans*. Dashed arrows, N-oxide transport; straight arrows, e^--transport. SDH, succinate dehydrogenase; NDH, NADH dehydrogenase; Q, quinone; bc_1, cytochrome bc_1 complex; c_{550}, cytochrome *c*; paz, pseudoazurin; NAR, membrane-bound NO_3^--reductase; NAP, periplasmic NO_3^--reductase (not normally involved in denitrification in this organism); NIR, cd_1-type NO_2^--reductase; NOR, bc-type NO-reductase; NOS, N_2O-reductase; NarK, NO_3^-/NO_2^- antiporter. (See Colour Plate Section in back of this volume)

Electrons are delivered to cytochrome cd_1 by a mono-heme *c*-type cytochrome, cytochrome c_{550}, or a cupredoxin protein called pseudoazurin. These two periplasmic and water-soluble proteins are reduced by an integral membrane complex called the cytochrome bc_1 complex, which in turn is reduced by ubiquinol. This complex is not specific for denitrification; it occurs in diverse respiratory chain systems in bacteria as well as in the mitochondrial e^--transfer chain (colour Figure A).

The NO generated by NO_2^--reductase must be restricted to low concentrations because of its potential toxicity, but nonetheless it is a definite free intermediate of denitrification. The NO-reductase is an integral membrane protein (Chapter 4) which is believed but not experimentally established to be supplied with e^- by either pseudoazurin or cytochrome c_{550} in common with the NO_2^--reductase.

The final step of the denitrification pathway is catalysed by N_2O-reductase, another periplasmic enzyme. Again it is assumed that pseudoazurin and cytochrome c_{550} are the immediate e^--donor proteins. N_2O-reductase is a Cu-enzyme (Chapter 5).

1.2.2 Pseudomonas stutzeri

The proteins of the denitrification system in *P. stutzeri*, a member of the gamma-group of organisms in contrast to *P. denitrificans*, which belongs to the α-group, are nevertheless similar to those in *P. denitrificans*. The main differences are that the putative NO_3^-/NO_2^- transport system is not organized as two fused proteins, and that the periplasmic cupredoxins and cytochromes *c* that transfer e^- from the cytochrome bc_1 complex to the terminal reductases for NO_2^-, NO and N_2O have not been identified.

1.2.3 Other Gram-negative bacteria

There are two substantial variations recognized in the enzymology of denitrification relative to *P. denitrificans* and *P. stutzeri*. The first is that a periplasmic NO_3^--reductase (Chapter 2) is often used. Use of this enzyme avoids any need for NO_3^- and NO_2^- transport across the cytoplasmic membrane but also sacrifices generation of H^+-motive force as e^- reach the enzyme from ubiquinol. *P. denitrificans* contains the periplasmic enzyme, but this is not normally used in denitrification.

The second major difference is that the cytochrome cd_1 NO_2^--reductase is often replaced by a Cu-containing enzyme (Chapter 3). There is no understandable basis for the presence of one of these enzymes rather than the other. Both types are almost never found in a given organism but there is no species-specific distribution; the genera *Pseudomonas* and *Azospirillum* contain both types of reductase.

1.2.4 Gram-positive bacteria

In Gram-negative bacteria, the NO_2^-- and N_2O-reductases are nearly always water-soluble proteins in the periplasm and periplasmic *c*-type cytochromes, together with cupredoxins, play important e^--transfer roles in denitrification. Since Gram-positive organisms do not have a periplasm, they have their counterpart anchored to the cytoplasmic membrane but with their globular domains still oriented towards the exterior of the cell. Furthermore, the NO-reductase is often a variant in which quinol is the direct reductant, i.e the cytochrome *bc*1 complex is bypassed. For reasons that should become clear from the discussion of the bioenergetics of denitrification (1.3), this means that there is no H^+-translocation as e^- flow from quinol to NO [6].

1.2.5 Archaeal and fungal denitrification

Denitrifying enzymes have been reported in some Archae, eg extreme halophiles and in a hyperthermophile, *Pyrobaculum aerophilum*. Recently, a Cu-containing dissimilatory NO_2^--reductase (CuNiR) was purified from denitrifying cells of a halophilic archaeon, *Haloarcula marismortui*. In addition, genes encoding putative NO_3^- transporters, NO_3^-- and NO_2^--reductases have been found in the sequenced genomes of members of both crenarchaeota and euryarchaeota. At least in some cases, archaea have a variant of the Nar-type of NO_3^--reductase (Chapter 2), which has an opposite orientation in the cell membrane. However, in most cases, either functional or sequence data, but not both, are available. Because the archaea are predominant microbial populations in extreme environments like saltwater lakes and hot springs, denitrifying archaea sustain the N-cycle under such hypersaline or hot conditions [7, 8].

Denitrifying fungi including some yeast, may contain NO_3^--, NO_2^-- and NO-reductases, but N_2O-reductases in these species have not yet been described. The NO_3^-- and NO_2^--reductases are counterparts of the bacterial dissimilatory NO_3^--reductases and Cu-type NO_2^--reductases, respectively. NO-reductase is the third enzyme in fungal denitrification. This unique enzyme, referred to as P450nor, resembles bacterial P450 cytochromes and they might be evolutionarily related. P450nor from *Fusarium oxysporum* exhibited a potent NO-reductase activity to form N_2O with NADH as the sole and direct e^- donor. The dissimilatory NO_3^-- and NO_2^--reductases of *F. oxysporum* are present in its mitochondria and thus they can be supplied with e^- from respiratory substrates such as succinate, formate or malate combined with pyruvate. The association with formate respiration means that the fungal Nar system is in this respect more similar to that involved in NO_3^- respiration by *E. coli* than that in the bacterial denitrifying system.

1.3 Bioenergetics of denitrification

Denitrification is a form of respiration and as such must generate an H^+-motive force across the cytoplasmic membrane of the cell. An important parameter is the number of net positive charges that are moved outwards from the cytoplasm to the periplasm for each pair of e^- moving from an ubiquinol to a terminal e^--acceptor (Table 1-1).

As explained earlier, the organization of the Nar NO_3^--reductase means that net H^+-translocation occurs with a stoichiometry of $2H^+/2e$. The periplasmic NO_2^-- and N_2O-reductases cannot themselves be involved in H^+-movement across the membrane. Although the NO-reductase is an integral membrane protein, several lines of evidence show that H^+ required for reduction of NO are taken from the periplasm and thus no net charge movement occurs. However, e^- transfer from ubiquinol to NO_2^-, NO or N_2O is coupled to H^+-translocation across the membrane because this is catalysed by the cytochrome bc_1 complex. The effective translocation stoichiometry – the

Table 1-1 Electron transfer mediated charge separations

Electron acceptors	Terminal oxidoreductases		q/10e$^-$	
			Electron donor	
			NADH	Succinate
2.5 O_2-> 5 H_2O	*aa*$_3$		50	30
	ba$_3$		40	20
2 NO_3^- -> N_2	Nar, Nir, Nos	cNor	30	10
	Nar, Nir, Nos	qNor	28	8
	Nap, Nir, Nos	cNor	26	6
	Nap, Nir, Nos	qNor	24	4

Note: q/10e$^-$, charge separations per 10e$^-$ transferred from donor to acceptor. The table indicates the number of charges moved per 10e$^-$ flowing from either NADH or succinate. 10e$^-$ are required to reduce two NO_3^- ions to one molecule of N_2. Figures are shown for e$^-$ flow via various oxidases to O_2 as well as via different types of protein to NO_3^-, NO_2^-, NO and N_2O.
aa_3, aa_3-type cytochrome *c* oxidase; ba_3, ba_3-type quinol oxidase; Nar, membrane-bound NO_3^--reductase; Nap, periplasmic NO_3^--reductase; Nir, NO_2^--reductase; Nos, N_2O-reductase; cNor and qNor, cytochrome c and quinol oxidizing NO-reductases respectively.

mechanism involving a quinone cycle [3] is beyond the scope of this chapter – is 2H$^+$/2e$^-$, i.e. the same as for the NO_3^--reductase reaction (Table 1-1). It is sometimes said that the periplasmic H$^+$-consumption by the NO_2^--, NO- and N_2O-reductases will cancel out the H$^+$-translocation activity of the cytochrome *bc*$_1$ complex. However, this is not the case as periplasmic H$^+$ consumption does not have any implication for the net charge moved across the membrane; the H$^+$ consumption is accompanied by equal e$^-$-uptake by the periplasmic reductases and so there is no change to the net charge translocation into the periplasm. Finally, it is notable that although a 'longer' e$^-$-transfer chain is used for e$^-$-delivery to NO_2^-, NO and N_2O, the H$^+$-translocation stoichiometry per 2e$^-$ is the same as for NO_3^-.

As mentioned earlier, O_2 is usually the preferred e$^-$-acceptor for denitrifying bacteria but the exact biochemical mechanism that ensures preferential e$^-$-flow to O_2 is not known. However, the energetic basis for the preferential case of O_2 can be identified. Thus, although the steady state H$^+$-motive force is likely to be of similar magnitude during aerobic respiration and denitrification, the H$^+$-translocation stoichiometry per 2e$^-$ will be higher by approximately 40% when O_2 is the e$^-$-acceptor (Table 1-1). This means that the ATP yield per 2e$^-$ flowing to a terminal acceptor from any given reductant available to the cells will be higher for aerobic respiration, thus explaining why denitrifying bacteria prefer aerobic growth conditions.

The preference for O_2 over NO_3^-, NO_2^-, NO and N_2O is also reflected in the regulation of expression of denitrification-specific genes. Current knowledge about *P. denitrificans* and *P. stutzeri* as paradigms is described below.

1.4 Genes coding for enzymes of denitrification

The subunits of membrane-bound NO_3^--reductase (Nar) are encoded by genes of a *narGHJI* operon. The organization of this operon is conserved in most species that express Nar. The *narGHI* genes encode the structural subunits, and *narJ* encodes a dedicated chaperone required for the proper maturation and membrane insertion of Nar. In many species, a *narK* gene encoding a NO_3^-/NO_2^- transporter precedes the *narGHJI* genes. A set of *narXL* genes encoding a two-component regulatory system, which modulates the expression of Nar in response to the NO_3^-/NO_2^- concentration, is found upstream of the *nar* gene cluster from denitrifying *Pseudomonads*. The α-proteobacteria have the *narXL* genes replaced by an *fnr*-like gene encoding an FNR homologue designated NarR. This protein is a transcriptional regulator that modulates the expression of the *nar* operon most likely in response to the intracellular NO_3^- or NO_2^- concentration.

Most of the operons that encode the subunits of periplasmic NO_3^--reductase (Nap) comprise the *napABCD* genes. The *napABC* genes encode the structural subunits, and *napD* has a likely role as chaperone in the correct assemblage of Nap. In addition to *napABCD*, *nap* operons may include one or more of *napKEFGH* genes, but their occurrence and position in the operons differ in species. The NapFG proteins might constitute an alternative e^--transfer pathway to this type of NO_3^--reductase. It is evident that the distinction in *nap* operon organization does not show a phylogenetic pattern of the species in which they are found; this is suggestive of lateral transfer of the *nap* operons.

The gene encoding the structural monomers of cd_1-type NO_2^--reductase, *nirS*, makes part of a *nir* gene cluster. The number and organization of *nir* genes in these clusters differ in different species. However, they share a number of genes that are proposed to encode a multimeric and multifunctional enzyme complex involved in the maturation and insertion of a specific heme d_1 into NO_2^--reductase.

The Cu-type NO_2^--reductase is encoded by the *nirK* gene. In some species, the *nirK* gene clusters with a downstream-located *nirV* gene, encoding a protein of unknown function.

The minimum genetic potential to express NO-reductase (Nor) appears to be a *norCBQD* operon, where *norCB* genes encode the structural subunits I and II. The NorQ and NorD proteins are essential for activation of Nor. Some more specialized denitrifiers have additional *norEF* genes, the products of which are involved in maturation and/or stability of Nor activity.

The *nos* gene clusters encoding N_2O-reductase (Nos) are highly conserved in denitrifiers and consist of at least seven genes located in the same transcriptional direction and in the order *nosRZDFYLX*. The *nosZ* gene encodes the N_2O-reductase protein. The *nosDFYL* genes encode proteins that are apparently required for Cu assemblage into Nos, although their specific role remains to be established. The NosRX proteins have roles in transcription regulation, activation and Cu assemblage of Nos.

1.5 Regulation of transcription of the denitrification genes

1.5.1 General description

Many, if not all, denitrifying bacteria are facultative aerobic chemo-organotrophic organisms. Their respiratory networks are designed such that there is preferential e^- flow to O_2 under aerobic growth conditions, as it is the most efficient route with respect to free energy transduction. Only when O_2 becomes limiting, and nitrate and/or nitrite is available, are the denitrification enzymes expressed [4, 9]. When a suspension of cells grown under anaerobic, denitrifying conditions is exposed to O_2 there is usually an immediate inhibition of any of the enzymes of denitrification, and the e^- flowing through the ubiquinone/ubiquinol couple are diverted to constitute oxidase activities. As far as is known, all four of the reductases from *P. denitrificans* are capable of functioning in the presence of O_2. There is some evidence that the presence of O_2 blocks the transport of NO_3^- across the cytoplasmic membrane and thus restricts the activity of the membrane-bound NO_3^--reductase [10].

The fact that the reduction steps during denitrification are sequential implies that the product of the one enzyme is a substrate for the next one. Fine-tuned regulation of the concentration and activity of the denitrification enzymes is, therefore, required in order to keep the free concentrations of NO_2^- and NO below cytotoxic levels. NO_3^--reductase is the first enzyme that is induced in response to O_2 depletion and NO_3^- availability. The activity of this enzyme gives rise to an initial increase in the intra- and extracellular NO_2^- concentration. Next, the NO_2^-- and NO-reductases are coordinately expressed; their concentrations and activities are tuned such that the free NO_2^- concentration drops to the micromolar range, while NO is kept in the nanomolar range. The induction profile of N_2O-reductase resembles that of the NO_2^-- and NO-reductases to a large extent, suggesting that the regulation of expression of these three denitrification enzymes involves the same transcriptional regulator. An intriguing difference is that Nos is expressed at a later stage in the growth phase than the other two reductases indicating that specific regulators other than the shared one control the sequential temporal expression of the denitrifying enzymes. The molecular basis of the regulatory networks in denitrifying species that control the oxic–anoxic shift as well as the fine-tuning of transcriptional activation is beginning to emerge. Not surprisingly, some of the key molecules that act as signals to these regulation pathways are O_2, NO_3^-, NO_2^- and NO. It has now become evident that more than one type of regulatory protein is involved in sensing each of these key molecules. Analogous systems for detecting similar signals in denitrification are used in different combinations in denitrifying species. In addition to the regulatory proteins that can monitor these key molecules, regulation of the on-set and fine-tuning of denitrification in some denitrifiers involves redox sensing mechanisms and the NirI and NosR proteins [9].

Regardless of the fact that there is no such thing as a universal mode of regulation of denitrification, these analogous networks in different denitrifiers operate such that homeostasis of NO_2^- and NO is realized.

1.5.2 Oxygen sensors

The two most important types of O_2 sensor involved in regulation of denitrification are FixL and FNR.

FixL is a membrane-bound O_2 sensor found in rhizobia like *Bradyrhizobium japonicum* and related species. Together with its cognate response regulator FixJ, these proteins belong to the group of two-component regulatory systems. The N-terminal domain of FixL contains a PAS domain with an O_2 responsive heme group. This is an independent domain found in the PAS (Per-ARNT-Sim) superfamily of proteins of about 100 amino acids, suitable for transmitting a signal from the receptor site to other domains of the same protein or to partner proteins through proper conformational changes. Despite the low sequence similarity of PAS domains, a high conservation in the fold and topology of the known structures suggests a strong evolutionary conservation of some functional features. The C-terminal domain of FixL has histidine kinase activity for autophosphorylation and subsequent P-group transfer to FixJ, which is the response regulator that activates transcription of its target genes. Binding of O_2 to the heme in FixL switches the iron from high to low spin, resulting in a conformational change, which inactivates the kinase activity and terminates the cascade.

FNR is an O_2 responsive transcription regulator that belongs to an expanding family of Crp/FNR like transcriptional activators [11]. All these regulators contain a signal sensing domain, a dimerization domain, a helix-turn-helix DNA-binding domain and up to three sites that are involved in contacting RNA polymerase. FNR of *E. coli* contains a sensory domain with four conserved cysteine residues that ligate an O_2 sensitive [4Fe–4S] cluster. This cluster disintegrates by direct interaction with O_2 rendering FNR inactive. Once the O_2 concentration drops below threshold levels, the cluster is assembled. This event triggers a conformational change, resulting in dimerization of FNR. The dimer binds with its DNA binding helix-turn-helix motif to a conserved target sequence TTGAT-N4-ATCAA (FNR-box). The DNA-binding motif contains conserved glutamate and serine residues, which make contact with the thymidine and guanine residues of the FNR-box half-site, 5′-TTGAT-3′, at positions 1 and 3, respectively. The FNR-box is usually located directly upstream of the RNA-polymerase binding site to facilitate contact between FNR and RNA polymerase. The physical contact between the dimer and RNA-polymerase on target promoters initiates a change in the geometry of the ternary transcription initiation complex, which results in transcription of the downstream gene. Homologues of FNR are widespread in Nature and have been encountered in a variety of prokaryotes that have to cope with changes in O_2 availability, including pathogens. Only some of these protein members have cysteine signatures diagnostic for binding an O_2 sensitive 4Fe–4S cluster. Most of them lack the cluster, indicating that these FNR-homologues respond to signals other than O_2.

1.5.3 Nitrate and nitrite sensors

Three types of NO_3^-/NO_2^- sensing systems have been characterized in denitrifying species, NarXL, NarQP and NarR. NarXL and NarQP are members of two-component regulatory systems [12]. The NarX and NarQ proteins are the signal sensors, which both respond to NO_3^- and NO_2^- although with different affinities. NarX is more specific for NO_3^- and NarQ for NO_2^-. The NarL and NarP proteins are their cognate response regulators respectively. In *E. coli*, they bind DNA to control induction of the genes encoding membrane-bound and periplasmic NO_3^--reductases and repression of genes encoding alternate anaerobic respiratory enzymes. The NarX and NarQ proteins communicate with both the NarP and NarL proteins in *E. coli*. Denitrifiers from the β- and γ-proteobacteria have genes encoding orthologues of NarXL. Their products regulate a NO_3^--induced expression of Nar. Expression of the *Neisseria* NO_2^-- and NO-reductase genes (1.6.8) is subjected to a NO_2^- response mediated by a NarQP couple.

NarR is a member of the FNR family of transcriptional activators, but it lacks the cysteines to incorporate a [4Fe–4S] cluster. NarR of *Paracoccus pantotrophus* and *P. denitrificans* is specifically required for transcription of the *narKGHJI* genes most likely in response to NO_3^- or NO_2^-. The mechanism of the response is not clear, but it is notable that NarR can also be activated by azide, which normally binds to metal centres, raising the possibility that NarR is a metalloprotein. Genes encoding NarR are found in the α-proteobacteria *Brucella suis, B. melitensis, P. denitrificans* and *P. pantotrophus* upstream of their *narGHJI* gene clusters. There are no indications that these organisms have counterparts of *narXL*. NarR might substitute for the NarXL system in the α-proteobacteria.

1.5.4 Nitric oxide sensors

The first clue for the identification of specific NO sensors came from the observation that mutants disturbed in NO_2^- reduction were unable to activate the transcription from *nir* and *nor* promoters, apparently due to the lack of NO-formation. NO was suggested to be required as an inducer for its own reductase. This view turned out to be correct as judged by the identification and characterization of NO-responsive transcriptional activators [13]. As yet, two different types of these activators are characterized in denitrifying species, NNR and NorR.

NNR (NO_2^- and NO gene regulator) is also a member of the FNR family of transcriptional activators, but, just like NarR, it lacks the cysteines to incorporate a [4Fe–4S] cluster. NNR orthologues are responsive to NO, but there are no indications on the mechanism. An alignment of NNR-like proteins does not reveal protein motifs diagnostic for ligating an NO-binding metal centre. Except for *Nitrosomonas europaea, Neisseria* and *Ralstonia* species, NNR orthologues are found in virtually all species that have Nir and Nor. These orthologues, sometimes referred to as NnrR, DNR or DnrD, have a dedicated task in the coordinate transcription of the *nir* and *nor* gene clusters in response to NO. Without exception, the *nnr* gene in these species is located in close proximity of its target operons. The promoters of these operons contain NNR binding sites that resemble the consensus FNR-box to a large extent. The NNR orthologues split in two phylogenetically distinct subgroups. Species from the first group possess genes encoding the *cd*1-type NO_2^--reductase, those from the second group a gene encoding a Cu-type NO_2^--reductase [13].

NorR is another protein involved in NO-responsive transcriptional regulation, and first identified in *Ralstonia eutropha*. NO reduction in this bacterium is catalysed by the quinol-dependent NO-reductase NorB (qNor). The *norB* gene and the adjacent *norA* form an operon that is controlled by the σ^{54}-dependent transcriptional activator NorR in response to NO. NorR is a member of the NtrC family of response regulators. The absence of possible phosphorylation sites as well as the presence of a conserved GAF domain indicative for signal perception suggests that the protein belongs to a sub-family of response regulators that sense their signal themselves rather than via a cognate signal sensor [14]. Similar motifs are present in several proteobacteria upstream of genes encoding proteins of NO metabolism, including NO-reductase (NorB), flavorubredoxin (NorV), NO dioxygenase (Hmp) and hybrid cluster protein (Hcp). In *E. coli*, NorR activates the transcription of the *norVW* genes encoding a flavorubredoxin (FlRd) and an associated flavoprotein respectively, which together have NADH-dependent NO-reductase activity. The regulatory domain of NorR contains a mononuclear non-heme iron centre, which reversibly binds NO. Binding of NO stimulates the ATPase activity of NorR, enabling the activation of transcription by RNA polymerase [15].

1.5.5 Redox sensors

The Reg regulon from *Rhodobacter capsulatus* and *R. sphaeroides* encodes proteins involved in numerous energy-generating and energy-utilizing processes such as photosynthetic CO_2-fixation, N_2-fixation, H_2-utilization, aerobic and anaerobic respiration, denitrification, e^--transport and aerotaxis. Amongst these proteins is the NirK-type NO_2^--reductase involved in denitrification. The expression of these proteins is controlled by the RegAB 2-component regulatory system. The redox signal that is detected by the membrane-bound sensor kinase, RegB, appears to originate from the aerobic respiratory chain, given that mutations in cytochrome *c* oxidase result in constitutive RegB autophosphorylation. Regulation of RegB autophosphorylation also involves a redox-active cysteine that is present in the cytosolic region of RegB [16].

1.5.6 NirI, NosR

NosR is a membrane-bound Fe–S flavoprotein. The flavin cofactor is presumably bound covalently to a large periplasmic domain, which is held in position by the first two helices. As such, the protein has redox centres positioned at opposite sides of the cytoplasmic membrane [17]. NosR seems to have a dual functionality. Apart from having a role in in vivo N_2O respiration, the protein is essential for expression of the *nos* gene cluster encoding N_2O-reductase in *P. stutzeri*, *S. meliloti* and *P. denitrificans*.

The latter species also expresses a homologue of NosR, NirI, which has the same structural features but has a specific role in transcription of the *nir* gene cluster encoding NO_2^--reductase in response to O_2 limitation and the presence of N-oxides [18]. The *nirI* gene is in a two-gene operon together with *nirX* just upstream of the *nirS* gene cluster. Transcription of the *nirIX* gene cluster is controlled by NNR. NirI mutants are unable to express Nir. Remarkably, attempts to complement a NirI mutation were successful only when the *nirI* gene was reintegrated into the chromosome of the NirI-deficient mutant via homologous recombination in such a way that the wild-type *nirI* gene was present directly upstream of the *nir* operon. This suggests a transcriptional and translational coupling of NirI and its target promoter. Interestingly, downstream-located mutations in NirI do not affect transcription but do yield inactive Nir, suggesting that the N-terminal domain of NirI is involved in the regulatory pathway whilst the C-terminal has a role in maturation of or metal insertion into NirS.

1.6 Regulatory networks in denitrifiers

1.6.1 General

Despite the variations in the make-up of the regulatory networks, homeostasis of NO_2^- and NO in all denitrifiers is realized according to a more or less general concept where the extent of transcriptional regulation depends on the concentration and activity of each of the regulators (Figure B,

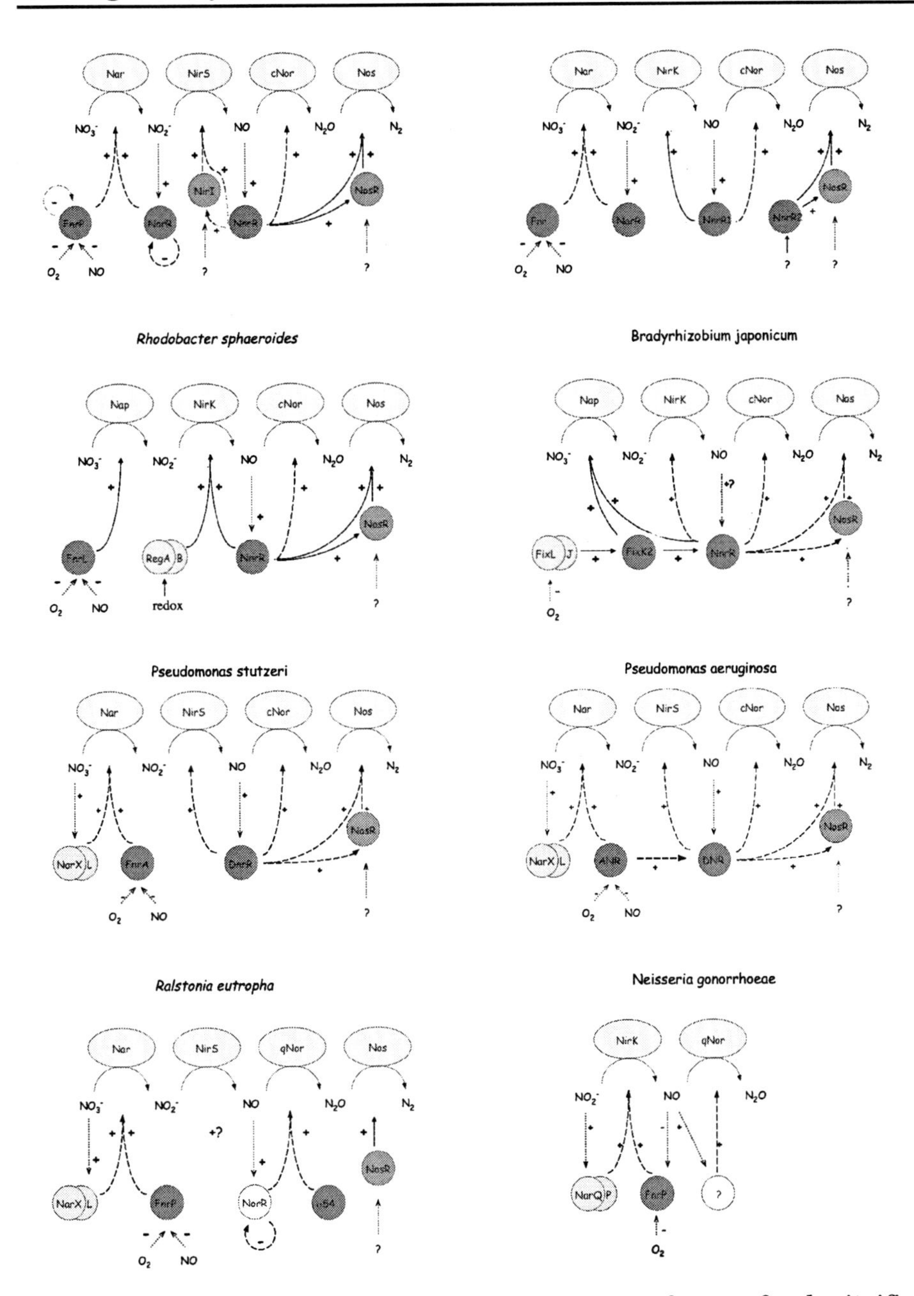

Figure B Schemes for the regulation of expression of genes for denitrification in some denitrifiers. Nar, NO_3^--reductase; NirS or NirK, cd_1-type or Cu-type NO_2^--reductases respectively; cNor or qNor, *bc*-type or quinol dependent NO-reductases respectively; Nos, N_2O-reductase. Dotted lines, activating (+) or inactivating (−) signals; dashed lines, positive (+) or negative (−) transcription regulation of the corresponding genes by the regulators. (See Colour Plate Section in back of this volume)

see Color Plate Section). Little is known about their concentrations, but at least some of the regulators are subject to autoregulation. Their activities depend on the free concentrations of their signalling molecules, which vary at the different phases of denitrification.

1.6.2 Paracoccus denitrificans

Three FNR-type transcriptional regulators FnrP (an O_2 sensor), NarR (a NO_2^- sensor) and NNR (a NO sensor) orchestrate the transcriptional changes during the shift from aerobic respiration to denitrification in *P. denitrificans*. Upon O_2 depletion, FnrP is activated and induces transcription of the *nar* operon in an unknown interplay with NarR. Both FnrP and NarR are subject to negative autoregulation so as to balance their concentrations. In a subsequent NO response, NNR activates expression of Nir, Nor and Nos. Since Nir expression is also subject to a product induction mechanism, the concentration of the enzyme progressively increases when more NO molecules become available from its NO_2^- reducing activities. A fine-tuned regulation of the concentration of Nir is accomplished by a mechanism involving NirI, for which expression is also under control of NNR. It is not yet known which signals are fed into the NirI pathway. As a result of Nir and Nor expression, NO_2^- and NO are consumed. Perhaps increasing concentrations of NO may also result in inactivation of FnrP and subsequent lowering of *nar* gene expression. Altogether, these systems realize homeostasis of NO_2^- and NO. Although the three FNR orthologues are highly homologous to one another, they display a remarkable degree of specificity towards their target promoters. In spite of these specificities, their binding sites on the DNA resemble each other to a large extent and are sometimes even identical [19]. These sites are correctly positioned relative to the RNA-polymerase binding site as judged from transcript start analyses. Perhaps subtle differences in the binding strength between the regulator and the DNA underlie the observed specificity. Alternatively, the FNR family members may interact with specific yet unknown transcription factors.

1.6.3 Brucella melitensis

The regulation of denitrification in *B. melitensis* is unknown, but the genetic make-up of this bacterium shows regulatory genes, which are quite similar to the *P. denitrificans* counterparts involved in regulation of denitrification. Notably, it has two copies of the *nnr* gene, one close to the *nir–nor* locus and the other adjacent to the *nos* gene cluster. Both proteins share a high degree of identity. It may well be that they are specifically involved in expression of the *nir/nor* genes and *nos* gene cluster respectively.

1.6.4 Rhodobacter sphaeroides

Nap is the enzyme responsible for denitrification in *R. sphaeroides* as judged by the fact that strains with mutations in the *nap* gene cluster lost NO_3^--reductase activity as well as the ability to grow with nitrate under anaerobic-dark conditions. This agrees well with the preliminary finding that *nar* gene sequences are absent from the partial genome sequence of *R. sphaeroides*. The onset of nitrate reduction by Nap requires O_2 depletion and NO_3/NO_2^- availability. The O_2 response is regulated by FnrL, but the proteins involved in NO_3^-/NO_2^- sensing are not known. The *R. sphaeroides* genome does not contain sequences that translate into NarXL or NarR-like proteins. In addition, the *R. sphaeroides* RegB/RegA system (PrrB/PrrA) was shown to control the expression of NO_2^--reductase. *nir* expression is also regulated through the transcription factor NnrR. Just like in *P. denitrificans*, NnrR regulates expression of the NO_2^-- and NO-reductases in response to sensing NO. Thus, RegA presumably acts in concert with NnrR to coordinate *nirK* expression [16].

1.6.5 Bradyrhizobium japonicum

Expression of genes involved in anaerobic metabolism in *B. japonicum* is controlled by a regulatory cascade initiated by the O_2 responsive FixLJ two-component regulators. FixJ activates the expression of *fixK*2, which is a member of the FNR family but lacks the O_2-sensing 4Fe–4S cluster. The FixK2 protein then activates a number of target genes or operons, including a regulatory gene presumably encoding an NnrR orthologue. This protein controls the expression of the NO_2^-- and NO-reductases, but it is not clear yet if this occurs in response to NO. Recent data suggest that the FixLJ–FixK2–NnrR cascade is also responsible for *napEDABC* gene expression encoding periplasmic NO_3^--reductase, which is the enzyme responsible for anaerobic growth of *B. japonicum* under NO_3^- respiring conditions.

1.6.6 Pseudomonads

Denitrification in *Pseudomonads* is initiated with transcription of the *narGHJI* gene cluster under control of an O_2 responsive FNR orthologue (FnrA or ANR) and the NO_3^- responsive NarXL two-component regulatory proteins. Nar expression is, therefore, subjected to a substrate induction mechanism rather than a product induction mechanism operating in the alpha proteobacteria. In *P. aeruginosa*, ANR also controls the expression of DNR in a cascade-like fashion. DNR is the NO-sensing orthologue of NNR, which regulates the expression of Nir and Nor, and perhaps also of Nos. Nos regulation also requires NosR. The regulation of denitrification is likely to be similar in *P. stutzeri*, although it is not clear if expression of DNR in this bacterium is under the control of FnrA as well.

1.6.7 Ralstonia eutropha

Regulation of *nar* expression in *R. eutropha* (now *Wautersia eutropha*) is similar to that in *Pseudomonads* in that it requires an O_2 responsive FNR orthologue and the NO_3^- responsive NarXL two-component regulatory proteins. It is not known which proteins regulate the expression of Nir and Nos. Expression of qNor requires NO, the NorR protein and sigma54. NorR represses its own synthesis via negative autoregulation. The protein has no role in expression of the *nir*, *nos* and *nar* gene clusters in *R. eutropha*.

1.6.8 Neisseria gonorrhoeae

This pathogenic species has genes encoding NirK and qNor, making up an incomplete denitrification pathway perhaps to cope with NO_2^- and NO challenges in the host. NO contributes to mammalian host defence by direct microbicidal activity and as a signalling molecule of innate immune responses. Macrophages produce NO via the inducible NO synthase (iNOS). Alternatively, the denitrification enzymes may be used in part for bioenergetic purposes. During the course of disease and host colonization, the bacterium has to withstand limited O_2 availability. N-oxide and N-oxyanions are thought to be present, which may constitute an alternative sink for e^- from its respiratory chain. An O_2 responsive FNR orthologue and the NO_2^- responsive NarQP two-component regulatory proteins control expression of NirK. The expression of the *norB* gene was not regulated by FNR or NarP but required anaerobic conditions, the presence of NO_2^- as well as a functional NO_2^--reductase encoded by the gonococcal *aniA* gene. NO might be the inducer for qNor expression in this bacterium.

1.6.9 Gram-positive organisms, archaea and fungi

Gram-positive organisms contain an Fnr-type system but otherwise nothing is currently known about how the expression of genes for denitrification proteins is controlled in these organisms.

1.7 Concluding remarks

Specialized denitrifiers such as *P. denitrificans* and the denitrifying *Pseudomonads* contain more than 40 genes, which encode the proteins that make up a full denitrification pathway. They include the structural genes for the enzymes and e^--donors, their regulators as well as many accessory genes required for assembly, cofactor synthesis and insertion into the enzymes. In contrast, some denitrifiers can only carry out the two central reactions of the pathway and use these activities to support growth, but the cost of maintaining this capability is a very small amount of genome space. Yet other, mostly pathogenic, organisms have the potential to express a special NO-reductase apparently to detoxify NO released in their environment. Recent years have seen the emergence of a great deal of genome information that is allowing a much better insight into the phylogenetic distribution, evolution and the propensity of the systems for lateral gene transfer of denitrification genes and gene clusters. In addition, it is providing insight into the regulation of gene expression and the way in which some denitrification enzymes play different roles in bacteria.

References

[1] B.C. Berks, S.J. Ferguson, J.W.B. Moir, D.J. Richardson, Enzymes and associated electron transport systems that catalyse the respiratory reduction of nitrogen oxides and oxyanions, Biochem. Biophys. Acta 1232 (1995) 97–173.

[2] O. Einsle, P.M. Kroneck, Structural basis of denitrification. Biol. Chem. 385 (2004) 875–883.

[3] D.G. Nicholls, S.J. Ferguson, Bioenergetics 3, Academic Press, London/San Diego, 2002.

[4] S.C. Baker, S.J. Ferguson, B. Ludwig, M.D. Page, O.M.H. Richter, R.J.M. Van Spanning, Molecular genetics of the genus *Paracoccus*: metabolically versatile bacteria with bioenergetic flexibility. Microbiol. Mol. Biol. Rev. 62 (1998) 1046–1078.

[5] J.W. Moir, N.J. Wood, Nitrate and nitrite transport in bacteria. Cell Mol. Life Sci. 58 (2001) 215–224.

[6] Suharti, S. de Vries, Membrane-bound denitrification in the Gram-positive bacterium *Bacillus azotoformans*, Biochem. Soc. Trans. 33 (2005) 130–133.

[7] P. Cabello, M.D. Roldan, C. Moreno-Vivian, Nitrate reduction and the nitrogen cycle in archaea, Microbiology 150 (2004) 3527–3546.

[8] L. Philippot, Denitrifying genes in bacterial and Archaeal genomes, Biochim. Biophys. Acta 1577 (2002) 355–376.

[9] W.G. Zumft, Cell biology and molecular basis of denitrification, Microbiol. Mol. Biol. Rev. 61 (1997) 533–616.

[10] S.J. Ferguson, Denitrification and its control, Anton Leeuwenhoek Int. J. Gen. M. 66 (1994) 89–110.

[11] S. Spiro, The FNR family of transcriptional regulators, Anton Leeuwenhoek Int. J. Gen. M. 66 (1994) 23–36.

[12] G. Unden, S. Becker, J. Bongaerts, G. Holighaus, J. Schirawski, S. Six, O_2-sensing and O_2-dependent gene regulation in facultatively anaerobic bacteria, Arch. Microbiol. 164 (1995) 81–90.

[13] W.G. Zumft, Nitric oxide signaling and NO dependent transcriptional control in bacterial denitrification by members of the FNR-CRP regulator family, J. Mol. Microbiol. Biotechnol. 4 (2002) 277–286.

[14] A. Busch, K. Strube, B. Friedrich, R. Cramm, Transcriptional regulation of nitric oxide reduction in Ralstonia eutropha H16, Biochem. Soc. Trans. 33 (2005) 193–194.
[15] B. D'Autreaux, N.P. Tucker, R. Dixon, S. Spiro, A non-heme iron centre in the transcription factor NorR senses nitric oxide, Nature 437 (2005) 769–772.
[16] S. Elsen, L.R. Swem, D.L. Swem, C.E. Bauer, RegB/RegA, a highly conserved redox-responding global two-component regulatory system. Microbiol. Mol. Biol. Rev. 68 (2004) 263–279.
[17] P. Wünsch, W.G. Zumft, Functional domains of NosR, a novel transmembrane iron-sulfur flavoprotein necessary for nitrous oxide respiration, J. Bacteriol. 187 (2005) 1992–2001.
[18] N.F. Saunders, E.N. Houben, S. Koefoed, S. de Weert, W.N. Reijnders, H.V. Westerhoff, A.P. De Boer, R.J.M. Van Spanning, Transcription regulation of the *nir* gene cluster encoding nitrite reductase of *Paracoccus denitrificans* involves NNR and NirI, a novel type of membrane protein, Mol. Microbiol. 34 (1999) 24–36.
[19] R.J.M. Van Spanning, A.P.N. De Boer, W.N.M. Reijnders, H.V. Westerhoff, A.H. Stouthamer, J. Van der Oost, FnrP and NNR of *Paracoccus denitrificans* are both members of the FNR family of transcriptional activators but have distinct roles in respiratory adaptation in response to oxygen limitation, Mol. Microbiol. 23 (1997) 893–907.

Chapter 2

The Prokaryotic Nitrate Reductases

David J. Richardson, Rob J.M. van Spanning and
Stuart J. Ferguson

2.1 Introduction

There are three kinds of NO_3^--reductase that can be used to initiate the respiratory denitrification or ammonification processes in prokaryotes, all of which bind a Mo-*bis*-molybdopterin guanine dinucleotide (Mo-*bis*-MGD) cofactor (Figure 2-1) and at least one 4Fe–4S cluster [1]. The best known is the integral membrane protein of eubacteria that is usually called NarGHI (Figure 2-2A). This is a molybdoenzyme, of known structure, and for which the active site faces the cytoplasm. An evolutionary precursor of this enzyme has the opposite orientation in denitrifying Archaea (Figure 2-2B). The Archaeal equivalent of the NarG subunit, which carries the active site molybdenum and Mo-*bis*-MGD cofactor, is synthesized as a precursor with a double arginine sequence motif which indicates export of the prefolded metalloprotein from the cytoplasm via the twin arginine-dependent translocase (TAT). In some Archaea it is clear that Mo or W can be coordinated by the Mo-*bis*-MGD at the active site. The

Biology of the Nitrogen Cycle
Edited by H. Bothe, S.J. Ferguson and W.E. Newton

Molybdopterin (MPT)

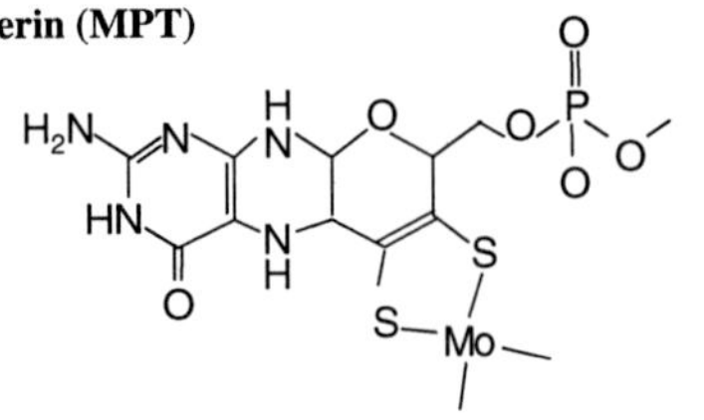

Molybdopterin Guanine Dinucleotide (MGD)

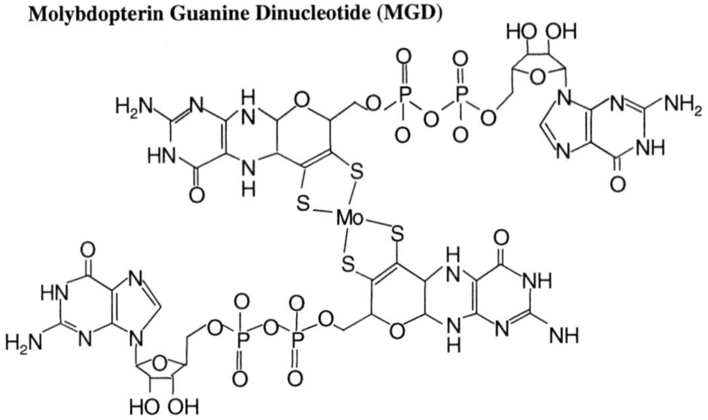

Figure 2-1 The structure of the molybdenum cofactor of prokaryotic (MGD) and eukaryotic (MPT) nitrate reductases.

third type of respiratory NO_3^--reductase is a water-soluble, periplasmically located protein known as Nap, which is also a substrate for the TAT export pathway (Figure 2-3A). The evolutionary precursor of this enzyme is a cytoplasmic enzyme that is involved in the assimilation of NO_3^- N into cellular N and which is synthesized without a signal peptide (Figure 2-3B). The respiratory and assimilatory prokaryotic NO_3^--reductases are quite distinct from the assimilatory NO_3^--reductase of eukaryotes (plants and fungi), which binds a simpler molybdopterin form of the molybdenum cofactor at the active site. The location of the active site of NarGHI and Nas at the cytoplasmic side of the cytoplasmic membrane immediately raises the problem as to how the negatively charged NO_3^- ion gains access to the active site, especially in the context that the cytoplasm is polarized at approximately −180 mV relative to the exterior of the cell. In this chapter, the structural organization and bioenergetics of the four prokaryotic NO_3^--reductases and the eukaryotic enzyme will be reviewed and the possible mechanisms of NO_3^- transport will be explored.

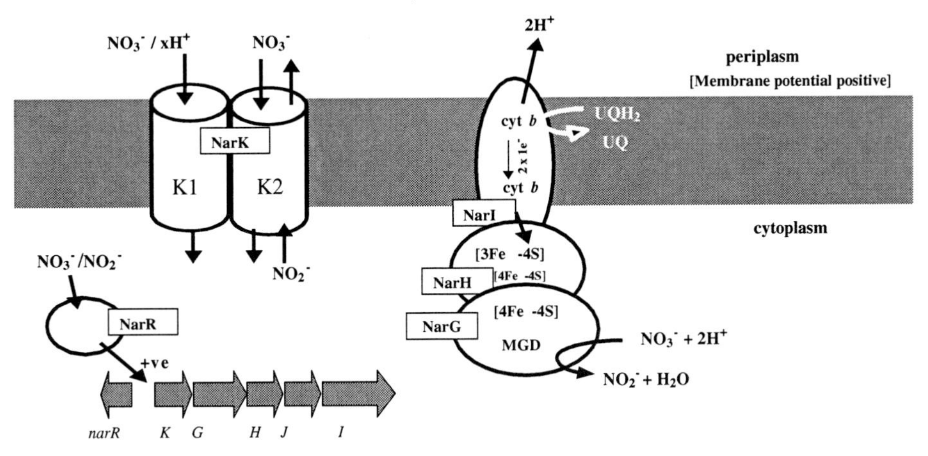

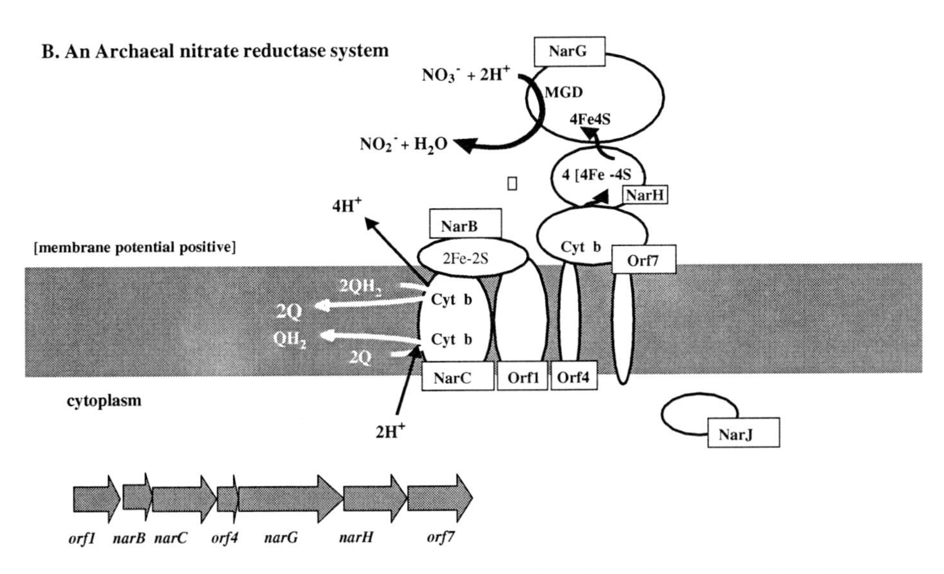

Figure 2-2 The organization of D-group nitrate reductases. (A) nNar, *P. denitrificans*; (B) pNar, *H. mediterranei*.

2.2 The membrane-bound NO_3^--reductase (NarGHI)

2.2.1 General characteristics

The membrane-bound NO_3^--reductase with the active site facing the cytoplasm is usually a three-subunit enzyme composed of NarGHI. The three-dimensional structure is only known for *Escherichia coli*, a non-denitrifier, but can be readily extrapolated to its counterpart in

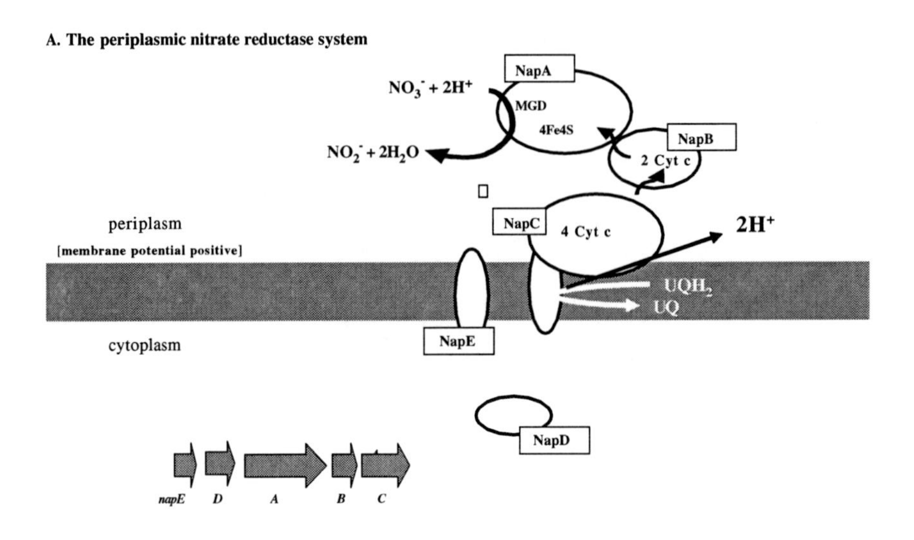

Figure 2-3 The organization of C-group nitrate reductases. (A) Nap, *P. denitrificans*; (B) Nas, *P. denitrificans*.

denitrifiers [2, 3]. NarG, the α-subunit of about 140 kDa, contains the Mo-*bis*-MGD cofactor at its catalytic site and a 4Fe–4S cluster. NarH, the β-subunit of about 60 kDa, contains four additional Fe–S centres: one 3Fe–4S and three 4Fe–4S. NarG and NarH are located in the cytoplasm and associate with NarI, the γ-subunit. The contact between NarGH and NarI is mediated solely by the C-terminal of NarH. NarI is an integral membrane protein of about 25 kDa with five transmembrane helices and the N-terminus facing the periplasm. This subunit carries two hemes *b*, one with low potential located at the periplasmic site and one with high potential located at the cytoplasmic side. Nar receives electrons from quinol, usually ubiquinol in denitrifiers, and is therefore linked to respiratory electron transfer. Oxidation of quinol occurs at the periplasmic side of NarI where the H^+ are released and the two electrons are moved from the outside low-potential heme b_l to the inside high-potential heme b_h. This charge separation makes the enzyme electrogenic in that it contributes to the generation of an H^+ electrochemical gradient across the membrane ($2q/2e^-$). Electrons from heme b_h of NarI are donated to the 3Fe–4S cluster of NarH. From there they flow via the Fe–S clusters in NarH to the one in NarG, which is the direct electron donor to the Mo-*bis*-MGD cofactor containing catalytic site in NarG where NO_3^- is reduced to NO_2^-.

The structure of the *E. coli* enzyme shows that electrons from quinol donated at the periplasmic side are transferred down a 90 Å 'ladder' of eight redox centres and ultimately reduce NO_3^- at the N-face of the cytoplasmic membrane (Figure C, see Color Plate Section). This redox ladder comprises two hemes, five Fe–S clusters and the Mo-*bis*-MGD. Each is within 12 Å of its nearest neighbour, thereby ensuring rapid electron transfer. In the NarH subunit, the four Fe–S centres (all 4Fe–4S) are arranged in two pairs in each of two domains. The very low redox potentials of some of these centres had raised the possibility that they were not directly involved in electron transfer between quinol and NO_3^-. However, consideration of the structure leaves no doubt that all four centres participate in mediating electron transfer and thus NarGHI provides an example of an electron transfer system in which not only there is a mixture of endergonic and exergonic electron transfer steps between the electron donor and electron acceptor, but also the overall ΔE favours unidirectional electron transport from quinol to NO_3^-.

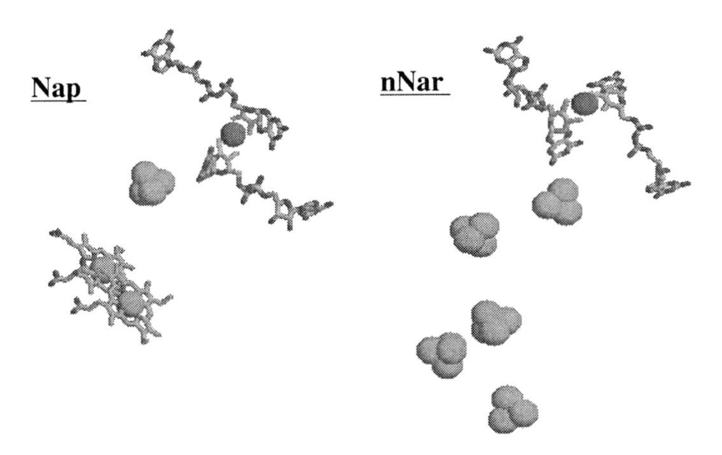

Figure C The organization of the Nap and nNar redox centres in the soluble NapAB and NarGH subunits (figure kindly provided by Dr. C. Butler). (See Colour Plate Section in back of this volume)

2.2.2 *Molybdenum coordination and the mechanism of nitrate reduction*

The Mo ion of NarG is coordinated by an aspartate ligand provided by the polypeptide chain. Different X-ray crystal structures of NarG show that this aspartate is coordinated as either a bidentate di-O ligand or as monodentate O ligand with the second O–H bonded to an active site histidine residue [2, 3]. The bidentate ligand does not present an obvious vacant coordination site of NO_3^- binding and its catalytic relevance is uncertain. In the case of the monodentate structure, the sixth coordination position is occupied by a Mo=O/H_2O group. This will be lost as water on reduction generating a vacant coordination site for NO_3^- binding. Studies in which the *E. coli* and *Paracoccus denitrificans* Nar enzymes have been interrogated using protein film voltammetry showed that at low nitrate concentrations optimum activity is observed when the enzyme is in a partially reduced rather than in a fully reduced state [4]. One explanation for this could be that NO_3^- binds with higher affinity to Mo(V) rather than with the fully reduced Mo(IV) state (Figure 2-4). Such a model might also account for similar observations with the assimilatory nitrate reductase [14].

2.2.3 *Nitrate transport to NarG*

Adjacent to the structural genes for NarGHI in many denitrifying bacteria are one or two members of genes encoding transport proteins known generally as NarK family proteins. NarK-type proteins are implicated in

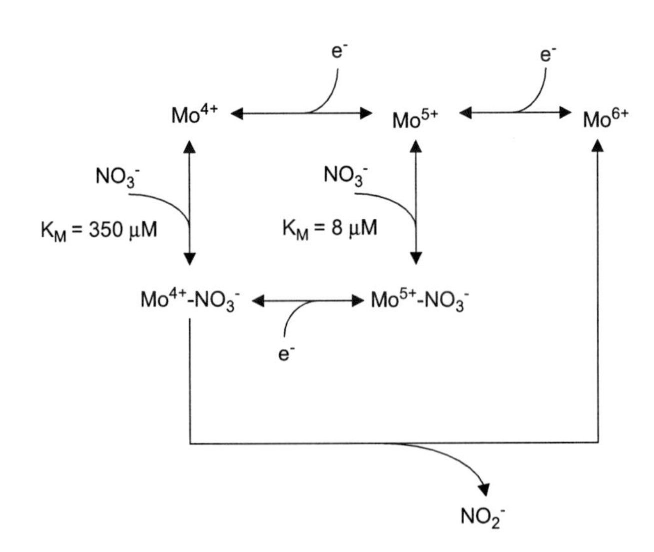

Figure 2-4 The Mo(V) and Mo(IV) cycles for nitrate reduction. Adpated from Ref. [4], but the K_M and reduction potential values are taken for the assimilatory nitrate reductase [14].

NO_2^- and/or NO_3^- transport in several non-denitrifying organisms including *E. coli* and several plant species where the role is related to uptake of NO_3^- or NO_2^- as cellular N-sources. Twelve transmembrane helices are predicted on the basis of hydrophobicity analysis for these proteins. However, the *narK* gene adjacent to *narG* in *Paracoccus* and *Brucella* codes for 24 putative transmembrane helices with two related NarK domains being present [5]. The two domains have been termed NarK1 and NarK2 (Figure 2-2A) and each shows a distinct phylogenetic relationship with other NarK proteins, which can also be subgrouped into NarK1 and NarK2 phylogenies. A variety of genetic and biochemical experiments suggest that NarK1 acts as NO_3^-/H^+ symport through which NO_3^- respiration could be initiated, while NarK2 is a NO_3^-/NO_2^- antiporter which sustains steady-state respiration (Figure 2-2A). It seems likely that NarK1 initiates NO_3^--reduction but NarK2 dominates in the steady state and thus ensures that the steady-state activity of NarK1 that will lead to net NO_2^- accumulation in the cytoplasm is minimized. The operation of NarK2 in steady state also serves to provide a supply of NO_2^- in the periplasm for use by periplasmic NO_2^--reductases. The number of *narK1* and *narK2* genes varies amongst NO_3^--reducing bacteria; for example, *E. coli* has two *narK2*-like genes and it cannot be assumed that all *narK1* genes are NO_3^-/H^+ symports or that all NarK2 proteins are NO_3^-/NO_2^- antiports. Indeed the evidence from *E. coli* is that this is not the case. However, the role of NarK proteins in the movement of N-oxyanions across the membrane to meet a particular organism's physiological requirement seems assured.

2.3 The Archaeal Nar system

Where respiratory NO_3^- reduction has been identified in Archaea, it is predicted to take place in a catalytic subunit with a TAT signal peptide, which may serve to export folded redox proteins across the cytoplasmic membrane. Examples of Archaea that have this type of NO_3^--reductase include *Haloferax mediterranei* and *Pyrobaculum aerophilum*. Some Eubacteria, for example *Symbiobacterium thermophilum*, also appear to have this system. There is a high percentage of sequence similarity between this NO_3^--reductase and the Nar with a cytoplasmic active site (see above) and there is confusion in the literature caused by these distinct enzymes having the same genetic nomenclature. Both types are predicted to have Mo-aspartate ligands and are hence termed D-group nitrate reductases. In bioenergetic terms, the membrane is energized in living cells by a transmembrane H^+ or Na^+ electrochemical gradient that is positive with respect to the outside of the membrane. This H^+ or Na^+ motive force has a magnitude of some 200 mV. On this basis, we define the Nar on the outside of the membrane as pNar (membrane potential *positive* side of membrane) and Nar on the inside of the membrane as nNar (membrane potential *negative* side). Since pNar and nNar are found in both Archaea and Bacteria and nNar is only found in Bacteria, it seems likely that pNar evolved first and evolutionary mutation of the TAT signal peptide after the time of the last universal ancestor would have led to a location of the enzyme at the cytoplasmic side of the membrane in some bacteria. Indeed, a remnant, or vestigial, TAT signal peptide can be identified at the N-terminal region of some nNarG in which one of the two critical arginines required for transport has been lost. The evolution of a functional nNar from a functional pNar is not straightforward. Simple re-location of the active site subunit does not yield an active cellular system. The newly located subunit requires a transport system for the substrate and it needs to latch on to a new means of extracting electrons from the quinol pool. In fact, for the 'young' nNar system this is well understood with the X-ray crystal structure of a quinol-oxidising complex having emerged and a transporter for NO_3^- having been identified as discussed above. Paradoxically, for the ancestor pNar we know very little about the enzyme, biochemically and bioenergetically. Thus, at present it is not known how electrons move from the Q-pool to pNar. However, the genetic context of the *H. mediterranei* pNar provides some interesting possibilities [6]. One of the genes encodes a protein that has some sequence similarity with the di-heme subunits of quinol-cytochrome *c* reductases, such as the well-studied cytochrome bc_1 complex of mitochondria. This subunit is predicted to bind two hemes stacked across the membrane. Next to this is a gene encoding a Rieske Fe–S protein also of the type found in the cytochrome bc_1 complex. There is also a gene that appears to encode a soluble *b*-type cytochrome. Thus, these three proteins are predicted to form a quinol-dependent electron transfer system that could be coupled to free energy transduction by a Q-cycle mechanism (Figure 2-2). This system would then be energetically equivalent to the nNar that operates as an H^+ motive Q-loop mechanism. It should be stressed that not all pNar proteins have the *H. mediterranei* genetic context, so they may not all be coupled in this way, but direct experimental evidence for a Q-cycle mechanism of quinol oxidation by *H. mediterranei* pNar and confirmation of it's location on the p side of the membrane is now important to establish.

2.4 The periplasmic nitrate reductase (Nap)

2.4.1 General characteristics

Periplasmic NO_3^--reductases (Nap) are also linked to quinol oxidation in respiratory electron transport chains, but do not transduce the free energy in the QH_2–NO_3^- couple into an H^+ motive force [7]. NO_3^--reduction via Nap can only be coupled to free energy transduction if the primary quinone reductase, for example NADH dehydrogenase or formate dehydrogenase, generates a H^+ electrochemical gradient. Thus, Nap systems have a range of physiological functions that include the disposal of reducing equivalents during aerobic growth on reduced carbon substrates [11] and anaerobic NO_3^- respiration as part of bacterial ammonification or denitrification pathways [12]. In Nap, e^- from quinol are generally passed through one or two cytochrome *c*-containing proteins (NapC and NapB) to the catalytic subunit, NapA, that contains a Mo-*bis*-MGD cofactor and a 4Fe–4S cluster. However, some Nap enzymes are now emerging in which the NapB subunit is absent. The Nap and membrane-bound NO_3^--reductases are structurally quite distinct. For example, in the periplasmic enzymes, cysteine provides a thiol ligand to the Mo ion (a C-group nitrate reductase), whereas aspartate provides one or two O ligands to the Mo ion in the pNar and nNar enzymes D-group nitrate reductases. In evolutionary terms the Nap and Nas enzymes are most closely related and this is reflected in similar spectroscopic properties between the two enzymes, although the redox properties of the Mo centres have become tuned for their different metabolic functions.

Bioinformatic analyses reveal that the Nap is phylogenetically widespread in proteobacteria [7, 12], but detailed biochemical and spectroscopic studies have been restricted to enzymes from relatively few species. The catalytic subunit (NapA) from the α-proteobacteria *P. denitrificans* and *Rhodobacter sphaeroides* form a very tight complex with a di-*c*-heme subunit (NapB) and a K_D of 0.5 nM has been determined for the NapAB complex of *R. sphaeroides* [8]. A 3.2 Å resolution structure suggests that tight complex formation is contributed to by two loops at the N- and C-terminal extremities of NapB that adopt an extended conformation and embrace the NapA subunit. However, another structure of NapA, the 1.9 Å resolution structure from *Desulfovibrio desulfuricans*, is that of a monomer [9]. *D. desulfuricans* lacks a *napB* gene and this has led to a subclassification of Nap into NapB-dependent and NapB-independent systems. In the NapB-dependent systems, the strength of interaction between NapB and NapA varies from species to species. It is very weak in *E. coli* where the NapA protein structure has also been solved as a monomer.

2.4.2 The molybdenum coordination of E. coli and D. desulfuricans NapA and a possible catalytic cycle

The coordination sphere around the Mo-*bis*-MGD includes four thiol ligands from the two MGD groups, one sulfur ligand from Cys143 and one O ligand (Figure 2-4B). In *E. coli* NapA, the Mo–O bond length is 2.6 Å and suggestive of H_2O ligation, whilst in *D. desulfuricans* NapA it is 2.1 Å, indicative of OH^- ligation. An oxo-transfer mechanism that requires involvement of an $Mo^{6+}{=}O$ species is widely envisaged for

NO_3^--reductases and other Mo-*bis*-MGD-dependent reductases such as the DMSO-reductase. On the basis of this assertion, a tentative catalytic cycle can be forwarded (Figure 2-5). In the absence of added NO_3^- only a two-electron reduced enzyme is thermodynamically accessible with physiological electron donors (e.g. electrons originating from reduced quinol in the electron transport chain) because of the very low E_m (<−400 mV) of the $Mo^{5+/4+}$ couple. This reduced enzyme will have the redox state $4Fe4S^{1+}/Mo^{5+}$ (Figure 2-5, state 1). The water molecule is lost on reduction to Mo^{5+}, vacating a coordination site for NO_3^- binding (Figure 2-5, state 2). Binding of NO_3^- to the Mo^{5+} (Figure 2-5, state 3) will raise the potential such that the Mo^{5+}–NO_3^- can be reduced by an e^- arising from the 4Fe–$4S^{1+}$ centre. The resulting Mo^{4+}–NO_3^- now holds the two electrons required for NO_3^--reduction to NO_2^- (Figure 2-5, state 4), which will yield a Mo^{6+}=O species (Figure 2-5, state 5) through an oxo-transferase reaction. This species must then protonate to give the stable Mo^{6+}–OH or Mo^{6+}–OH_2 (Figure 2-5, state 6) states resolved in the crystal structures of the *E. Coli* and *D. desulfuricans* NapA structures [9].

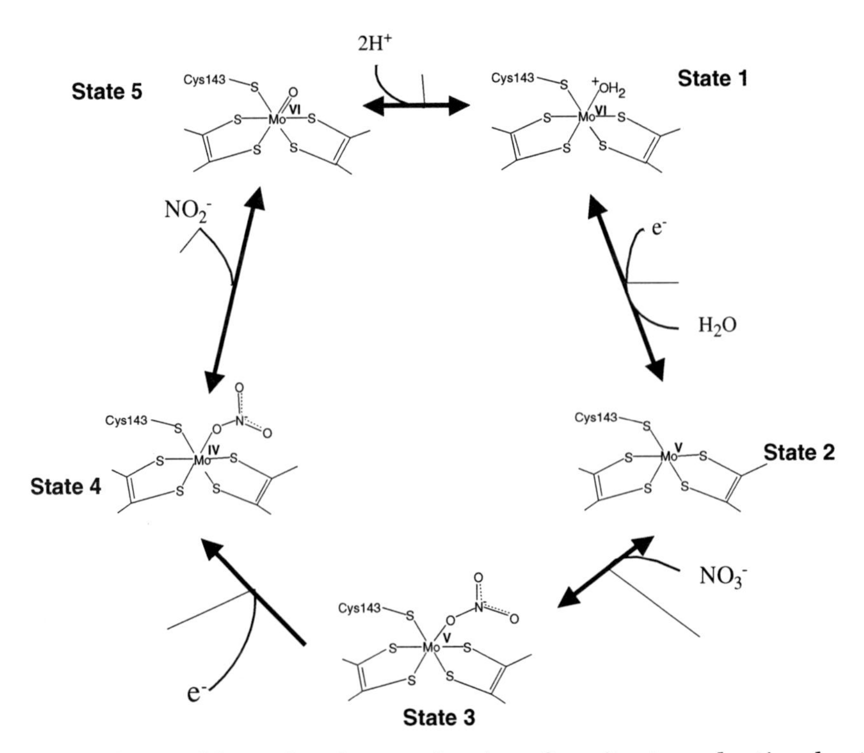

Figure 2-5 A possible molecular mechanism for nitrate reduction by the periplasmic nitrate reductase of *E. Coli* and *D. desulfuricans* (figure kindly provided by Dr. B. Jepson).

2.4.3 Quinol oxidation by the Nap system

In many Nap systems, quinol oxidation is catalysed by NapC (25 kDa), which has a single N-terminal transmembrane helix that anchors a globular domain with four *c*-type hemes on to the periplasmic surface of the cytoplasmic membrane [10]. All four hemes have relatively low E_0' (~ −50 to −250 mV), are low-spin bis-histidinyl axial ligated and possibly form two di-heme pairs. NapC belongs to a large family of bacterial tetra-heme and penta-heme cytochromes that have been proposed to participate in electron transfer between the quinol/quinone pool and periplasmic redox enzymes such as the reductases for trimethylamine N-oxide-, dimethylsulfoxide-, fumarate-, NO_2^--reductase and hydroxylamine oxidoreductase. On the basis of these features, the electron pathway from quinol to NO_3^- through Nap can be deduced. The enzyme receives electrons from membrane-embedded benzoquinols or napthoquinols, which are donated to the tetra-heme periplasmic domain of NapC. From there they flow via the hemes in NapB and the 4Fe–4S cluster in NapA to the Mo-*bis*-MGD containing catalytic site of NapA where the $2e^-$ reduction of NO_3^- to NO_2^- is catalysed. As discussed above for Nar, this electron transfer system will also contain a mixture of endergonic and exergonic e^--transfer steps, but it is expected that none of the eight redox centres involved will be more than 14 Å apart, thus ensuring rapid electron transfer. Electron transfer through NapCAB is non-electrogenic since e^- and H^+ required for the reduction of NO_3^- to NO_2^- are both taken up at the same side of the membrane, i.e. the periplasm.

In some cases, a second quinol oxidation system is encoded by the *nap* gene cluster that involves NapH. This is an integral membrane protein, which is predicted to bind cytoplasmic Fe–S clusters and additional as yet uncharacterized redox centres. In some *nap* gene clusters, NapC is actually absent and so NapH may be the sole means of quinol oxidation. However, this protein has never been purified and the nature of the cofactors bound and mechanism of interaction with quinols remain unknown [13].

2.5 The assimilatory nitrate reductase (Nas)

The assimilatory NO_3^-/NO_2^- system is quite distinct from the respiratory NO_3^-/NO_2^--reductase systems of denitrifiers and ammonifiers and is regulated in a different way. Although current understanding is incomplete, the broad picture is that, with the exception of Nap, the enzymes of denitrification are only synthesized and active under anaerobic conditions in the presence of NO_3^- whilst the presence of the latter and the absence of ammonia are the conditions that promote functioning/formation of the assimilatory system under aerobic or anaerobic conditions. There is no typical *nas* gene cluster amongst bacteria and Archaea. However, many common genes are present [13]. In *P. denitrificans* the gene cluster that encodes the NO_3^-/ NO_2^- assimilatory apparatus (Figure 2-3B) is (from 5′ to 3′) *nasTSADE*, *nirC*, *nasC*, all transcribed in the same direction. NasC is predicted to encode a Nas with a 4Fe–4S and Mo cofactor similar to a cyanobacterial type that has been biochemically characterized, but with a possible C-terminal extension that may bind an additional 2Fe–2S centre (as also speculated for the *Klebsiella oxytoca* Nas). NasA is a predicted member of the NarK family with 12 transmembrane helices. It has 71% similarity (58% identity) to the *P. denitrificans* NarK1, but only 40% similarity (25% identity) to NarK2. Thus, it that can be placed phylogenetically in the NarK1 H^+/NO_3^- symporter (an activity defined only for plant and fungal proteins [8, 9]) subgroup of the NarK family. This is very probably a pathway of NO_3^- uptake for delivery to NasC (Figure 2-3B). NasD is predicted to be a typical siroheme-dependent cytoplasmic NO_2^--reductase and NasE is likely to be a 2Fe–2S ferredoxin. The NO_2^--reductase, NasD, can be serviced intracellularly by NO_2^- generated from NO_3^- reduction by NasC. However, bioinformatic analysis also shows the presence of a putative NO_2^- transporter NirC of a family for which the prototype was first identified in *E. coli* where it operates in anaerobic dissimilatory NO_2^--reduction. The expression of the *nas–nir* gene cluster is likely to be under the control of NasS and NasT proteins. NasS is homologous to NO_3^-/NO_2^- periplasmic binding proteins of the ATP-dependent Nrt transport systems found in cyanobacteria. However, NasS has no signal peptide and so must be a cytoplasmic NO_3^- binding protein. NasT has the character of an anti-terminator of RNA translation and so together with NasST might mediate NO_3^-/NO_2^--dependent regulation of the *nas–nir* genes at a post-transcriptional level, as suggested for homologous genes in *Azotobacter vinelandii*. The presence of the NasS sensor shows that, in addition to having multiple genes for NO_3^-/NO_2^- transport, *P. denitrificans* also has distinct sensing systems for these oxyanions associated with the *nar* and *nas* gene clusters.

Integration of data from the *E. coli* monomeric NapA, the *R. sphaeroides* heterodimeric NapAB, *D. desulfuricans* monomeric NapA and the cytoplasmic Nas enzymes together give insight into the evolution of the NapB-independent and NapB-dependent periplasmic NO_3^--reductases. The current consensus of opinion is that NapA evolved from cytoplasmic eubacterial assimilatory NO_3^--reductases, which are ferredoxin or flavodoxin dependent and so are tuned to work at lower potentials than NapA, but also appear to bind nitrate in the Mo(V) state [14] (Figure 3-5). During evolution, these cytoplasmic enzymes picked up a signal peptide and were translocated to the periplasmic compartment. Once in the periplasmic compartment, the NO_3^--reductase would need a new electron donor that could allow it to derive electrons from the respiratory electron transport chain. The monomeric NapA of the *D. sulfuricans* is currently cited as the closest known relative of Nas and as yet there is no evidence for this enzyme being NapB dependent. The recently published genome of *S. thermophilum*, an uncultivable Gram-positive bacterium that depends on microbial commensalisms, reveals that it has the genetic potential to encode a *D. desulfuricans* type of NapA. This is the first example of Nap in an organism outside of the proteobacteria even in a bacterium without a periplasm. The gene cluster encoding *S. thermophilum* NapA, like that of *D. desulfuricans*, does not encode a NapB and no *napB* gene can be found elsewhere on the genome. The mature forms of both the *D. desulfuricans* and *S. thermophilum* NapA proteins are characterized by being around 720 amino acids in length, similar to the ferredoxin-dependent Nas (NarB) of *Synechococcus* sp. At some point in evolution, NapA must have recruited the di-heme NapB as a redox partner, but the molecular interaction with the primordial NapB would have intially been weak, a situation perhaps still reflected by the transient NapA–NapB interaction in *E. coli*. As the NapA and NapB proteins co-evolved, the interaction would have become stronger such that a tight heterodimeric NapAB complex, typified in the α-proteobacteria *R. sphaeroides* and *P. pantotrophus*, was the final outcome. These heterodimeric and monomeric subclasses of NapB-dependent NapA from *R. sphaeroides* and *P. pantotrophus* can both be distinguished from the monomeric NapB-independent NapA from *D. desulfuricans* and *S. thermophilum* on the basis of their larger size, since the mature proteins are around 800 amino acids, rather than 720 amino acids.

One notable gene, *napG*, is a common feature in the *D. sulfuricans* and *S. thermophilum nap* clusters. This gene encodes a putative 4[4Fe–4S] ferredoxin with a signal peptide predicted to direct export to the periplasmic compartment. It is notable that 4[4Fe–4S] ferredoxin proteins commonly donate electrons directly to the Mo-*bis*-MGD subunits of respiratory reductases, such as the DMSO-, tetrathionate-reductase or Nar, and so it is reasonable that NapG is the periplasmic electron donor to NapA in *D. desulfuricans* and *S. thermophilum*. It is then notable that many *nap* gene clusters that also encode *napB* have retained *napG*, including that of *E. coli* [9]. The weak association of NapB with NapA in this organism may allow for some promiscuity of NapA in also accepting electrons from NapG and some genetic evidence supports the operation of dual electron transport pathways. However, this is not the case for the *R. sphaeroides* or *P. pantotrophus* where the tight association between NapA and NapB precludes promiscuity and in which, accordingly, *napG* is deleted from the *nap* gene cluster. In *R. sphaeroides* and *P. pantotrophus*, the *napA* and *napB* genes overlap in a way that is indicative of translational coupling to yield a 1:1 ratio of translated polypeptide. By contrast, in *E. coli* the *napA* and *napB* genes are non-contiguous, being separated by the *napGH* genes and so these genes are not translationally coupled.

2.6 The eukaryotic nitrate reductases

Plants and fungi have the ability to assimilate NO_3^- N into biomass and possess NADH-, NADPH- or NADH/NADPH-dependent NO_3^--reductases. There is a Mo enzyme, but the Mo ion is bound within the Molybdopterin (MPT) form of the cofactor which is simpler than Mo-*bis*-MGD in that it is a monopterin cofactor that lacks the guanine di-nucleotide extension Figure 2-1. The enzyme is most closely related to eukaryotic SO_3^{2-} oxidases and no prokaryotic counterpart of the eukaryotic Nas has yet convincingly been identified. In addition to MPT, the enzyme binds FAD and a *b*-heme. Structures have been determined for the cytochrome and MPT domains [15]. On the basis of the structure and associated spectroscopic evidence, the Mo ion has been argued to be a dioxo-species in the Mo(VI) state and with reduction to Mo(IV) via the FAD and heme domains yielding a two-electron reduced mono-oxo Mo(IV) state with a vacant coordination site to which NO_3^- binds. It should also be noted that some fungi have the capacity to reduce NO_3^- as part of a denitrification process and here the NO_3^--reductase is located in the mitochondrial membrane and is likely to emerge as being a prokaryotic pNar or nNar type.

References

[1] S.J. Ferguson, D.J. Richardson, Electron transport systems for reduction of nitrogen oxides and nitrogen oxyanions, in: D. Zannoni (Ed.) Respiration in Archaea and Bacteria, vol. 2: Diversity of Prokaryotic Respiratory Systems, Kluwer Academic Publishers, The Netherlands, 2004, pp. 169–206.

[2] M.G. Bertero, R.A. Rothery, N. Boroumand, M. Palak, F. Blasco, N. Ginet, J.H. Weiner, N.C. Strynadka, Insights into the respiratory electron transfer pathway from the structure of nitrate reductase A, Nat. Struct. Biol. 10 (2003) 681–687.

[3] M. Jormakka, D.J Richardson, B. Byrne, S. Iwata, The architecture of NarGH reveals a structural classification of MGD enzymes, Structure 12 (2004) 95–104.

[4] L. Anderson, D.J. Richardson, J.N. Butt, Protein film electrochemistry of the membrane-bound nitrate reductase, Biochemistry 40 (2001) 11294–11307.

[5] N. Wood, T. Alizadeh, D.J. Richardson, S.J. Ferguson, J.W.B. Moir, Two domains of a dual function NarK protein are required for nitrate uptake, the first step of denitrification in *Paracoccus pantotrophus*, Mol. Microbiol. 44 (2002) 157–170.

[6] B. Lledo, R.M. Martinez-Espinosa, F.C. Marhuenda-Egea, M.J. Bonete, Respiratory nitrate reductase from haloarchaeon *Haloferax*

mediterranei: biochemical and genetic analysis, Biochim. Biophys. Acta 1674 (2004) 50–59.
[7] D.J. Richardson, The 1999 Fleming Lecture. Bacterial respiration: a flexible process for a changing environment, Microbiology 146 (2000) 551–571.
[8] P. Arnoux, M. Sabaty, J. Alric, B. Frangioni, B. Guigliarelli, J.M. Adriano, D. Pignol, Structural and redox plasticity in the heterodimeric periplasmic nitrate reductase. Nat. Struc. 10 (2003) 928–934
[9] J.M. Dias, M.E. Than, A. Humm, R. Huber, G.P. Bourenkov, H.D. Bartunik, S. Bursakov J. Calvete, J. Caldeira, C. Carneiro, J.J. Moura, I. Moura, M.J. Romao. Crystal structure of the first dissimilatory nitrate reductase at 1.9 A solved by MAD methods. Structure (1999) 7 65-79.
[10] M.D. Roldan, H.J. Sears, S.J. Ferguson, M.R Cheesman, A.J. Thomson, B.C. Berks, D.J. Richardson, Spectroscopic characterization of a novel tetra-heme c-type cytochrome widely implicated in bacterial electron transport, J. Biol. Chem. 273 (1998) 28785–28790.
[11] M.J.K. Ellington, K.K. Bhakoo, G. Sawers, D.J. Richardson, S.J. Ferguson, Hierarchy of carbon source selection in *Paracoccus pantotrophus*: Strict correlation between reduction state of the carbon substrate and aerobic expression of the nap operon, J. Bact. 184 (2002) 4767–4774.
[12] D.J. Richardson, B.C. Berks, S. Spiro, D.A. Russell, C.J. Taylor, The biochemical, genetic and physiological diversity of prokaryotic nitrate reduction, Cell. Mol. Life Sci. 58 (2001) 165–178.
[13] T.H.C. Brondijk, D. Fiegen, D.J. Richardson, J.A. Cole, The roles of NapF, NapG and NapH gene products in electron transfer from ubiquinol to the periplasmic nitrate reductase of *Escherichia coli*, Mol. Microb. 44 (2002) 245–255.
[14] B. N. Jepson, L. Anderson, L. Rubio, C. Taylor, C. Butler, A. Herrero, E. Flores, J.N. Butt, D.J. Richardson, Tuning a nitrate reductase for function: the first spectropotentiometric characterization of a bacterial assimilatory nitrate reductase reveals novel redox properties, J. Biol. Chem. 279 (2004) 32212–32218.
[15] K. Fischer, G.G. Barbier, H.J. Hecht, R.R. Mendel, W.H. Campbell, G. Schwarz, Structural basis of eukaryotic nitrate reduction: crystal structures of the nitrate reductase active site, Plant Cell 17 (2005) 1167–1179.

Chapter 3

Nitrite Reductases in Denitrification

Serena Rinaldo and Francesca Cutruzzolà

3.1 Introduction

Nitrite reductase (NIR) is a key enzyme in the dissimilatory denitrification chain, catalysing the reduction of NO_2^- to NO [1]. Although this has been a matter of debate for a long time, NO is now accepted as a product of NO_2^--reduction and an obligatory intermediate in most denitrifiers; it is further reduced to N_2O by NO reductase.

Purification and characterization of NIR from several bacterial sources have shown that there are two distinct classes of dissimilatory NIRs that yield NO as the main reaction product, containing either copper (CuNIR) or heme (cd_1NIR) as cofactor, the heme-containing enzyme occurring more frequently. The genes coding for the CuNIR and cd_1NIR apoproteins are called *nirK* and *nirS*, respectively. Besides the species from which the enzyme has been purified, several others were shown to contain one of the two types of NIR on the basis of DNA hybridization and/or inhibition of denitrification by the Cu chelator diethyldithiocarbamate (DDC). The enzymes containing Cu and heme never coexist within the same bacterial species. Functional complementation of a cd_1NIR-deficient

Biology of the Nitrogen Cycle
Edited by H. Bothe, S.J. Ferguson and W.E. Newton

strain of *Pseudomonas stutzeri* with the Cu NIR from *Pseudomonas aureofaciens* indicates that the two enzymes fulfil the same role in vivo.

This chapter mainly focuses on the structure–function relationships in the two classes of dissimilatory NIRs and special attention is paid to recent structural information on enzymes from different sources, which have different structures and catalyse the reduction of NO_2^- to NO via different mechanisms. The reader is referred to recent reviews on cd_1NIR [2–4] and CuNIR [2, 5, 6] with a more extensive bibliography.

3.2 Cd_1 nitrite reductase

3.2.1 General properties and structure

Heme NO_2^--reductases are generally dimers of two identical subunits, each containing one heme c and one unique heme d_1 (Figure 3-1). Up to now, cd_1 enzymes have been purified from *Alcaligenes faecalis, Azospirillum brasilense* SP7, *Magnetospirillum magnetotacticum, Pseudomonas aeruginosa, P. nautica, P. stutzeri, Paracoccus denitrificans, P. halodenitrificans, P. pantotrophus, Ralstonia eutropha, Roseobacter denitrificans* and *Thiobacillus denitrificans*.

The cd_1 enzymes are periplasmic soluble proteins and involved in respiratory NO_2^--reduction, apart from those from *R. denitrificans* and *M. magnetotacticum*, which have been assigned an O_2-reductase and a Fe(II): NO_2^--oxidoreductase activity, respectively.

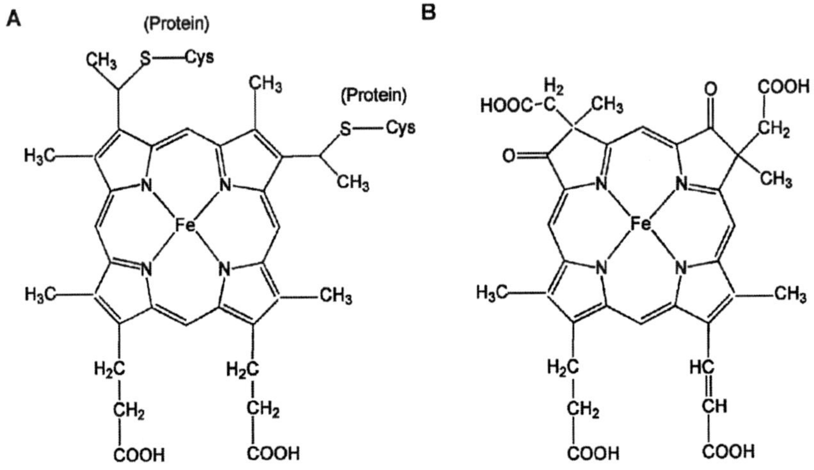

Figure 3-1 Chemical structure of the hemes of cd_1NIR. (A) Structure of c heme showing the covalent attachment to the protein. (B) Chemical structure of d_1 heme: notice that this cofactor is more oxidized than a classical b-type heme.

The first enzyme that was characterized is that from *P. aeruginosa* (E.C. 1.9.3.2) discovered by Horio and coworkers and initially studied for its O_2-reductase activity, which is inhibited by both CO and CN^- and produces water. Later, the work of Yamanaka showed that the enzyme is also capable of catalysing the reduction of NO_2^-, an activity which is now accepted to be the only physiological role. This assignment is based both on kinetic and equilibrium results with the two substrates (Table 3-1) and on genetic evidence showing that strains of *P. stutzeri* and *P. aeruginosa* in which *nirS*, the gene coding for cd_1NIR, has been selectively inactivated cannot grow on N-oxides. NIR activity is inhibited by CN^- but insensitive to CO and its reaction product is NO. It is also pH-dependent in vitro with an optimum between 5.8 and 7. Interestingly, a pH dependence of cd_1NIR activity with accumulation of NO_2^- has been observed in vivo in cultures of *P. denitrificans* grown at suboptimal pH (6.8), indicating that an inactivation may occur at low pH.

The peculiarity of cd_1NiR in catalysing both the monoelectronic reduction of NO_2^- to NO and the tetraelectronic reduction of O_2 to H_2O, which are mechanistically very different, is a very intriguing feature of this enzyme. Other, less relevant, activities have been attributed to cd_1NiR such as the oxidation of CO to CO_2 and the reduction of NH_2OH to NH_3.

The spectroscopic, redox and catalytic properties of the cd_1NIRs from *P. aeruginosa*, *P. stutzeri*, *P. nautica*, *T. denitrificans* and *P. pantotrophus* are summarized in Table 3-1. In the optical visible spectrum of the oxidized protein from most species the c heme is characterized by absorption maxima at 520 and 411 nm, while the d_1 heme shows a broad shoulder around 470 nm and a band centred at 640 nm. Two bands are, however, seen in the region between 640 and 710 for the *P. pantotrophus* and *R. denitrificans* cd_1 NIRs. The reduced enzyme absorption maxima for the c heme are at 548, 520 and 417 nm, and for the d_1 heme at 650 and 456 nm. The midpoint redox potentials (E_0') of the two hemes under various conditions, known for the *P. aeruginosa* and *P. nautica* enzymes, range between +200 and +300 mV (Table 3-1). It is noteworthy that E_0' of the c heme is very sensitive to the ligation state of the d_1 heme, being more negative when CO is bound, and more positive when NO is bound (Table 3-1). Moreover, in the semiapoprotein of *P. aeruginosa* containing only the c heme, E_0' of the c heme is more positive than that measured in the holoprotein. These findings, together with data on specific ligands, suggest the existence of redox interactions among the different hemes.

Electron donors to cd_1NiRs have been identified in vitro with a survey of soluble electron carriers, either *c*-type cytochromes (cyt c_{550}, cyt c_{551}, cyt c_{554}) or Cu-proteins, like azurin or pseudoazurin. In a few cases the physiological electron donor could be identified in vivo, for example, for the *P. aeruginosa* and *P. denitrificans* enzymes. However, the results do not

Table 3-1 Properties of cd_1NIR.

Property	*Pseudomonas aeruginosa*	*Pseudomonas stutzeri*	*Paracoccus pantotrophus*	*Pseudomonas nautica*	*Thiobacillus denitrificans*
Molecular mass (kDa)(homodimer)	119–121	119–134	120	131	118
VIS absorbance (nm) (oxidized form)	411, 520 640	411, 525 641	406, 525, 644, 702	409, 521, 636	407, 525, 642
(reduced form)	418, 460, 521, 549, 554, 625–655	417, 460, 522, 548, 554, 625–655	418, 460, 521, 547, 553, 625–655	416, 460, 521, 548, 552, 625–655	418, 460, 523, 549, 553, 625–650
EPR parameters (*g* values)					
Heme d1	2.51, 2.43, 1.71	2.56, 2.42, 1.84	2.52, 2.19, 1.84	2.51, 2.33, 1.67	2.50, 2.43, 1.70
Heme c	3.01, 2.29, 1.40	2.97, 2.24, 1.58	3.05	2.92, 2.35	3.6
$E_{m,7}$ (mV)					
Heme c	+288[a] (>290[b]) (<214[c])		Hysteretic redox behaviour[d]	+234 (pH7.6)	

Heme d1	+287			+199 (pH7.6)
Electron donor(s)	Cyt c_{551}/azurin	Cyt c_{551}	Cyt c_{550} / pseudoazurin	Cyt c_{552}
Km for NO_2^- (μM) (electron donor)	53 (hydroxyquinone) 6 (azurin)	1.8 (horse heart Cyt c)	6.7–12[e] (Cyt c_{550})	
Km for O_2 (μM) (oxygen electrode)	28		80	
$\boldsymbol{k}_{\mathbf{cat}}$ (s^{-1})				
with NO_2^-	8	8	2.1–74[e]	
with O_2	0.6–6.4		3–6	

[a] When d_1 heme is unbound.
[b] When d_1 heme is NO bound.
[c] When d_1 heme is CO bound.
[d] See Ref. [3].
[e] These two values refer to "as isolated " and "pre-activated" states of the enzyme, respectively.

give a unifying picture, since in the former case only cyt c_{551} is involved whereas in the latter both cyt c_{550} and pseudoazurin are implicated [7].

3.2.2 *Structure*

The primary structure of the enzyme from several species reveals that NIR is synthesized as a pre-protein, with a leader peptide responsible for the periplasmic export. The mature subunit is over 500 amino acids long; it contains only two Cys residues at the N-terminal region, which are covalently bound to the vinyl groups of the c heme (Figure 3-1A). The degree of homology between cd_1NIRs is much higher for the C-terminal d_1 heme domain than for the N-terminal c heme domain, and is particularly low at the N-terminus. For the entire protein a 52.7% identity and 70% homology can be calculated.

The total number of amino acid sequences available for cd_1NIR (and also for CuNiR) in public databases is rapidly growing since oligonucleotide primers amplifying the d_1-heme domain of *nirS* (or the whole gene of *nirK*) are often used to monitor the presence of denitrifiers in the environment.

Cd_1NIR is a homodimer (MW = 120 kDa) and each subunit contains one c heme and one d_1 heme; thus the native dimeric protein carries four metal centres. The high-resolution structure has been determined for several derivatives of the enzyme from *P. pantotrophus* (hereinafter Pp-cd_1NIR) and *P. aeruginosa* (hereinafter Pa-cd_1NIR). It shows a conserved overall architecture: each monomer is organized into two structurally distinct domains, one carrying the c heme and the other the d_1 heme (Figure 3-2). The former, which is the electron acceptor pole of the molecule, contains mainly α-helices, whereas the latter has an eight-bladed β-propeller structure common to other proteins. The distances between the hemes within a subunit are about 11 Å edge to edge and 20 Å iron to iron; those between subunits are much larger (>35 Å). As discussed below, the d_1 heme is the site where substrate binds and catalysis occurs. With regard to quaternary structure, non-covalent interactions between monomers in the dimer are very strong, since dissociation is not observed even in 3 M NaCl or 6 M urea, but only at extreme pH (>11), or after extensive succinylation.

A closer look at the structure reveals significant differences in the c and d_1 domains between the two species; some of these can be related to different primary sequences and are worthy of discussion since they may be relevant to the interpretation of the reaction mechanism.

In Pa-cd_1NIR the α-helical c heme domain has a topology similar to that of class I cytochromes c (Figure 3-2), whereas a different connectivity of the helices is observed in Pp-cd_1NIR. In both proteins a N-terminal segment from the c-domain extends towards the d_1-domain and is inserted in the active site pocket. In the Pp-cd_1NIR this exchange occurs within the

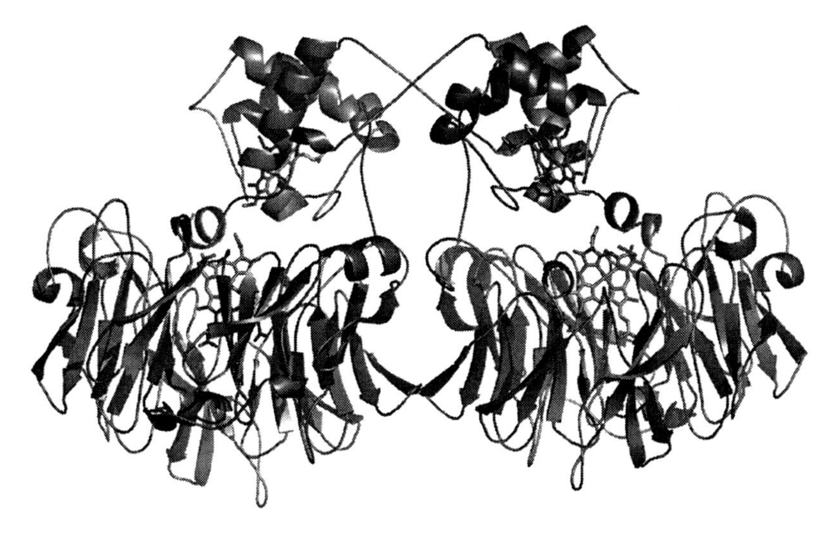

Figure 3-2 *P. aeruginosa* cd_1NIR. Three-dimensional structure of the dimer in the oxidized form (PDB code 1NIR). Notice the swapping of the N-terminal arm between neighbouring monomers. The α-helical c heme domain is the electron acceptor pole, whereas the β-propeller domain contains the d_1 heme active site.

same monomer, but in the *P. aeruginosa* enzyme a "domain swapping" occurs, since the tail of one monomer contacts the d_1 site of the neighbouring monomer (Figure 3-2). The ligands of the c heme in the oxidized form of the two enzymes are strikingly different: His^{51}-Met^{88} in Pa-cd_1NIR (Figure 3-3A) and His^{17}-His^{69} for Pp-cd_1NIR. Spectroscopic studies (MCD, NMR, EPR) available on the oxidized form of both enzymes confirm that this pattern of c heme ligation is the same also in solution; the His-Met type ligation is also expected for the cd_1NIR of *P. stutzeri* on the basis of EPR and MCD studies. In Pp-cd_1NIR the axial ligands change to His^{69}-Met^{106} upon reduction, triggering a large conformational change, which involves also the d_1 domain (see below). In the *P. aeruginosa* enzyme the c heme ligands are unchanged upon reduction. The c heme is thus hexacoordinated and low spin in both reduction states; nevertheless it may form a complex in Pa-cd_1NIR with several heme ligands (NO, CN^-) under conditions in which eukaryotic cytochrome *c* does not.

The c-domain is the electron-accepting site of the molecule and the interaction with the electron donors is mainly electrostatic in nature. Several negatively charged residues (mainly glutamates) that could be involved in electron transfer complex formation were identified by analysis of the c-domain structure in both Pa-cd_1NIR and Pp-cd_1NIR. As for other systems, the formation of the electron transfer complex is mainly driven by the anisotropic distribution of surface charges, which leads to

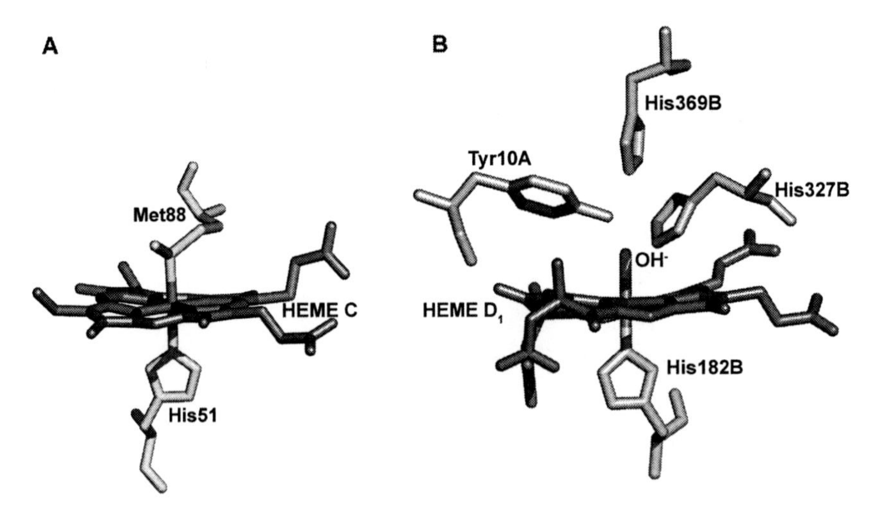

Figure 3-3 Structure of the c heme (A) and the d_1 heme (B) sites in the oxidized *P. aeruginosa* cd_1NIR (PDB code 1NIR). In (B), the residues are labelled with the sequence number and a letter, which refers to the different subunits (A,B).

strong dipolar moments in the redox partners. However, some cross-reactivity between electron donors and reductases from different species was observed and can be related to the relatively "loose" specificity of recognition of the interacting surfaces in the complex.

The d_1 heme is unique to denitrifiers containing the cd_1NIR, where it is synthesized starting from δ-aminolevulinic acid and via uroporphyrinogen III. The genes responsible for the d_1 heme-specific reactions of the pathway are mostly found in the same operon downstream to the *nirS* gene in *P. aeruginosa*, *P. stutzeri* and *P. denitrificans* and thus are also induced by low O_2 tension and presence of N-oxides.

Heme d_1 has been unequivocally identified as a 3,8-dioxo-17-acrylate-porphyrindione (Figure 3-1B); since it is so unique, it was suggested that it could be responsible for some catalytic features of the enzyme. The presence of the electronegative oxo groups confers to the macrocycle distinct redox properties, rendering it harder to oxidize than the corresponding isobacteriochlorines, and thus shifting the E_0' of the Fe to more positive values. The chemical nature of heme d_1 has been confirmed by inspection of the crystal structures of all cd_1NIRs: in the oxidized and reduced/NO-bound Pp-cd_1NIR, heme d_1 adopts a planar conformation, whereas in the others a more distorted, saddle-like conformation is preferred.

The d_1 heme binds the physiological oxidants (O_2, NO_2^-) as well as other heme ligands. A semiapoprotein containing only the c heme can be obtained either by acidic acetone treatment or by recombinant DNA techniques. Reconstitution of both apoproteins with purified or chemically synthetized d_1 heme is possible, with good recovery of the spectroscopic properties of the holoenzyme and of the O_2- and NIR-activities.

On the basis of EPR, NMR and MCD measurements, the d_1 heme in Pa-cd_1NIR has been assigned as a low-spin hexacoordinated species in the ferric state, and high spin pentacoordinated in the ferrous state. One axial ligand is always provided by a histidine. The sixth ligand in the low-spin ferric form is a hydroxide ion in the oxidized Pa-cd_1NIR structure (Figure 3-3B). The ligation is different in the oxidized Pp-cd_1NIR structure where the sixth ligand is provided by the phenolate of Tyr^{25}, a residue belonging to the N-terminal segment and thus to the c-domain. The Tyr^{25}-coordinated enzyme is catalytically inert and needs to be activated by reduction (either with sodium dithionite or with the protein NapC [8]). Substitution of Tyr^{25} with Ser has no effect on the catalytic properties of the enzyme [8, 9]; it is possible that the Tyr^{25}-coordinated structure may represent a resting form of the enzyme, consistent with the fact that this residue is poorly conserved among the cd_1NIRs from different species. Tyr^{25}-coordination could be protective towards unwanted reaction of the d_1-heme with harmful reactive oxygen species (ROS) that could be present in a partially aerobic environment; this hypothesis is still to be confirmed experimentally. In Pa-cd_1NIR a Tyr is found in the d_1 heme pocket, but is located further away from the iron: its mutation to Phe does not alter the catalytic and optical properties of the enzyme. Spectroscopic studies in solution have either directly confirmed, or are consistent with these different ligation patterns of the two enzymes, in either oxidation state. A low-spin ferric d_1 heme-associated α-band in the 636–644 nm region is common to all NIRs and may indicate a common His/OH or His/Tyr ligation: however, whether the latter form is *on* or *off* the catalytic pathway still remains unclear (see below).

Among the other residues found in close proximity of the d_1 heme, an important role in catalysis is played by two His residues which are H-bonded to a H_2O molecule in the Pp-cd_1NIR oxidized structure or directly to the ligand hydroxide in the Pa-cd_1NIR oxidized structure (Figure 3-3B). These residues are good candidates to be involved in the protonation and dehydration of NO_2^- (see below).

3.2.3 Catalytic mechanism

The monoelectronic reduction of NO_2^- to NO is the main activity of cd_1NIR in vivo. The mechanism of NO_2^--reduction involves chemically complex steps such as transfer of a reducing equivalent into an electron-rich species, e.g. the NO_2^- anion, possibly followed by protonation and dehydration. Product inhibition may occur, since the NO produced at the catalytic site forms stable complexes with the ferrous d_1 heme and, at acidic pH, also with the c heme. A possible reaction scheme for cd_1NIRs is shown in Figure 3-4, panel A.

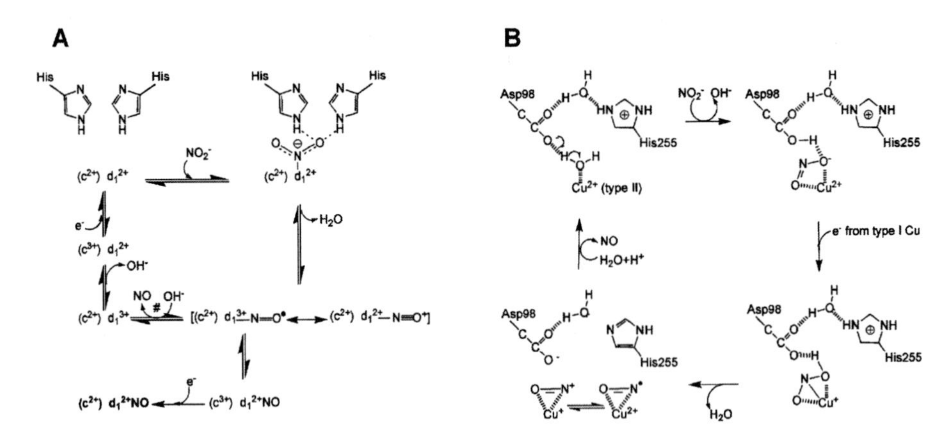

Figure 3-4 Reaction mechanism of NIRs with NO_2^-. (A) Proposed mechanism of cd_1NIR, modified from [13]; (#) the histidine residues in the active site are involved in the product (NO) release by stabilizing the hydroxide (OH^-) ligand of the oxidized d_1 heme iron. Lack of this stabilization favours the formation of the dead-end state (reduced NO-bound adduct). (B) Proposed mechanism of CuNIR, modified from [5, 16]; in this model the "side-on" product (NO) binding is also included.

Catalysis occurs at the d_1 heme with the substrate (NO_2^-) and product (NO) bound to the Fe via the N-atom, as seen in the crystal structure of the NO_2^--bound transient species of Pp-cd_1NIR and NO-bound reduced derivative of both Pa-cd_1NIR and Pp-cd_1NIR. A recent proposal from a theoretical study suggests an alternative O-NO_2^- ligation [10], but still awaits experimental support. A nitrosyl intermediate (d_1-$Fe^{2+}NO^+$) is formed during catalysis, as shown in $^{18}O/^{15}N$ exchange experiments. Spectroscopic evidence for this species, which is EPR silent, has also been obtained by FTIR on the oxidized, NO reacted *P. stutzeri* enzyme and by stopped-flow FTIR on Pp-cd_1NIR [11]. The nitrosyl intermediate (d_1-$Fe^{2+}NO^+$), which is formally equivalent to d_1-$Fe^{3+}NO$, is chemically unstable and rapidly decays to d_1-Fe^{3+} plus NO (Figure 3-4A). Complete reduction of the transient species leads to the formation in vitro of a paramagnetic (d_1-$Fe^{2+}NO$) stable adduct with slow NO dissociation rates, which has been observed by EPR. Data on the NO complexes of d_1 heme synthetic models show that the formation of a ferrous π-cation radical with a bent Fe-NO bond is also possible. Formation of this species is facilitated by the oxo groups of the macrocycle that increase the positive charge on the iron; under these conditions, a weakening of the Fe-NO bond is observed and NO dissociation occurs readily in the presence of a nucleophile. Whereas formation of NO can be monitored at steady state (e.g. using an NO-sensitive electrode), rapid reaction studies have failed to detect kinetically competent NO release from cd_1NIRs.

An important feature of the reduction of NO_2^- by cd_1NIR is that for the enzyme to accomplish a productive turnover, a balance between product release and re-reduction of the d_1 heme is critical to avoid product inhibitory effects (Figure 3-4A). Product release in the periplasmic space might be enhanced by the efficient scavenging activity of NO reductase. The internal electron transfer rate, relevant for the catalytic mechanism, might be under allosteric control in cd_1NIRs. In *P. stutzeri* cd_1NIR and Pa-cd_1NIR a marked decline in the intramolecular electron transfer rate is seen with increasing enzyme reduction [12]. Whether a tertiary or a quaternary conformational change is involved in this allosteric control is still under investigation; it is known that in a crystal form of Pa-cd_1NIR, reduction of the c heme alone does not trigger a conformational change in the d_1 heme environment. Noteworthy, large conformational changes (of unknown meaning) have been seen after mutagenesis of Pa-cd_1NIR [13] and in a crystal form of reduced Pp-cd_1NIR [14].

Although there is agreement on the chemical species populated along the reaction pathway (Figure 3-4A), the relative rates of the single steps are different in the enzymes characterized so far, in agreement with the different turnover numbers observed at steady state (Table 3-1, [3]). For Pa-cd_1NIR at pH = 8, the binding and dehydration step is very fast, almost diffusion-limited ($k = 10^8\ M^{-1}\ s^{-1}$), while the internal electron transfer is slow ($1\ s^{-1}$) and the final species is the d_1-$Fe^{2+}NO$ complex. In the reaction with NO_2^- of Pp-cd_1NIR at pH = 7, the internal electron transfer rate is much faster (in agreement with pulse radiolysis experiments on the same enzyme), but no release of NO could be detected [11]. Although the origin of these differences is unclear, they may arise from variations in sequence and structure between the Pp-cd_1NIR and Pa-cd_1NIR enzymes that result in substantially different electron transfer kinetics.

Insight into the molecular mechanism of catalysis comes from the results on the mutants of the two conserved histidines (His327 and His369 in Pa-cd_1NIR) located near the active centre. Substitution of either of the two His with Ala has a dramatic effect on NO_2^--reduction [13]. The two His are not equally important in controlling the affinity for the anionic substrate NO_2^-: His369 was shown to be essential. A common feature of the two His mutants in the reaction with NO_2^- is that they are both trapped in the reduced-NO bound species faster than the wild type Pa-cd_1NIR. Given that the intramolecular electron transfer rate was found to be unchanged in these mutants, the increased probability of trapping may originate in a decreased rate of NO dissociation from the ferric d_1 heme. Reduction of the positive potential in the d_1 heme pocket by substitution of either one of the two invariant active site His with Ala may lead to the loss of the hydroxyl coordinated to the ferric d_1 heme iron in the wt enzyme, which should assist NO dissociation (Figure 3-4A). This hypothesis is fully consistent with the unusually high affinity of the reduced protein for anions, both NO_2^- (the physiological substrate) and other anionic ligands as cyanide or hydroxide, which is controlled by the His side chains.

Several key points of the reaction mechanism of cd_1NIRs have been clarified with the contribution of functional and structural information, and can be summarized as follows: (a) the oxidized enzyme has both hemes in the low-spin, hexacoordinated state; (b) the c heme is the electron entry site, whereas the d_1-heme is the catalytic site; (c) large conformational changes are seen upon reduction, leading to a pentacoordinated d_1 heme iron and a catalytic site open for the binding of ligands; (d) catalysis occurs at the d_1 heme with the substrate (NO_2^-) and product (NO) bound to the iron via the N-atom; (e) the His residues in the distal pocket of the d_1 heme are involved in substrate binding and protonation; (f) binding of anions to the reduced enzyme is facilitated by the positive potential of the distal pocket and by the presence of the specialized d_1 heme; (g) a nitrosyl intermediate is formed during catalysis; (h) in Pa-cd_1NIR dissociation of the product (NO) is assisted by the active site His residues which stabilize the OH^- bound to the oxidized enzyme.

There are still open questions, among which the most relevant during catalysis are the role of the d_1 heme and the role of the redox state of the c heme. The latter may be involved in triggering the conformational changes seen upon reduction and related to the low rate of c-to-d_1 electron transfer measured for the Pa-cd_1NIR and *P. stutzeri* cd_1NIR. Biochemical and structural chararacterization of cd_1NIRs from other species will shed more light on the reaction mechanism of this puzzling enzyme.

3.3 Copper nitrite reductase

3.3.1 General properties and structure

Cu-containing NO_2^--reductase (CuNIR) (E.C. 1.7.99.3) is present in both Gram-negative and Gram-positive (*Bacillus* sp.) eubacteria and in Archaea (*Haloarcula* sp.). The better studied enzymes from the Gram-negative group comprises those from *Achromobacter cycloclastes*, *Alcaligenes faecalis* S-6 and *Alcaligenes xylosoxidans*, *Rhodobacter sphaeroides* and *Pseudomonas aureofaciens*.

The CuNIR class is more heterogeneous than the cd_1NIR one in terms of molecular properties (Table 3-2) although the primary structure is well conserved within each subclass, ranging from 60% to 80% protein identity in different species. The proteins are synthesized as longer precursors, with a leader peptide responsible for the periplasmic export. Common features are the trimeric quaternary structure in which a monomer (~40 kDa) contains two distinct Cu-centres, one belonging to the type I Cu subclass and the other being a type II Cu. To date the only known exception was

Table 3-2 Properties of CuNIR.

Property	*Alcaligenes xylosoxidans*	*Achromobacter cycloclastes*	*Alcaligenes faecalis S-6*	*Rhodobacter sphaeroides*
Molecular mass (kDa) homotrimer	103	108	100.4–119	140
VIS absorbance (nm)	593, 770	400, 458, 585, 700	400, 457, 587, 700	390, 457, 589, 700, 810
No. of Cu atoms/holoenzyme	3.5 + 0.8	4.6	4.5	
EPR parameters				
Type I Cu, g	2.212	2.195	2.19	2.19
Type I Cu A (mT)	6.3	7.3	2.30	7.8
Type II Cu, g	2.29	2.262	2.34	
Type II Cu, A (mT)	14.2	17.5	16.3	
Electron donor(s)	Cyt c_{553}	Pseudoazurin/ cyt c	Pseudoazurin	Cytochrome c_2
$\boldsymbol{E_m}$ (mV)				
Type I Cu	+260	+240		+247
Type II Cu		+260		< +200
Activity (μmol min^{-1} mg^{-1})				
NO_2^- —> NO	240	280	380	40
(electron donor)[a]	(MV)	(PMS-Asc)	(MV)	(Yeast Cyt c)
$\boldsymbol{K_m}$ **for** $\boldsymbol{NO_2^-}$ (μM)[b]	230	500	74	14 (Yeast Cyt c) 35 (BV)

[a] MV, methyl viologen; PMS-Asc, phenazine methosulphate plus ascorbate; BV, benzyl viologen.
[b] Values obtained with the same e^- donors as for the nitrite reductase activity, unless otherwise stated.

found in *Hyphomicrobium denitrificans* A3151 (HdNIR) [15], where the CuNIR is composed of six identical subunits.

Spectroscopic and site-directed mutagenesis studies on the two Cu-centres showed that the type I site is the redox active site from which the electrons necessary for catalysis are transferred to the type II site, where substrate binding occurs. The primary product of CuNIR is NO; however, the enzyme can also yield a small amount (3–6%) of N_2O if NO is allowed to accumulate. Other reactions such as the conversion of NO_2^- to NH_4^+ or reduction of O_2 have also been described, although at much lower rates.

The total number of Cu atoms found in enzymes from different sources may differ considerably from the ideal six, depending on purification and storage, but six Cu atoms have been found in all crystal structures determined to date. The three-dimensional structure of different forms of the enzyme from *A. cycloclastes* [16], *A. faecalis* S-6 [2], *A. xylosoxidans* [17], *R. sphaeroides* [18] has been determined.

3.3.2 Structure

The overall architecture of the enzyme is organized as a homotrimer where the three identical subunits are tightly associated around a central channel of 5–6 Å in a threefold axis symmetry. Each monomer is composed of two distinct domains (Figure 3-5A), each consisting of a Greek key β-barrel fold similar to that of cupredoxins. An extensive network of H-bonds, both within and between the subunits of the trimer, confers considerable rigidity to the molecule. The protein conformation does not experience large shifts in atom positions neither at different pH values nor in the different redox- or ligand-bound states.

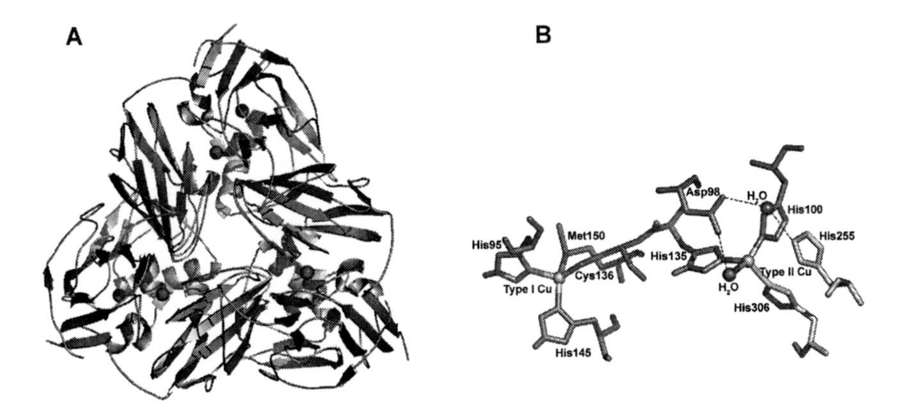

Figure 3-5 Structure of the *A. faecalis* CuNIR in the oxidized form (PDB code 1AS7). (A) Overall view of the homotrimer; the copper atoms are represented as filled circles. (B) Type I and type II Cu sites in the oxidized protein. Residues His^{255} and His^{306} come from the adjacent subunit.

As outlined above, CuNIRs contain both type I and type II Cu centres. In the three-dimensional structure of CuNIR purified from *A. cycloclastes* it was shown for the first time that the type I Cu sites (two His, one Cys and one Met ligands) are located within each monomer, while the type II sites are at the interface between monomers in the trimer, with ligands (three His and one solvent molecule) provided by two different subunits. The two Cu sites are linked via a His-Cys bridge (Figure 3-5B) with a Cu–Cu distance of 12.5–13 Å. The type II Cu is not essential for subunit stability; consistently, only minor structural changes are observed in the type II Cu-depleted enzyme from *A. cycloclastes*.

According to the spectroscopic properties of the type I Cu, these proteins can be classified as blue and green CuNIRs (Table 3-2). The blue CuNIRs from *A. xylosoxidans* (NCIB 11015 and GIFU 1051 strains) share a 593 nm absorption band caused by the charge-transfer transition of type I Cu ($S_{(Cys)} \rightarrow Cu^{2+}$), which confers an EPR signal of this metal centre with axial symmetry. The green CuNIRs (*A. cycloclastes* and *A. faecalis*) show two bands at 460 nm and 584 nm also due to a $S_{(Cys)} \rightarrow Cu^{2+}$ transfer transition of type I Cu but with a rhombic EPR signal.

The E_0' of the type I Cu is between +240 and +260 mV in enzymes from different species (Table 3-2). Small copper proteins, like azurins and pseudoazurins, and bacterial c-type cytochromes can donate electrons to CuNIR, the hemeproteins being less frequently involved. The interaction involves electrostatically complementary surfaces on both redox partners, as shown by site directed mutagenesis on CuNIR and by NMR studies on a pseudoazurin–CuNIR complex [19]. In this model the intermolecular electron transfer involves formation of a complex between the pseudoazurin as electron donor and the type I Cu-containing domain on CuNIR in a well-defined mode [19]. These experimental and theoretical data are supported by the recent studies on the HdNIR where each subunit is composed of one plastocyanin-like domain and one green CuNIR-like domain connected to each other, resembling the complex of green CuNIR and pseudoazurin [15].

Type II Cu, bound at the interface of two subunits lies at the bottom of a 12–13 Å deep solvent channel, is the substrate-binding and reduction site. Looking at the primary and 3D structure, one of the three histidines involved in the type II Cu binding is located immediately before of the cysteine residue involved in the type I Cu binding (Figure 3-5B), suggesting that this stretch of amino acids can act as electron carrier during the catalysis. An E_0' of +260 mV was measured for type II Cu in *A. cycloclastes* NO_2^--reductase, but this value may vary in the presence of NO_2^- and can be as low as 200 mV, as is the case for *R. sphaeroides* enzyme. The NO_2^--dependent modulation of the active site E_0' suggests that the intramolecular electron transfer is not energetically favoured in the absence of the substrate; the first-order electron transfer rate measured by pulse radiolysis was found to be $1.4 \times 10^3\ s^{-1}$, slower than reduction of type I Cu by external electron donors.

3.3.3 Catalytic mechanism

A mechanism for the monoelectronic reduction of NO_2^- from CuNIR was initially proposed which resembled that already described for cd_1NIR, involving a Cu nitrosyl intermediate (Cu^+–NO^+) formed at the type II Cu site. The presence of this species was inferred from isotope exchange experiments in which CuNIR was shown to produce N_2O in the presence of N_3^- and NH_2OH. The active site in the absence of substrate shows a type II Cu with nearly perfect tetrahedral geometry, with three out of four coordination bonds occupied by the Nε of three His, and the fourth by an O-atom from a water molecule (Figure 3-5B). The water ligand forms a hydrogen bond with the Carboxylate group of the Asp^{98}, close to the active site; furthermore, this residue together with the His^{255} residue bridge a second H_2O molecule to form a H-bonding network.

Spectroscopic evidence from electron nuclear double resonance (ENDOR) studies on *A. xylosoxidans* CuNIR shows that NO_2^- displaces a bound water ligand; no relevant coupling was, however, seen when $^{15}NO_2^-$ instead of $^{14}NO_2^-$ was used, raising the possibility that NO_2^- binds to the catalytic Cu via an O-atom. The structure of the type II Cu site in the absence and in the presence of NO_2^- has been elucidated for both oxidized (from *A. cycloclastes*, *A. faecalis* and *A. xylosoxidans*) and reduced CuNIR (from *A. faecalis* and *R. sphaeroides* -strain 2.4.3- at high pH).

Analysis of NO_2^--soaked crystals show that NO_2^- is coordinated to oxidized type II Cu in an asymmetric bidentate fashion through two O-atoms instead of the solvent molecule (Figure 3-4B). The N-atom in NO_2^- is more bent away from the Cu in the *A. faecalis* structure than in the *A. cycloclastes* one. A lower occupancy (i.e. a weaker bond) of the NO_2^- molecule is found in the reduced form, with a less ordered H-bonding network. This evidence, together with the absence of a fourth ligand in the fully reduced *A. faecalis* CuNIR structure, has been used to support the hypothesis that *O*-coordinated binding of NO_2^- to oxidized Cu is the first event in catalysis, followed by reduction of the type II Cu. This mechanistic view is also supported by observations on the *R. sphaeroides* CuNIR, which show that upon NO_2^- binding a decreased covalency of the Cu–N_ε(His) bond is observed. This may explain the NO_2^--induced increase in type II Cu E_0', and the increased probability of electron transfer from type I Cu.

The active site residues are more hydrophilic on one side and more hydrophobic on the other, suggesting a possible route for NO escape. The Asp^{98} and the His^{255} residues are both involved in the catalysis as proton donors and in the control of internal electron transfer. Leu^{106} has also been shown to be important in the control of the Asp^{98} position in the active site pocket [16]. Mutants of Asp^{98} or His^{255} with Ala show, respectively, an increased and reduced intramolecular electron transfer rate constant (k_{ET}), indicating that both residues around the type II Cu control the internal electron transfer by the H-bonding network. The pH dependence of the catalytic activity, similarly to that of k_{ET}, shows an optimum around 5.5; the only ionizable and conserved residues around the type II Cu are the Asp^{98} and the His^{255}. All the Asp^{98} mutants show a decrease in the apparent rate constant, as in the case of the His^{255} mutants, and an increased K_M value, a parameter which is unchanged in the His mutants. Therefore, His^{255} is not directly involved in the substrate binding but could control the rate of the catalysis either by positioning the Asp^{98} residue through the network of hydrogen bonds and/or by acting as a proton donor.

A possible reaction mechanism is summarized in Figure 3-4B, in agreement with structure analysis of native and NO_2^--soaked CuNIRs and kinetic analysis of mutants. In the first step, the H_2O molecule bound to type II Cu is displaced by the substrate and presumably is released as OH^-; upon NO_2^- binding, type II Cu reduction occurs via intramolecular electron transfer. This step was proposed to be irreversible [16]. NO_2^- protonation and electron transfer from type II Cu allow the N-O bond cleavage and release of a H_2O molecule. Both Asp^{98} and His^{255} act as acid–base catalysts giving the two H^+ required for the NO_2^--reductase activity.

As previously shown for cd_1NIR, knowledge of structural data on different forms of CuNIR has helped considerably to unravel the mechanism of reduction of NO_2^- to NO. Also in the CuNIR system, some details still are unclear: among these, the apparent paradox between the implicit N-coordinated binding of the productive nitrosyl intermediate during catalysis and the observed O-coordinated substrate binding. The recent crystal structure determination of type II Cu-nitrosyl complex [20] may shed light on this apparent paradox. The crystals obtained using reduced CuNIR in NO-saturated solution reveal a surprising "side-on" type II Cu-nitrosyl complex, suggesting a revised catalytic mechanism. NO_2^- binds as an O-coordinated species to the oxidized type II Cu site, displacing the water ligand, in the H^+-form due presumably to the Asp^{98} residue. The internal electron transfer from type I Cu site reduces the active site triggering a rearrangement of NO_2^- to release water and form a Cu^+-NO^+ intermediate (Figure 3-4B). The model leads to the hypothesis that the Cu^+-NO^+ intermediate could be "side-on" bound according with the crystallographic data on the Cu-nitrosyl derivative. Furthermore, the proposed intermediate may be stabilized by the negative charge of Asp^{98} residue. Finally, NO is displaced by water to form the resting state of the enzyme [20].

The resemblance of the CuNIR active site to that of carbonic anhydrase and other Zn-containing enzymes has been noted previously, suggesting that a common catalytic strategy might be operative in dehydration reactions driven by metalloenzymes. In both classes of enzymes, binding of the substrate occurs asymmetrically with one O-bound to the catalytic metal (Cu or Zn) and a H-bond formed to a H^+-abstracting group (carboxylate or hydroxide). Release of the product leaves a hydroxide or a water molecule bound to the metal and hydrogen bonded to the same ionizable group.

References

[1] W.G. Zumft, Cell biology and molecular basis of denitrification, Microbiol. Mol. Biol. Rev. 61 (1997) 533–616.

[2] F. Cutruzzolà, Bacterial nitric oxide synthesis, Biochim. Biophys. Acta 1411 (1999) 231–249.

[3] J.W.A. Allen, S.J. Ferguson, V. Fulop, Cytochrome cd_1 nitrite reductase, in: A. Messerschmidt, R. Huber, T. Poulos, K. Wieghardt (Eds.) Handbook of Metalloproteins, Wiley, New York, 2001, pp. 424–439.

[4] F. Cutruzzolà, S. Rinaldo, F. Centola, M. Brunori, NO production by *Pseudomonas aeruginosa* cd_1 nitrite reductase, IUBMB Life 55 (2003) 617–621.

[5] S. Suzuki, K. Kataoka, K. Yamaguchi, Metal coordination and mechanism of multicopper nitrite reductase, Acc. Chem. Res. 33 (2000) 728–735.

[6] E.T. Adman and M.E.P. Murphy, Copper nitrite reductase, in: A. Messerschmidt, R. Huber, T. Poulos, K. Wieghardt (Eds.) Handbook of Metalloproteins, Wiley, New York, 2001, pp. 1381–1389.

[7] I.V. Pearson, M.D. Page, R.J.M. van Spanning, S.J. Ferguson, A mutant of *Paracoccus denitrificans* with disrupted genes coding for cytochrome c_{550} and pseudoazurin establishes these two proteins as the in vivo electron donors to cytochrome cd_1 nitrite reductase, J. Biol. Chem. 185 (2003) 6308–6315.

[8] R.S. Zajicek and S.J. Ferguson, The enigma of *Paracoccus pantotrophus* cytochrome cd_1 activation, Biochem. Soc. Trans. 33 (2005) 147–148.

[9] R.S. Zajicek, M.R. Cheesman, E.H.J. Gordon, S.J. Ferguson, Y25S variant of *Paracoccus pantotrophus* cytochrome cd_1 provides insight into anion binding by d_1 heme and a rare example of a critical difference between solution and crystal structures, J. Biol. Chem. 280 (2005) 26073–26079.

[10] R. Silaghi-Dumitrescu, Linkage isomerism in nitrite reduction by cytochrome cd_1 nitrite reductase, Inorg. Chem. 43 (2004) 3715–3718.

[11] S.J. George, J.W. Allen, S.J. Ferguson, R.N. Thorneley, Time-resolved infrared spectroscopy reveals a stable. ferric heme-NO intermediate in the reaction of *Paracoccus pantotrophus* cytochrome cd_1 nitrite reductase with nitrite, J. Biol. Chem. 275 (2000) 33231–33237.

[12] S. Wherland, O. Farver, I. Pecht, Intramolecular electron transfer in nitrite reductases, Hemphyschem. 6 (2005) 805–812.
[13] F. Cutruzzolà, K. Brown, E.K. Wilson, A. Bellelli, M. Arese, M. Tegoni, C. Cambillau, M. Brunori, The nitrite reductase from *Pseudomonas aeruginosa*: essential role of two active-site histidines in the catalytic and structural properties, Proc. Natl. Acad. Sci. U S A 98 (2001) 2232–2237.
[14] T. Sjogren and J. Hajdu, The Structure of an alternative form of *Paracoccus pantotrophus* cytochrome cd_1 nitrite reductase, J. Biol. Chem. 276 (2001) 29450–29455.
[15] K. Yamaguchi, K. Kataoka, M. Kobayashi, K. Itoh, A. Fukui, S. Suzuki, Characterization of two type 1 Cu sites of *Hyphomicrobium denitrificans* nitrite reductase: a new class of copper-containing nitrite reductases, Biochemistry 43 (2004) 14180–14188.
[16] S.V. Antonyuk, R.W. Strange, G. Sawers, R.R. Eady, S.S. Hasnain, Atomic resolution structures of resting-state, substrate- and product-complexed Cu-nitrite reductase provide insight into catalytic mechanism, Proc. Natl. Acad. Sci. U S A, 102 (2005) 12041–12046.
[17] F.E. Dodd, J. Van Beeumen, R.R. Eady, S.S. Hasnain, X-ray structure of a blue-copper nitrite reductase in two crystal forms. The nature of the copper sites, mode of substrate binding and recognition by redox partner, J. Mol. Biol. 282 (1998) 369–382.
[18] F. Jacobson, H. Guo, K. Olesen, M. Okvist, R. Neutze, L. Sjolin, Structures of the oxidized and reduced forms of nitrite reductase from *Rhodobacter sphaeroides* 2.4.3 at high pH: changes in the interactions of the type 2 copper, Acta Crystallogr. D. Biol. Crystallogr. 61 (2005) 1190–1198.
[19] A. Impagliazzo, L. Krippahl, M. Ubbink, Pseudoazurin-nitrite reductase interactions, Chembiochem. 6 (2005) 1648–1653.
[20] E.I. Tocheva, F.I. Rosell, A.G. Mauk, M.E. Murphy, Side-on copper-nitrosyl coordination by nitrite reductase, Science 304 (2004) 867–870.

Chapter 4

Nitric Oxide Reductase: Structural Variations and Catalytic Mechanism

Simon de Vries, Suharti and Laurice A.M. Pouvreau

4.1 Introduction

Genetic and biochemical analyses have shown that nitric oxide reductase (NOR) and cytochrome oxidase (CcO) have evolved from a common ancestor, assumed to be an anaerobic, NO-reducing enzyme [1–3]. Its archaic function might have been mainly to rid cells of the toxic NO – formed by lightning – rather than catalyzing part of the complete denitrification from nitrate to dinitrogen, which serves bioenergetic purposes.

The denitrifying NO reductases and the respiratory heme–copper CcOs are members of the superfamily of heme–copper oxidases, which all are complex integral membrane metallo-enzymes. CcOs are present in bacteria, archaea and eukaryotes [2]; more specifically, in the mitochondria of eukaryotes CcOs occur as cytochrome aa_3 oxidase. CcOs reduce O_2 to H_2O in a highly complex reaction in which protons are pumped across the mitochondrial membrane. The pumped protons are used by the F_oF_1-ATP synthase to produce ATP in a bioenergetic process called 'oxidative phosphorylation' [4].

Biology of the Nitrogen Cycle
Edited by H. Bothe, S. J. Ferguson and W.E. Newton

At present no crystal structures are available for NORs, but we have quite detailed structural and mechanistic information on NORs from biochemical and biophysical studies (kinetics, electron paramagnetic resonance (EPR), resonance Raman (RR), fourier transform infrared (FTIR), and magnetic circular dichroism (MCD), from sequence comparison between NORs and CcOs and from the crystal structures of CcOs [5–7 and refs. therein]. The close evolutionary and functional relation between NORs and CcOs is illustrated by the occurrence in both branches of cytochrome *c*- and quinol-linked activities and by the observation that NORs can reduce O_2 and CcOs NO. NO reductases (specifically cNOR) are structurally and functionally most closely related to the cbb_3-type oxidases. Presently three different bacterial NORs have been characterized, cNOR, qNOR and qCu_ANOR [3, 8, 9].

The best studied NORs are the cytochrome *bc*-complexes (cNOR) from Gram-negative bacteria e.g. *Pseudomonas stutzeri*, *Paracoccus denitrificans* and *Paracoccus halodenitrificans* [8–12]. In general, cNORs use membrane or soluble *c*-type cytochromes or small soluble blue copper proteins (azurin, pseudoazurin) as physiological electron donors (Figure 4-1). The qNOR, uses ubihydroquinone (QH_2) or menahydroquinone (MQH_2) as electron donor and is found in denitrifying archaea and soil bacteria, but also in pathogenic microorganisms such as *Neisseria meningitides*, *Neisseria gonorrhoea* and *Corynebacterium diphteriae*, which do not denitrify (Figure 4-1) [1]. In the latter organisms and in, for example, the cyanobacterium *Synechocystis* the qNOR has a detoxifying role, in particular in pathogenic species likely where qNOR attacks NO produced as a defense by the infected host. A third type of NOR has so far been found only in the Gram-positive bacterium *Bacillus azotoformans*. This NOR – qCu_ANOR – is bifunctional using both MQH_2 and a specific *c*-cytochrome (c_{551}) as electron donor (Figure 4-1) [13]. It was suggested that the MQH_2-linked activity of qCu_ANOR serves detoxification (cf. qNOR) and the c_{551} pathway involving the cytochrome b_6f complex has a bioenergetic function (cf. cNOR) [14].

4.2 Structural variations in NORs

4.2.1 cNOR (bc-heme containing nitric oxide oxidereductase)

Purified cNORs consist of two subunits, a heme *c*-containing subunit, NorC (~17 kDa) and a large, heme *b*-containing subunit, NorB (~53 kDa) [8–12] (Figure 4-1). NorB is an extremely hydrophobic subunit with 12 putative transmembrane α-helices and homologous to the large subunit (SUI) of CcOs. The *nor* operons of various bacteria suggests that cNOR might in vivo consist of four subunits, NorE, homologuous to SUIII of CcOs and NorF.

The primary sequences of the large subunit of NORs and CcOs reveal the presence of six conserved histidine residues. In CcOs these histidines

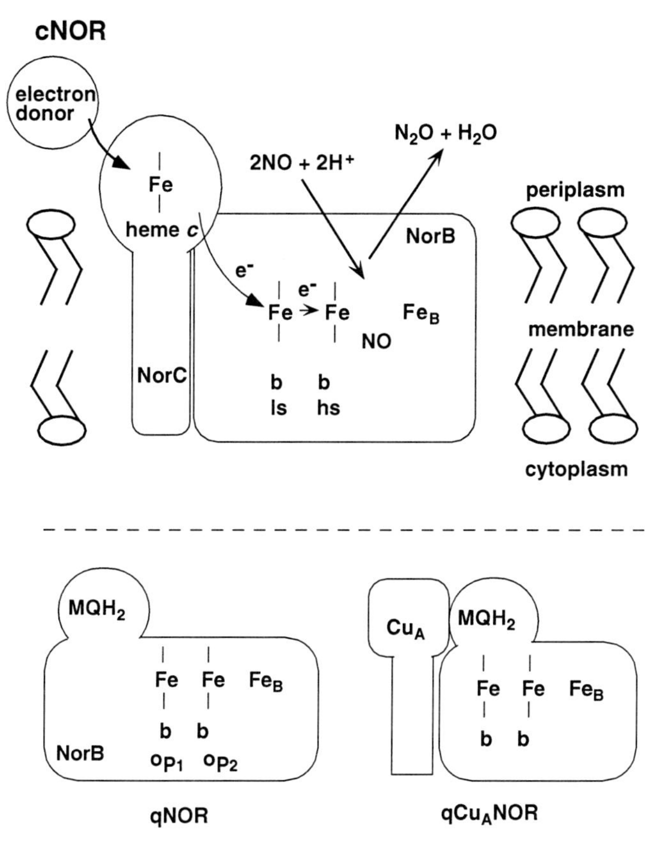

Figure 4-1 Schematic structure of the subunits and metallo-centers in NORs. The upper panel details the active site and the orientation and sidedness of electron transfer pathways, substrates including protons, and products in cNOR. See text for explanation of the different reaction steps. cNOR consists of two subunits (NorC and NorB), like the qCuANOR; qNOR consists of one subunit (NorB). Electron donors are membrane or soluble c-cytochromes, (pseudo)azurin (cNOR), (M)QH_2 (qNOR) or both MQH_2 and membrane-bound cytochrome c_{551} for qCu_ANOR. In qCu_ANOR the c_{551} and MQH_2 show non-competitive inhibition, suggesting different binding sites [13, 14, 19]; in fact c_{551} donates directly to the Cu_A center. Ls and hs indicate low spin and high spin and refer to the spin state of *heme b* and *heme* b_3, respectively. O_{p1} and O_{p2} are the modified hemes of the *P. aerophilum* qNOR. e^- denotes electron and arrows electron transfer pathways.

ligate the redox groups heme *a*, heme a_3 and Cu_B [1–8]. Heme *a* transfers electrons to the site where O_2 reduction takes place. This site is a dinuclear center consisting of heme a_3 and a copper center called Cu_B. In NORs the six conserved histidine residues coordinate heme *b* (*bis*-His) and heme b_3 (*mono*-His), and the non-heme iron Fe_B (*tris*-His) (Figure 4-1). The non-heme iron center Fe_B is structurally and functionally similar to Cu_B of CcO [9]. In fact, Fe_B and heme b_3 form a dinuclear center where NO binding and reduction to N_2O takes place [5–13].

The NorC subunit contains only a single, putative transmembrane α-helix. The low-spin *c*-heme is located in the periplasmic portion of the subunit. The dinuclear center in NOR is presumed to be located within the membrane at a similar position as the dinuclear heme a_3/Cu_B center in CcO. Electrons from, e.g., soluble or membrane bound *c*-cytochromes are first transferred to the heme *c* group of NorC and then via a low-spin heme *b* in NorB to the dinuclear Fe_B/heme b_3 active site where NO is reduced to N_2O (Figure 4-1). The overall reaction catalyzed by NOR does not lead to the generation of a proton electrochemical gradient – *in contrast to CcOs* [4] – because electrons and protons enter the catalytic center of the enzyme from the same, periplasmic side according to the overall reaction [15]:

$$2NO + 2cyt.c^{2+}_{peri} + 2H^{+}_{peri} \rightarrow N_2O + H_2O + 2cyt.c^{3+}_{peri}.$$

Electron paramagnetic resonance (EPR), magnetic circular dichroïsm (MCD), resonance (RR) and UV-Vis spectroscopy methods have been used to demonstrate that one heme *b* is low-spin (LS) while the other, heme b_3, is a penta-coordinated high-spin heme. In its reduced state heme b_3 is penta-coordinated, high-spin and *mono*-His ligated [10, 16]. However, upon binding of CO or NO it becomes low-spin and hexa-coordinated [16]. The high-spin heme b_3 and Fe_B are EPR silent in the oxidized enzyme and form an antiferromagnetically coupled oxo-bridged heme/non-heme di-iron center (Figure 4-2) [10]. RR has indicated Fe–Fe distance of 3.5 Å and an (Fe–O–Fe) angle of ~145° [16].

4.2.2 qNOR (b-heme containing nitric oxide oxidereductase)

qNORs have been purified from *Ralstonia eutropha* and the hyperthermophilic archaeon *Pyrobaculum aerophilum* [17, 18]. qNOR consists of a single subunit (~80 kDa). While the enzyme lacks heme *c*, it contains heme *b* and non-heme iron in a 2:1 ratio, arranged in a similar dinuclear center as in cNORs. qNORs are inactive with cytochrome *c* as electron donor, but can use QH_2 or MQH_2, thus classifying them as a qNOR. The most intriguing feature of qNOR is its N-terminal extension of about 280 amino acid residues that might fold into two transmembrane α-helices. These are separated by a large relatively hydrophilic domain, which may

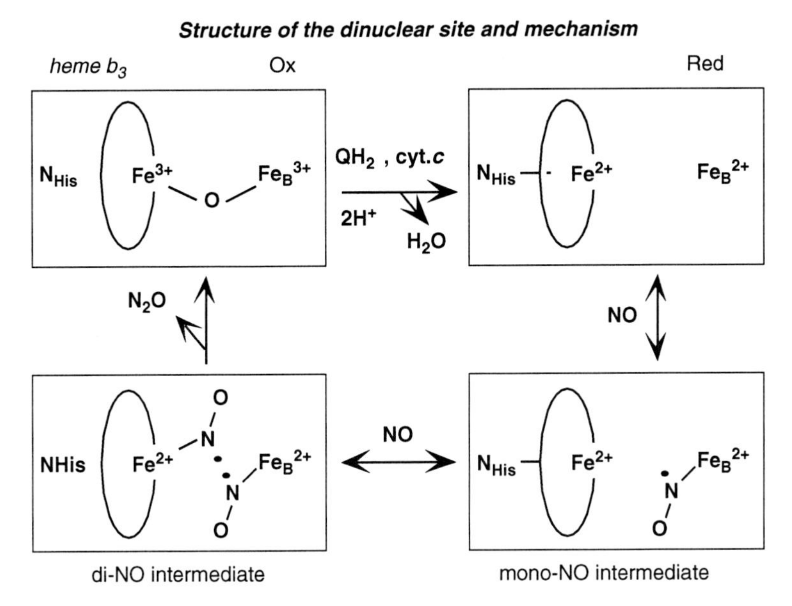

Figure 4-2 Structure of the active site of NOR and catalytic mechanism. See text for explanation of the different reaction steps and possible alternative mechanisms. Only NorB containing the dinuclear site is shown. N_{His} indicates the N-atom of a histidine residue, which is ligated to the heme iron in 'Red' but not in Ox; Fe_B is ligated by three histidine residues and possibly a glutamate (not shown in the figure). The heme Fe–Fe_B distance is 3.5 Å. Reduction is by a *c*-cytochrome (cNOR), by a (M)QH_2 (qNOR) or by either electron donor (qCu_ANOR) in a non-competitive manner.

contain the QH_2- or MQH_2-binding site [1, 17]. This domain has no sequence similarity to the quinol-oxidizing subunit of the terminal quinol oxidases, but a thorough analysis indicates sequence similarity to NorC of cNORs. Thus qNOR is apparently a fusion of NorC and NorB. During evolution this fused peptide could have lost the heme *c* center and obtained a quinol-binding site [1]. Alternatively, the opposite evolutionary scenario is possible, where qNOR could precede cNOR and the quinol-binding site evolved to a heme *c* anchoring domain.

The NO reductase from *P. aerophilum* is similar to that of *R. eutropha* but retains MQH_2: NO oxidoreductase activity at temperatures as high as 95 °C and can be boiled for 90 minutes losing only 50% activity [18]. The thermal stability can be understood by its relatively high content of branched amino acids and by the presence of modified hemes *b* (one of each O_{p1} and O_{p2}) containing a peripheral hydrophobic tail, ethenylgeranylgeranyl and hydroxyethylgeranylgeranyl derivatives of protoporphyrin IX.

The archaeal qNOR is the first example of a NOR, which contains modified hemes rather than heme *b*, reminiscent of bo_3 and aa_3 oxidases. Whether this modification indeed reflects an adaptation to life at high temperatures or yet serves another purpose remains to be established [18].

4.2.3 *qCu$_A$NOR (a bifunctional copper and b-heme containing nitric oxide oxidereductase)*

The first purification of a NO reductase from a Gram-positive bacterium *Bacillus azotoformans* was described recently [13, 14, 19]. The purified enzyme consists of two subunits and contains one Fe_B, two copper atoms and two *b*-type hemes per enzyme complex (Figure 4-1). Heme *c* was absent. One heme *b* is high-spin and forms a dinuclear center with Fe_B as in the other NORs. The other heme is a low-spin heme *b*, in which the two axial histidine imidazole planes are positioned at an angle of 60–70° in stead of the usual 'parallel over meso' position in all other NORs and CcOs [13]. Further, the enzyme contains the electron transfer center copper A thought to be unique to CcOs and nitrous oxide reductases. Like the qNOR, the qCu$_A$NOR enzyme uses MQH_2 as electron donor, but it also reacts with a membrane-bound specific cytochrome c_{551} [14, 19]. The observation that qCu$_A$NOR can react with two substrates each of which is bound to the membrane, is understandable in view of the absence of a periplasmic space in Gram-positive bacteria. Since a periplasmic space is also absent in archaea, denitrification is also membrane bound utilizing MQH_2 [3, 18]. Note that the soluble electron donors are located in the periplasm of Gram-negative bacteria. The qCu$_A$NOR is bifunctional regarding its electron donating substrates, and so far unique within the superfamily. The specific orientation of the LS heme *b* might be related to its bifunctionality [13].

4.3 Catalytic mechanism of NOR

The estimated midpoint potentials (E_m) at pH 7 of the metallo-centers are 310 and 345 mV for the heme *c* and LS heme *b*, and 114 and 320 mV for the heme b_3 and Fe_B, respectively [20, 21]. The relatively low E_m value of heme b_3 might prevent its initial reduction [21] and the formation of a very stable Fe(II)–NO complex, which might lead subsequently to complete inhibition [11, 22]. RR showed that in the fully reduced enzyme heme b_3 is penta-coordinated to a neutral histidine while this histidine ligand is lost in the oxidized state. Indeed, the ferric heme remains five-coordinated and its sole axial ligand is an oxo group bridging the two irons [16]. From the description of the oxidized and reduced state of the enzyme by various spectroscopic techniques, the relative binding affinities of heme b_3 and Fe_B for CO and pre-steady state kinetic analysis a (minimal) catalytic cycle is proposed [8, 10, 16, 23, 24] (Figure 4-2). Reduction of the di-iron site weakens the oxo-bridge and favors the formation of the heme Fe(II)–His bond. The enzyme can bind one NO molecule per ferrous iron, the first going to Fe_B. As observed in several hemoproteins, e.g., hemoglobin or soluble guanylyl cyclase, the heme Fe(II)–His bond is broken upon binding of NO to form a pentacoordinate heme nitrosyl complex. The close proximity of the two nitrosyl groups destabilizes these otherwise poorly reactive $\{FeNO\}^7$ species and promotes formation of the N–N bond. The product N_2O leaves the active site; in the second turnover or during steady state catalysis formation of the oxo-bridge (ox intermediate, Figure 4-2) may occur [8, 10, 16, 23] or apparently bypassed in case reduction to the 'Red' form of the enzyme is fast [24]. Alternative views on the mechanism are discussed in Refs. [11, 22, 25, 26].

The catalytic scheme of Figure 4-2 is consistent with steady-state kinetic analyses of NOR, which show a sigmoïdal relation between [NO] and activity and lack of inhibition by the product N_2O is [3, 10, 13, 18]. NORs do not obey simple Michaelis–Menten kinetics because the reaction is an obligatory two-substrate (i.e. 2NO) reaction. NORs also display the phenomenon of substrate (NO) inhibition at the non-physiological [NO]>10 μM [3, 10, 13, 18]. This inhibition occurs most likely due to binding of NO to the ferric heme b_3 [14, 19, 25] and is not indicated in Figure 4-2. A complete equation describing the steady-state kinetics has been derived in Ref. [10]. Maximal turnover rates of purified NOR amount to about 200 NO/s, a value approximately twofold lower than the steady-state rate in denitrifying cells. The use of non-physiological electron donors and the detergent extraction procedure are probable causes for the lower rates observed in assays of the purified enzymes. Consequently, the apparent K_m value for NO (or rather the [NO] at half-maximal activity) of 250–500 nM should be considered as an upper limit [10]. Since the free concentration of NO in, e.g. soil, is generally much lower ($<$10 nM) the true K_m for NO may indeed be closer to this latter value.

4.4 Comparison between catalysis by NOR and CcO

cNORs are like CcOs, capable of catalyzing the four-electron reduction of O_2 to water. The rate of this reaction (40 electrons per second per cNOR enzyme molecule) is relatively high, only about fivefold lower than that of NO reduction. However, the apparent K_m for O_2 is very high, ~20 μM yielding a specificity constant (k_{cat}/K_m) of about $2 \times 10^6\ M^{-1}s^{-1}$ in contrast to $> 5 \times 10^8\ M^{-1}s^{-1}$ for the NO reduction activity. Mitochondrial CcOs are capable of reduction of NO to N_2O but are usually (reversibly) inhibited after a single or few turnovers. The prokaryotic oxidases, which are more related to the NO reductases, can sustain catalysis, but the rate is only

0.5 NO/s at 50 μM of NO, i.e. giving a specificity constant of 10^4 $M^{-1}s^{-1}$ [27]. Thus, although NORs and CcOs each catalyze both NO and O_2 reduction, oxidases have evolved to efficiently catalyze O_2 reduction (250 O_2/s; K_m = 5–200 nM; specificity constant $>10^9$ $M^{-1}s^{-1}$), and NO reductases to efficiently catalyze the reduction of NO (specificity constant $>5 \times 10^8$ $M^{-1}s^{-1}$).

The primary sequence of NorB reveals several conserved glutamic acid residues which are not found in SUI of CcOs [1, 2, 8, 11]. The carboxylate groups of one or more of these glutamate residues may play a role in Fe_B binding, providing a possible fifth or sixth ligand. The vicinity of these charges is expected to lower the E_m value of Fe_B and heme b_3, e.g., with respect to heme a_3 and Cu_B in CcO. However, since the E_m values are in the same range as in CcO, which lacks analogous glutamate residues, they may play a role in proton binding [11] needed to form water as part of the overall reaction (Figure 4-2).

The heme Fe–Fe_B distance of approximately 3.5 Å in cNOR [16] is much larger that the 5 Å heme Fe–Cu_B in CcO [5, 6]. Instead of the diferric μ-oxo bridge observed in NOR, the dinuclear center of CcO contains hydroxy- and aqua-ligands between the heme a_3 and the Cu_B. In the ba_3-cytochrome *c* oxidase from *Thermus thermophilus* the electron density between heme a_3 and Cu_B is consistent with a single oxygen atom (oxo-, hydroxo-, or aqua) but the heme Fe–Cu_B distance of 4.4 Å is still larger than in NORs [6]. Despite the difference in metal–metal distance and metal composition, NOR and CcO are both capable of catalyzing NO and O_2 reduction to N_2O and H_2O, respectively, albeit with significantly different specificities (see above). There seems to be no a priori chemical reason why iron would be preferred over copper for NO reduction, or copper over iron for O_2 reduction. The major difference between NORs and CcOs is apparently in the conservation of redox energy by means of creating an electrochemical proton gradient [4]. While under 'natural conditions' – low [NO] – the •*G* for NO reduction as calculated from the Nernst equation, is similar to that of O_2 reduction (0.5–0.6 V), NORs do not conserve energy whatsoever, while CcOs have evolved to conserve (almost) all. This observation remains enigmatic from a biochemical and an evolutionary point of view.

References

[1] J. Hendriks, A. Oubrie, J. Castresana, A. Urbani, S. Gemeinhardt, M. Saraste, Nitric oxide reductases in bacteria, Biochim. Biophys. Acta 1459 (2000) 266–273.

[2] M.M. Pereira, M. Santana, M. Teixeira, A novel scenario for the evolution of haem–copper oxidases, Biochim. Biophys. Acta 1505 (2001) 185–208.

[3] S. de Vries, I. Schröder, Comparison between the nitric oxide reductase family and its aerobic relatives, the cytochrome oxidases, Biochem. Soc. Trans. 30 (2002) 662–667.
[4] D.G. Nicholls, S.J. Ferguson, Bioenergetics 3, Academic Press, London, 2001.
[5] C. Ostermeier, A. Harrenga, U. Ermler, H. Michel, Structure at 2.7 Å resolution of the *Paracoccus denitrificans* two-subunit cytochrome *c* oxidase complexed with an antibody FV fragment, Proc. Natl. Acad. Sci. USA 94 (1997) 10547–10553.
[6] T. Soulimane, G. Buse, G.P. Bourenkov, H.D. Bartunik, R. Huber, M. Than, Structure and mechanism of the aberrant *ba*(3)-cytochrome *c* oxidase from thermus thermophilus, EMBO. J. 19 (2000) 1766–1776.
[7] A.V. Cherepanov, S. de Vries, Microsecond freeze-hyperquenching (MHQ). Development of a new ultrafast micro-mixing and sampling technology and application to enzyme catalysis, Biochim. Biophys. Acta 1656 (2004) 1–31.
[8] I.M. Wasser, S. de Vries, P. Moenne-Loccoz, I. Schöder, K.D. Karlin, Nitric oxide in biological denitrification: Fe/Cu metalloenzyme and metal complex NO_x redox chemistry, Chem. Rev. 102 (2002) 1201–1234.
[9] W.G. Zumft, Nitric oxide reductases of prokaryotes with emphasis on the respiratory, heme–copper oxidase type, J. Inorg. Biochem. 99 (2005) 194–215.
[10] P. Girsch, S. de Vries, Purification and initial kinetic and spectroscopic characterization of NO reductase from *Paracoccus denitrificans*, Biochim. Biophys. Acta 1318 (1997) 202–216.
[11] G. Butland, S. Spiro, N.J. Watmough, D.J. Richardson, Two conserved glutamates in the bacterial nitric oxide reductase are essential for activity but not assembly of the enzyme, J. Bacteriol. 183 (2001) 189–199.
[12] J. Hendriks, A. Warne, U. Gohlke, T. Haltia, C. Ludovici, M. Lubben, M. Saraste, The active site of the bacterial nitric oxide reductase is a dinuclear iron center, Biochemistry 37 (1998) 13102–13109.
[13] Suharti, M.J.F. Strampraad, I. Schröder, S. de Vries, A novel copper A containing menaquinol NO reductase from *Bacillus azotoformans*, Biochemistry 40 (2001) 2632–2639.
[14] Suharti, S. de Vries, Membrane-bound denitrification in the gram-positive bacterium *Bacillus azotoformans*, Biochem. Soc. Trans. 33 (2005) 130–133.
[15] J.H. Hendriks, A. Jasaitis, M. Saraste, M.I. Verkhovsky, Proton and electron pathways in the bacterial nitric oxide reductase, Biochemistry 41 (2002) 2331–2340.
[16] P. Moënne-Loccoz, O.-M. H. Richter, H.-W. Huang, I.M. Wasser, R.A. Ghiladi, K.D. Karlin, S. de Vries, Nitric oxide reductase from *Paracoccus denitrificans* contains an oxo-bridged heme/non-heme diiron center, J. Am. Chem. Soc. 122 (2000) 9344–9345.

[17] R. Cramm, A. Pohlmann, B. Friedrich, Purification and characterization of the single-component nitric oxide reductase from *Ralstonia eutropha*, FEBS Lett. 460 (1999) 6–10.
[18] S. de Vries, M.J.F. Strampraad, S. Lu, P. Moënne-Loccoz, I. Schröder, Purification and characterization of the MQH_2: NO oxidoreductase (qNOR) from the hyperthermophilic archaeon *Pyrobaculum aerophilum*, J. Biol. Chem. 278 (2003) 35861–35868.
[19] Suharti, H.A. Heering, S. de Vries, NO reductase from *Bacillus azotoformans* is a bifunctional enzyme accepting electrons from menaquinol and a specific endogenous membrane-bound cytochrome c_{551}, Biochemistry 43 (2004) 13487–13495.
[20] K.L.C. Gronberg, M.D. Roldan, L. Prior, G. Butland, M.R. Cheesman, D.J. Richardson, S. Spiro, A.J. Thomson, N.J. Watmough, A low-redox potential heme in the dinuclear center of bacterial nitric oxide reductase: implications for the evolution of energy-conserving heme–copper oxidases, Biochemistry 38 (1999) 13780–13786.
[21] H.A. Heering, F.G. Wiertz, C. Dekker, S. de Vries, Direct immobilization of native yeast iso -1- cytochrome *c* on bare gold: fast electron relay to redox enzymes and zeptomole protein-film voltammetry, J. Am. Chem. Soc. 126 (2004) 11103–11112.
[22] C.S. Butler, H.E. Seward, C. Greenwood, A.J. Thomson, Fast cytochrome *bo* from *Escherichia coli* binds two molecules of nitric oxide at CuB, Biochemistry 36 (1997) 16259–16266.
[23] S. Lu, Suharti, S. de Vries, P. Moënne-Loccoz, Two CO molecules can bind concomitantly at the diiron site of NO reductase from *Bacillus azotoformans*, J. Am. Chem. Soc. 126 (2004) 15332–15333.
[24] H. Kumita, K. Matsuura, T. Hino, S. Takahashi, H. Hori, Y. Fukumori, I. Morishima, Y.O. Shiro, Reduction by nitric oxide reductase from denitrifying bacterium *Pseudomonas aeruginosa*: characterization of reaction intermediates that appear in the single-turnover cycle, J. Biol. Chem. 279 (2004) 55247–55254.
[25] E. Pinakoulaki, S. Gemeinhardt, M. Saraste, C. Varotsis, Nitric-oxide reductase. Structure and properties of the catalytic site from resonance Raman scattering, J. Biol. Chem. 277 (2002) 23407–23413.
[26] E. Pinakoulaki, T. Ohta, T. Soulimane, T. Kitagawa, C. Varotsis, Simultaneous resonance Raman detection of the heme a 3 –Fe–CO and CuB –CO species in CO–bound ba3 –cytochrome *c* oxidase from *Thermus thermophilus*: evidence for a charge transfer Cu B –CO transition, J. Biol. Chem. 279 (2004) 22791–22794.
[27] A. Giuffre, G. Stubauer, P. Sarti, M. Brunori, W.G. Zumft, G. Buse, T. Soulimane, The heme–copper oxidases of *Thermus thermophilus* catalyze the reduction of nitric oxide: evolutionary implications, Proc. Natl. Acad. Sci. USA 96 (1999) 14718–14723.

Chapter 5

Nitrous Oxide Reductases

Walter G. Zumft and Heinz Körner

5.1 Introduction

Nitrous oxide (N_2O) is an inorganic metabolite of the prokaryotic cell during denitrification. Thus, denitrifiers comprise the principal group of N_2O producers, with roles played also by nitrifiers, methanotrophic bacteria, and fungi. Among them, only denitrifying prokaryotes have the ability to convert N_2O into N_2. Conversion of N_2O into N_2 is the last step of a complete nitrate denitrification process and is an autonomous form of respiration. N_2O is generated in the denitrifying cell by the activity of respiratory NO reductase [1]. N_2O respiration can sustain the bioenergetic needs of a denitrifying bacterium in the absence of O_2. The reaction is strongly exergonic, $N_2O + 2H^+ + 2e^- \rightarrow N_2 + H_2O$ [E_o' (pH 7.0) = +1.35 V; $\Delta G_o' = -339.5$ kJ·mol^{-1}], and provides an electron sink for an anaerobic electron transport chain. Charge separation and energy conservation are not catalyzed by the reductase but depend on coupling sites of the electron transport pathway to N_2O. An involvement of the cytochrome bc_1 complex in N_2O reduction has been demonstrated for various denitrifying bacteria. N_2O reductase (N_2OR) is not thought to interact directly with a

Biology of the Nitrogen Cycle
Edited by H. Bothe, S.J. Ferguson and W.E. Newton

respiratory complex, but with mobile carriers similar to what has been shown for respiratory nitrite reductase (see Chapter 3). In this chapter we describe the multicopper enzyme N_2OR that has been intensively studied in *Pseudomonas stutzeri* and *Paracoccus* species. Microbial inorganic N_2O metabolism has been covered in a recent review where further details on all aspects can be found [2].

5.2 Properties of N_2O reductase

N_2O respiration is found in numerous taxonomic groups and a wide variety of habitats. The range of organisms spans extremophilic archaea, phototrophic and chemoautotrophic bacteria, pathogens, marine organisms, piezophiles, sulfur- and halorespirers, thermo- and psychrophilic bacteria, as well as degrading organisms of contaminated soils and sediments. Genome sequences have uncovered *nos* (for nitrous oxide respiration) genes in several archaea and bacteria hitherto not known to metabolize N_2O [2]. Although bacteria can grow by N_2O respiration, early attempts to isolate N_2OR failed. An indirect strategy was successful in the discovery of N_2O reductase from the ZoBell strain of *P. stutzeri* by exploring the nutritional requirements of the N_2O-grown cells with respect to a trace metal. The observation that Cu specifically stimulates anaerobic growth under N_2O led to the finding of N_2OR among Cu-containing proteins. Depending on the bacterial host and the purification procedure, distinct forms of N_2OR are obtained, which differ with regard to their activity and spectroscopic properties. Nevertheless, all N_2ORs isolated are homodimers that contain two multinuclear Cu centers per monomer. We refer to this N_2OR as the Z-type paradigm because of its Cu_Z site and head-to-tail arrangement of its two subunits, and also to differentiate it from the *Wolinella*-type enzyme and a postulated new form in *Archaea*. N_2O reduction is inhibited by acetylene, the inhibition is non-competitive with a $K_i \leq 45$ μM and the mechanism of action is unknown.

A modified catalytic site, Cu_Z^*, is formed when N_2OR reacts with O_2 in cell extract. Cu_Z^* gives an electron paramagnetic resonance signal but does not show redox activity, neither with dithionite nor with ferricyanide. Its potential involvement in catalysis is unsolved, but Cu_Z^* is detectable in the cellular context of an N_2O-respiration negative, double mutant of the paralogous genes *nosX* and *nirX* of *Paracoccus denitrificans*.

5.3 Enzyme structure and Cu centers

The crystal structure of N_2OR has been solved (Figure 5-1). From X-ray scattering of N_2OR from *Alcaligenes xylosoxidans*, a three-dimensional

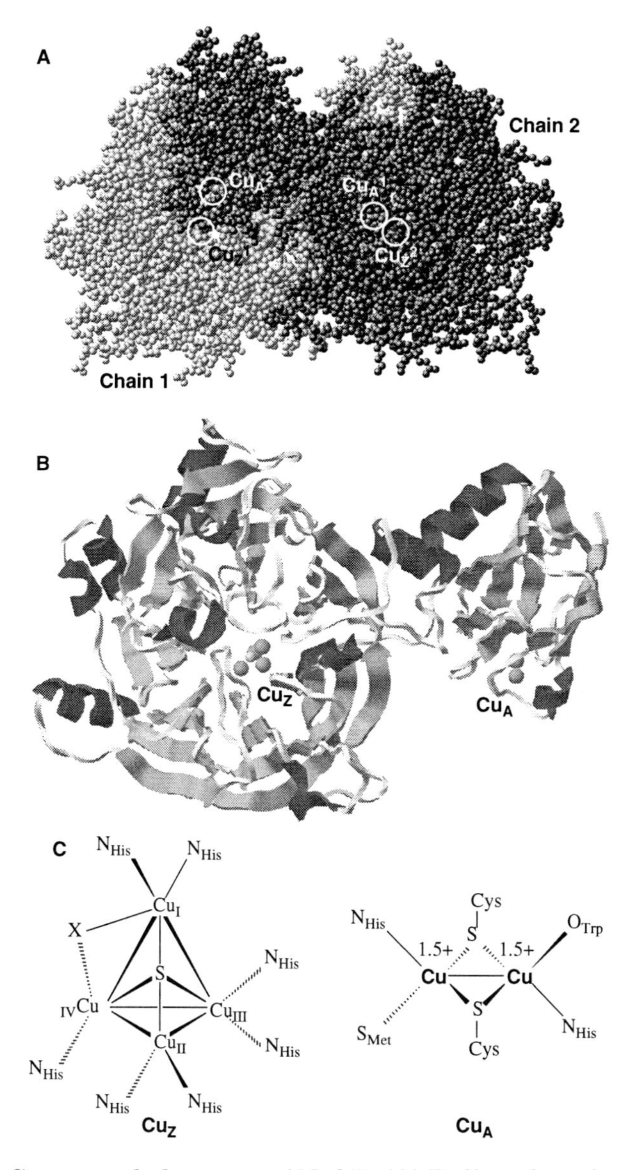

Figure 5-1 Structural elements of N_2OR. (A) Ball-and-stick representation of the N_2OR dimer; the positions of Cu_A and Cu_Z in the depth of the molecule are circled. Note the approximation of these centers from opposite subunits. Data for *Marinobacter hydrocarbonoclasticus* (formerly *Pseudomonas nautica*), Protein Data Bank file 1QNI. (B) Ribbon drawing by RasTop of the N_2OR monomer, showing the seven-bladed beta-propeller fold ligating Cu_Z and the cupredoxin fold with Cu_A. Orientation is with the subunit interface toward the viewer. (C) Structures and ligands of Cu_Z and Cu_A.

model at about 20 Å resolution was deduced [3]. It has the two subunits of N_2OR arranged in a twofold symmetry unit around a central axis, suggesting head-to-tail positioning, which was confirmed in the X-ray structure. Prior to obtaining the crystal structure site-directed mutagenesis was crucial to identify ligands of the Cu_A center [4].

Each N_2OR subunit is composed of two domains. The N-terminal domain hosts the catalytic tetranuclear μ_4-sulfide bridged Cu_Z site. It makes up the bulk of the protein and adopts a highly symmetrical seven-bladed beta-propeller. The C-terminal domain carries the Cu_A electron transfer center enveloped in a cupredoxin fold. It is a mixed-valent $[Cu_A(1.5+) \ldots Cu_A(1.5+)]$ electron transfer site with two bridging cysteine ligands and forms a highly spin-delocalized Cu_2S_2 rhomb (Figure 5-1). The Cu_A-binding consensus in the primary structure is shared between N_2OR and subunit II of cytochrome *c* oxidase. Unraveling the binuclearity of the Cu_A center in N_2OR was central for recognizing the same structure in cytochrome *c* oxidase [5]. The NO reductase of *Bacillus azotoformans* is currently the third example of a Cu_A-containing enzyme (see Chapter 4).

The Cu_A domain is important for the structural integrity of N_2OR. Deletion of this domain renders the enzyme unstable and subjects it to rapid degradation. The mixed-valence state of Cu_A is energetically very favorable. The oxidation to the Cu(II)–Cu(II) form has not been achieved for any Cu_A center to date. Remarkably, the Cu_A center is formed from apoprotein and Cu(II). Electron transfer mediated by Cu_A is very efficient. The reason for utilizing a binuclear electron transfer center in cytochrome *c* oxidase and N_2OR, instead of mononuclear type-1 Cu, is thought to be the unidirectional electron transfer through the site or the lower energy of reorganization. A distinct property of the Cu_A center is its function in electron transfer pathways crossing a subunit interface.

Elemental analysis and resonance Raman spectroscopy of isotopically labeled N_2OR conclusively demonstrated the presence of an acid-labile sulfur ligand [6]. The 1.6 Å-resolution structure of N_2OR from *P. denitrificans* confirmed the existence of one sulfide at the Cu_Z site. Together, this marks the discovery of a new type of biological metal center in respect to its Cu–S nature and the arrangement of the four Cu atoms in a 'butterfly' cluster, coordinated solely by histidine residues [7].

The beta-propeller domain is engaged in the formation of the active site and in protein–protein interactions that are essential for dimerization of the N_2OR monomer. The homodimer is stabilized by polar and non-polar, as well as chelating interactions. The shortest Cu_A–Cu_Z distance of $\approx$40 Å within the monomer is too large for an efficient electron transfer between the two sites. However, the corresponding distance of 10 Å across the subunit interface is well-suited for fast electron transfer (Figure 5-1). The dimeric architecture is essential for catalysis and the residues at the dimer interface tend to be highly conserved.

Of the four copper ions at the Cu_Z site (Figure 5-1), Cu_I appears to be the predominantly oxidized Cu atom, which is consistent with Cu_I having a four-coordinate structure while the other three Cu atoms show lower coordination numbers. The high-resolution structure of the enzyme from *P. denitrificans* shows a weakly bound ligand, *X*, most likely water-derived [7], whose nature has not been identified. N_2OR from different sources is activated in vitro by reduction with reduced viologens. This indicates that the substrate-reducing form of Cu_Z is the fully reduced $[Cu_4S]^{2+}$, $S = 0$ state. Incubation of the reductively activated N_2OR from *Achromobacter cycloclastes* with N_2O-saturated buffer leads to a partially oxidized enzyme as shown by optical and electron paramagnetic resonance (EPR) spectra, and converts N_2O into product [8]. Computational analysis suggests that N_2O binds in the lowest energy structure of the $Cu_Z(4Cu_I)$–N_2O complex at the Cu_I–Cu_{IV} edge in a bent, μ-1,3 bridging mode with the terminal N atom coordinating to Cu_I.

5.4 Metal center assembly

N_2OR is subjected to a catalyzed process of metal cluster assembly. On characterizing the *nos* gene locus it was found that the structural gene, *nosZ*, is a part of several genes whose mutagenesis affects the Cu content and properties of Cu sites in the enzyme. An essential requirement for homologous or heterologous expression of holoN_2OR is that the assembly system complements *nosZ*. Several lines of evidence support a periplasmic maturation process for N_2OR. (i) Export of NosZ proceeds undisturbed in accessory gene mutants. (ii) Cu deficiency does not affect the location of N_2OR, but lowers enzyme activity. The effect can be remedied by exogenous Cu. (iii) N_2OR assembly involves the periplasmic components NosL and NosD. (iv) Mutagenesis of the signal peptide or the Tat protein transporter retains NosZ in the cytoplasm. The topologically aberrant cytoplasmic enzyme is devoid of Cu, but a correctly folded NosZ is deduced from the possibility to incorporate Cu in vitro into the Cu_A site. (v) NosZ of *Wolinella succinogenes* is thought to be transported by the Sec system, which acts on unfolded proteins and requires subsequent periplasmic assembly and insertion of the Cu cofactors. The principal events in NosZ processing are shown in Figure 5-2.

The *nosDFY* genes encode an ABC transporter exhibiting clear signatures in the *nosF* product of an ABC-type ATPase. NosF has been purified and shown to have ATPase activity. Cytoplasmic NosF is complemented by the integral membrane protein, NosY, with five predicted transmembrane helices and the periplasmic component NosD. NosD is a member of the CASH protein family.

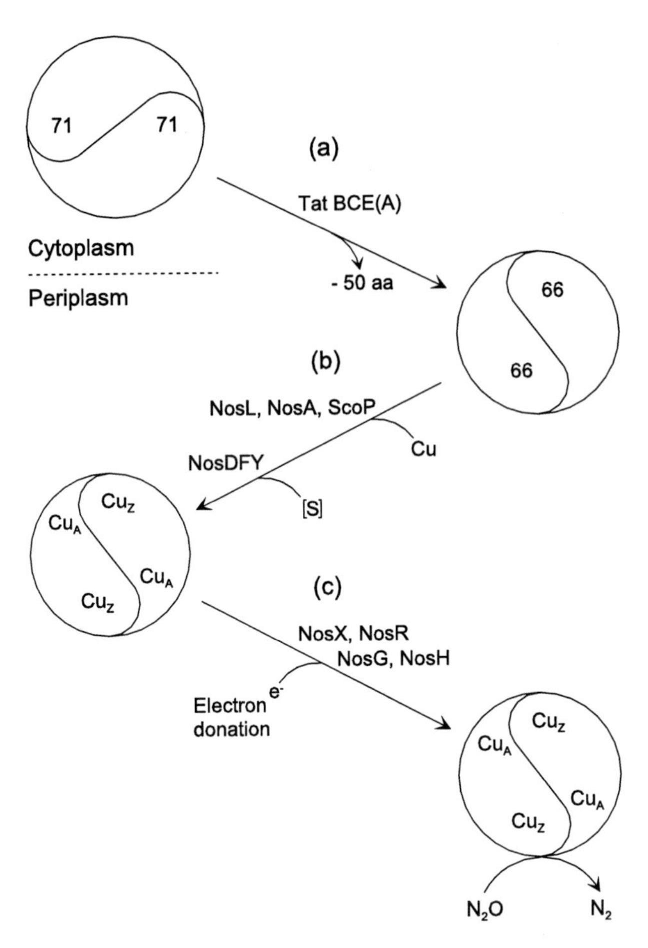

Figure 5-2 Maturation sequence of N_2OR and role of accessory components. (a) Transport of apoenzyme (142 kDa) to the periplasm by the Tat translocon; numbers given are for *P. stutzeri*. (b) Synthesis of Cu centers to assemble holoN_2OR. (c) Electron donation and conversion to or maintenance of catalytically active N_2OR. The indicated Nos proteins are not found in every N_2O-reducing bacterium. For explanation of gene products and alternative roles in distinct bacterial backgrounds see the text.

The biogenesis of Cu_Z requires sources for Cu and sulfide. Likely are independent functions for S^- and Cu provision. The most feasible concept is that NosDFY is a sulfur (sulfide?) transporter. If a sulfur-donor molecule is being transported, the question of how S^- is generated arises. Little is known about Cu uptake and cellular routing in N_2OR-harboring bacteria. NosA, NosL, and Sco1 homologs are presumed to process Cu [9]. The role of NosA may be that of an outer membrane pore for a Cu uptake system. The *nosL* product is predicted as lipoprotein of the outer membrane. The role of NosL may be to destine Cu in the periplasm for NosZ. The protein from *A. cycloclastes* has been purified and shown to bind cuprous and cupric ions with differential affinity to qualify as a Cu chaperone. The biogenesis of Cu_A in cytochrome *c* oxidase of yeast depends on Sco1. A catalyzed insertion of Cu into Cu_A is in line with the chaperone concept and routing of Cu along discrete pathways. However, none of the putative Cu processing factors generates a recognizable phenotype in *P. stutzeri* mutants indicating, thus, that the Cu supply route may draw from alternatives or that certain steps of metallation can occur spontaneously.

5.5 Gene patterns and dissemination of *nos* genes

The entire denitrification system is based on somewhat over 50 identified and classified genes, of which up to 10 are specific for N_2O utilization [2, 10]. The cluster *nosDFYL* is found in every N_2O-reducing prokaryote, and its presence is taken as evidence for a Cu−S center. The genes *nosR*, *nosX*, *nosC*, *nosG*, and *nosH* are not ubiquitous but are distributed according to taxonomic patterns (Figure 5-3).

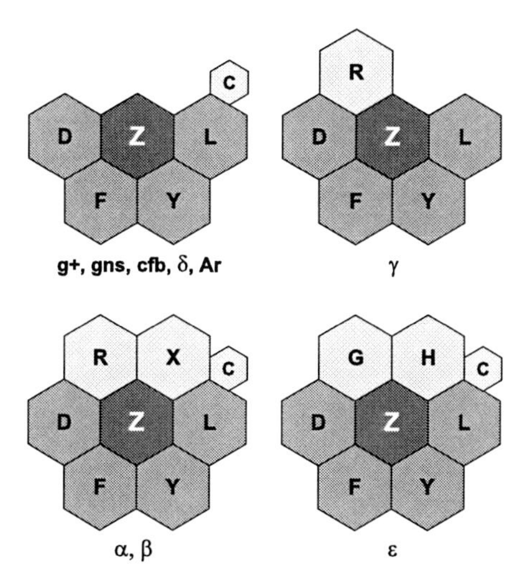

Figure 5-3 Cluster patterns of *nos* genes among distinct taxons. Capital letters represent the corresponding *nos* gene product. NosC is a putative cytochrome of irregular occurrence. Greek key letters denote the respective groups of proteobacteria; Ar, *Archaea*; g+, Gram-positive; cfb, cytophaga-flavobacter-bacteroides group; gns, green non-sulfur bacteria.

Analysis of *W. succinogenes* resulted in the identification of two homologs of *nap* genes, *nosH* and *nosG*, positioned within the *nos* gene cluster (NGC). NosG shows similarity to the C-terminal part of NosR. NosG and NosH are thought to be involved in electron donation to N_2OR [11]. Spectroscopic evidence and structural prediction indicates that the N_2OR from *W. succinogenes* exhibits the Cu_Z center and folding pattern of the Z-type enzyme. However, this N_2OR is modified by a C-terminal extension of about 150 amino acids, containing a *c*-type cytochrome domain. The cytochrome maturation system is presumed to be required for heme insertion in the C-terminal domain. The signal peptide of this enzyme is Sec specific. Overall, these modifications justify considering the N_2OR of *Wolinella* a variant (Table 5-1).

Archaea and *Bacteria* exhibit the same denitrification pathway with the same metalloenzymes, which suggests that the process existed in a common ancestor before separation of these domains. Gas formation from nitrate or nitrite is known for several halophilic archaea [12]. The 3.13-Mb chromosome I of *Haloarcula marismortui*, the larger one of two chromosomes, carries the *nos* genes, with *nosZ* being part of an NGC that includes homologs of *nosDFY* also. The deduced *nosZ* product exhibits the Cu_Z domain of N_2OR, with the set of conserved histidine residues, and the Cu_A domain carrying the canonical Cu ligands. Like its bacterial counterpart, this archaeal N_2OR is inhibited by C_2H_2.

In addition to the existence of the Z-type N_2OR in archaea, two independent observations suggest still another type of enzyme. The genome of

Table 5-1 Structural types of N_2O reductases.

Designation	Organisms	Catalytic site	Domain order	Accessory proteins
Z	*Paracoccus denitrificans* *Pseudomonas stutzeri*	Cu–S[a]	Tat signal, beta-propeller, cupredoxin	NosDFYL, NosR (X)
H	*Wolinella succinogenes* *Magnetospirillum magnetotacticum*	Cu–S[b]	Sec signal, beta-propeller, cupredoxin, C-domain	NosDFYL, NosGH (C)
A	*Haloferax volcanii*[c] *Pyrobaculum aerophilum*	Cu–S[d]	Unknown	NosDFYL

[a] From spectroscopic and analytic evidence or crystal structure.
[b] From spectroscopic evidence.
[c] Unpublished sequence information from COMB, University of Maryland.
[d] Predicted from extant accessory proteins.

Haloferax volcanii indicates that this archaeon is a denitrifier, as it reveals genes for the Cu-type nitrite reductase, NirK, the NorZ-type NO reductase, and genes *nosDFYL*. Although the latter are clearly indicative for a Cu–S center, no corresponding *nosZ* gene is evident. *Pyrobaculum aerophilum* represents a similar case. It produces N_2 from nitrite together with traces of N_2O and NO [13]. Although the physiological evidence indicates existence of an N_2OR, and the genome reveals the accessory genes *nosDFL* and *sco1*, the structural information for NosZ is not detectable in the genome. We suggest therefore a further type of N_2OR for archaea (Table 5-1). Because the accessory genes for the Z-type enzyme are present, a Cu–S center is assumed also for this variant.

The organization of the *nos* genes usually involves several transcriptional units. In *Pseudomonas aeruginosa*, the *nos* genes are arranged in a single hexacistronic *nosRZDFYL* operon. The *nos* genes are under the control of regulators that belong to the Crp–Fnr family. Although carrying different names, Dnr, DnrD, Nnr, NnrR, or $FixK_2$, they belong to the same type of regulator [14]. N_2O is only a weak inducer of *nosZ*. In contrast, NO as signal molecule strongly upregulates the *nosR*, *nosZ*, and *nosD* promoters. The NO signal is probably processed via the regulator but the mechanism of activation by NO is not resolved. Additional signals for gene activation may be required from the redox status or O_2 tension that are processed by an Fnr-type Fe–S protein of the Crp–Fnr family. For *Bradyrhizobium japonicum*, dependence of *nos* gene expression on the heme-based O_2 sensor FixLJ is suggested in cooperation with $FixK_2$.

5.6 Evolutionary aspects

Figure 5-4 shows the phylogenetic relationship among NosZ proteins across nine taxonomic lineages from organisms for which a complete NosZ sequence is known. The protein tree shows three well-populated clades consisting of α-, β-, and γ-proteobacteria. Other taxons are the archaeon *H. marismortui*, the Gram-positive bacterium *Desulfitobacterium hafniense*, *Salinibacter ruber* from the cytophaga-flavobacter-bacteroides superphylum, *Thermomicrobium roseum* from the green non-sulfur bacteria phylum, and the delta-proteobacterium *Anaeromyxobacter dehalogenans*. The epsilon-proteobacteria are represented by *W. succinogenes* and *Thiomicrospira denitrificans*.

The tree of NosZ proteins follows the phylogenetic pattern of the host taxa in a 16S rRNA tree remarkably well. The Nos trait seems to be an early evolutionary event that subsequently conformed to the phylogenetic trajectory of speciation. With the information about NGCs from genomes, it is also possible to reconstruct the phylogeny of the other Nos proteins in support of the above conclusion [2]. The argument of an autonomous mode

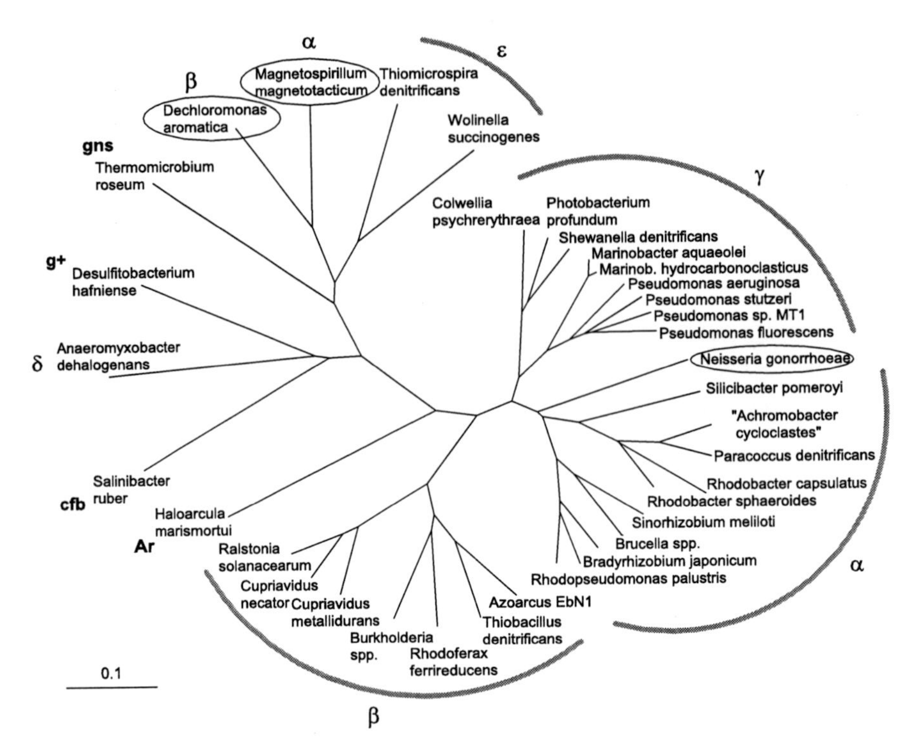

Figure 5-4 Phylogenetic relationship among NosZ proteins in an unrooted tree display, constructed on the basis of a ClustalX multiple alignment by means of the PHYLIP algorithms for protein distance matrix and neighbor-joining calculations. Tree branch lengths indicate phylogenetic distance in units of the scale shown. Abbreviations as in Figure 5-3. Encircled species denote cases of putative lateral gene transfer.

for N_2O respiration implies that an NGC evolved along a comparable path for the components making up the system. The extant congruous phylogenetic relationship indicates co-evolution of NosZ and the accessory Nos proteins due to selective pressure from a shared function. The few exceptions, where the NosZ and 16S rRNA trees do not coincide, may represent events of lateral gene transfer (Figure 5-4).

Occasionally, *nos* genes are encoded on a plasmid. The first analysis of such a situation was carried out for *Cupriavidus necator* H16 (formerly *Ralstonia eutropha*) whose capability for chemolithoautotrophy and denitrification resides on a 452-kb megaplasmid. In the diazotrophic *Sinorhizobium meliloti*, the *nos* genes form part of the pSymA plasmid. A 115-kb plasmid carries an NGC in the phototrophic bacterium *Rhodobacter sphaeroides* IL106. Genomic sequencing of *Silicibacter pomeroyi* has revealed the plasmid-encoded nature of its *nos* genes. *nosD* and *nosL* genes are found as separate loci on the plasmid pNG700 of *H. marismortui*, whereas the principal *nosZDFY* set is chromosomally encoded. The existence of plasmid-borne NGCs emphasizes the independence of the N_2O-respiratory system and the possibility to move it laterally. Notwithstanding, we notice in the plasmid-borne cases of *nos* genes a coincidence of the NosZ clades with the taxonomic grouping (Figure 5-4), and acquisition of any of the plasmid-borne NGCs is not attributable to lateral transfer across large phylogenetic distance.

Denitrification is found among several genera of Gram-positive bacteria. Although an N_2OR has not yet been isolated from any representative, the biochemical nature of the enzyme can be predicted via genome analysis of *D. hafniense*. This organism is of Gram-positive morphotype, forms endospores and belongs to the group of low G + C Gram-positive bacteria. Its genome reveals a *nosCZ·orf·nosDLFY* cluster that is suggested to encode the Z-type N_2OR of Gram-negative bacteria. All Cu_Z-ligating histidine residues are conserved in NosZ, so are the Cu_A domain and other structural elements.

5.7 Transport processes for Nos proteins

N_2OR is located in the periplasm where it functions as an electron sink for the anaerobic respiratory electron transfer chain. The twin arginine translocation pathway (Tat) and the Sec translocation systems have to cooperate to assemble a functional N_2O respiratory system in this compartment. NosZ and NosX are exported by the Tat system that is directed at folded proteins, whether or not they are associated with a non-proteinaceous cofactor [15]. N_2OR was among the very first proteins where a conserved sequence motif with an arginine pair was observed in the signal peptide. The importance of the conserved motif is seen in an Arg20Asp substitution of the signal peptide, which causes NosZ to accumulate in the cytoplasm without its Cu cofactors. For *Flexibacter canadensis*, *Thiobacillus denitrificans*, and *B. azotoformans* a membrane association of N_2OR has been described.

The cognate signal peptide of the Tat system carries a heptameric consensus motif (S/T)RRXFPLK. The *tat* locus of *P. stutzeri* consists of *tatABC* and an additional *tatE* gene as part of the NGC. *tatE* and *tatA* are exchangeable since a *tatE* mutant is not rendered transport incompetent for NosZ [16]. The Tat system of *P. stutzeri* or other N_2O-metabolizing bacteria is assumed to follow functionally the *Escherichia coli* paradigm. A specific conformation seems to be required for the NosZ substrate to explain the observations with certain recombinant forms. For instance, substitution of the Cu_A ligand Cys^{618} for valine leads to a strong rejection of NosZ by the Tat translocon. Most of NosZ is left in the unprocessed form. This and other amino acid substitutions show that the Tat system is constrained by distinct conformations in the folded form of NosZ, which indicates some control of correctness of folding. NosX carries a Tat specific signal sequence and is presumed to be a periplasmic component of the N_2O respiratory system.

The Sec system is the principal transport system for translocation of unfolded bacterial proteins. Cofactors, if present, are processed separately. NosZ of *W. succinogenes* and a few other NosZ proteins have Sec-type signal peptides and, in contrast to the usual Tat export pathway, seem to be exported by the Sec system [11]. The Nos components NosR and NosY are integral membrane proteins. Both have Sec-specific signal peptides. The periplasmic location of NosD is deduced from reporter gene fusions. NosL is predicted to be a component of the outer membrane since it carries the signal peptide of a lipoprotein with the lipobox consensus sequence LA(A/G)C. Lipoproteins cross the membrane by the Sec system and are targeted to the outer membrane by the Lol system [17].

5.8 Role of accessory flavoproteins

The *nosR* gene is found in the NGCs of many N_2O-respiring bacteria located mostly adjacent to and upstream of *nosZ*. Its inactivation in *P. stutzeri* results in a complete lack of *nos* gene expression since it affects the transcription of both *nosZ* and the *nosD* operon. A *nosR* mutant of *B. japonicum* is defective in the N_2O-reducing system, which manifests itself by the termination of nitrate denitrification with N_2O. In *P. aeruginosa*, disruption of *nosR* approximately halves transcriptional activity of the *nosR* promoter. Nonetheless, NosR proteins do not have properties of transcription factors and their apparent regulatory function is not understood.

The *nosR* gene product is a membrane-bound protein. It has a complex topology with different cofactors at either side of the cytoplasmic membrane and a central five-helix bundle (Figure 5-5). The N-terminal

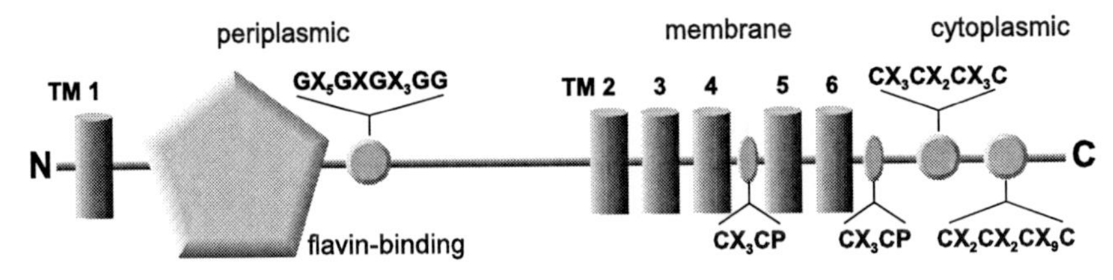

Figure 5-5 Domain structure of unprocessed NosR. Upright cylinders denote transmembrane helices as predicted from protein secondary structure analyses; TM1 is absent from mature NosR. The pentagon indicates the flavin-binding domain; buttons mark positions of conserved motifs.

domain, comprising about half of the protein, is located in the periplasm. The C-terminal domain contains two iron–sulfur clusters that are exposed to the cytoplasm. CX_3CP motifs, conserved in every NosR protein, follow transmembrane helices four and six; they may undergo reversible metal binding or −SH redox chemistry for positioning these helices in a functionally important interaction. The periplasmic domain of NosR has structural similarity with the FMN-binding domain of the NqrC subunit of the Na^+-translocating NADH:quinone oxidoreductase. NosR has been purified and shown to be a flavoprotein [18]. Following the flavin-binding region is a remarkably conserved glycine-rich region. A genuine NosR protein can be recognized from the assemblage of its distinctive structural elements (Figure 5-5). The distribution of NosR in only the alpha, beta, and gamma divisions of the proteobacteria limits the NosR function to a specific type of N_2O respiratory system (Figure 5-3).

Genetically engineered forms indicate that NosR is an indispensable factor in sustaining cellular N_2OR activity. Manipulations of the periplasmic domain or metal-binding centers of NosR cause the loss of whole-cell N_2O-reducing activity. But N_2OR proteins synthesized in the genetic background of a defective NosR protein show the UV-vis spectral signatures of Cu_A and Cu_Z [18]. When expressed as separate proteins, the N-terminal, periplasmic domain, and the transmembrane, C-terminal domains of NosR complement each other functionally. Certain recombinant modifications of NosR result in the transformation of the Cu_Z center to the Cu_Z^* state.

NosX is a 34-kD periplasmic flavoprotein that has been shown by mutational evidence to provide an essential function for a certain variant of N_2O respiration or type of N_2O-utilizing bacterium. The location and properties of NosX are consistent with a role directed at N_2OR. The *nosX* gene is found currently only in alpha- and beta-proteobacteria (Figure 5-3). Investigation of the NosX phenotype at the level of isolated N_2OR reveals that a catalytically active holoN_2OR with spectral evidence for Cu_A and Cu_Z^* is synthesized. Since Cu_Z^* becomes manifest upon *nosX* and *nosR* mutations, it must have physiological relevance.

References

[1] W.G. Zumft, Nitric oxide reductases of prokaryotes with emphasis on the respiratory, heme-copper oxidase type, J. Inorg. Biochem. 99 (2005) 194–215.

[2] W.G. Zumft, P.M.H. Kroneck, Respiratory transformation of nitrous oxide (N_2O) to dinitrogen by *Bacteria* and *Archaea*, Adv. Microbial. Physiol., 52 (2006) 109–230.

[3] S. Ferretti, J.G. Grossmann, S.S. Hasnain, R.R. Eady, B.E. Smith, Biochemical characterization and solution structure of nitrous oxide

reductase from *Alcaligenes xylosoxidans* (NCIMB 11015), Eur. J. Biochem. 259 (1999) 651–659.
[4] J.M. Charnock, A. Dreusch, H. Körner, F. Neese, J. Nelson, A. Kannt, H. Michel, C.D. Garner, P.M.H. Kroneck, W.G. Zumft, Structural investigations of the Cu_A centre of nitrous oxide reductase from *Pseudomonas stutzeri* by site-directed mutagenesis and X-ray absorption spectroscopy, Eur. J. Biochem. 267 (2000) 1368–1381.
[5] H. Beinert, Copper A of cytochrome *c* oxidase, a novel, long-embattled, biological electron-transfer site, Eur. J. Biochem. 245 (1997) 521–532.
[6] T. Rasmussen, B.C. Berks, J. Sanders-Loehr, D.M. Dooley, W.G. Zumft, A.J. Thomson, The catalytic center in nitrous oxide reductase, Cu_Z, is a copper sulfide cluster, Biochemistry 39 (2000) 12753–12756.
[7] T. Haltia, K. Brown, M. Tegoni, C. Cambillau, M. Saraste, K. Mattila, K. Djinovic-Carugo, Crystal structure of nitrous oxide reductase from *Paracoccus denitrificans* at 1.6 Å resolution, Biochem. J. 369 (2003) 77–88.
[8] J.M. Chan, J.A. Bollinger, C.L. Grewell, D.M. Dooley, Reductively activated nitrous oxide reductase reacts directly with substrate, J. Am. Chem. Soc. 126 (2004) 3030–3031.
[9] P. Wunsch, M. Herb, H. Wieland, U.M. Schiek, W.G. Zumft, Requirements for Cu_A and Cu−S center assembly of nitrous oxide reductase deduced from complete periplasmic enzyme maturation in the nondenitrifier *Pseudomonas putida*, J. Bacteriol. 185 (2003) 887–896.
[10] W.G. Zumft, Cell biology and molecular basis of denitrification, Microbiol. Mol. Biol. Rev. 61 (1997) 533–616.
[11] J. Simon, O. Einsle, P.M.H. Kroneck, W.G. Zumft, The unprecedented *nos* gene cluster of *Wolinella succinogenes* encodes a novel respiratory electron transfer pathway to cytochrome *c* nitrous oxide reductase, FEBS Lett. 569 (2004) 7–12.
[12] W.D. Grant, M. Kamekura, T.J. McGenity, A. Ventosa, Class III. Halobacteria *class. nov.*, in: D.R. Boone, R.W. Castenholz, G.M. Garrity (Eds.) Bergey's Manual of Systematic Bacteriology, 2nd ed., vol. 1: The *Archaea* and the Deeply Branching and Phototrophic *Bacteria*, Springer, New York, , 2001, pp. 294–334.
[13] P. Völkl, R. Huber, E. Drobner, R. Rachel, S. Burggraf, A. Trincone, K.O. Stetter, *Pyrobaculum aerophilum* sp. nov., a novel nitrate-reducing hyperthermophilic archaeum, Appl. Environ. Microbiol. 59 (1993) 2918–2926.
[14] H. Körner, H.J. Sofia, W.G. Zumft, Phylogeny of the bacterial superfamily of Crp−Fnr transcription regulators: exploiting the metabolic spectrum by controlling alternative gene programs, FEMS Microbiol. Rev. 27 (2003) 559–592.

[15] T. Palmer, F. Sargent, B.C. Berks, Export of complex cofactor-containing proteins by the bacterial Tat pathway, Trends Microbiol. 13 (2005) 175–180.

[16] M.P. Heikkilä, U. Honisch, P. Wunsch, W.G. Zumft, Role of the Tat transport system in nitrous oxide reductase translocation and cytochrome cd_1 biosynthesis in *Pseudomonas stutzeri*, J. Bacteriol. 183 (2001) 1663–1671.

[17] H. Tokuda, S. Matsuyama, Sorting of lipoproteins to the outer membrane in *E. coli*, Biochim. Biophys. Acta 1694 (2004) IN1–IN9.

[18] P. Wunsch, W.G. Zumft, Functional domains of NosR, a novel transmembrane iron-sulfur flavoprotein necessary for nitrous oxide respiration, J. Bacteriol. 187 (2005) 1992–2001.

Chapter 6

Denitrification in Rhizobia-Legume Symbiosis

María J. Delgado, Sergio Casella and Eulogio J. Bedmar

6.1 Introduction

The genera *Allorhizobium*, *Azorhizobium*, *Bradyrhizobium*, *Mesorhizobium*, *Rhizobium* and *Sinorhizobium*, collectively referred to as rhizobia, are members, among others, of the bacterial order *Rhizobiales* of the α-proteobacteria [1]. Rhizobia are soil, Gram-negative bacteria with the unique ability to establish a N_2-fixing symbiosis on legume roots and on the stems of some aquatic legumes. Symbiotic N_2 fixation (SNF) by the legume-rhizobia couple is a process of both ecological and agricultural importance. In the nodule, maintenance of nitrogenase activity is subject to a delicate equilibrium. A high rate of O_2-respiration is necessary to supply the energy demands of the N reduction process, but O_2 also irreversibly inactivates the nitrogenase complex. These conflicting demands are met by control of O_2 flux through a diffusion barrier in the cortex of nodules, which greatly limits permeability to O_2 [2]. Then, O_2 is delivered to the bacteroids by the plant O_2-carrier, leghemoglobin, which is present

Biology of the Nitrogen Cycle
Edited by H. Bothe, S.J. Ferguson and W.E. Newton

exclusively in the nodule [3]. To cope with the low-ambient O_2-concentration within the nodule, N_2-fixing bacteroids use a high-affinity cytochrome cbb_3-type oxidase encoded by the *fixNOQP* operon to produce ATP [4].

In recent years it has emerged that many rhizobia species have genes for enzymes of some or all of the four reductase reactions (see Chapter 1) for denitrification. In fact, denitrification can be readily observed in many rhizobia species, in their free-living form, in legume root nodules, or in isolated bacteroids [5]. In this chapter, we update progress on denitrification by free-living and symbiotic rhizobia.

6.2 Denitrification in free-living rhizobia

Although the ability to denitrify may enhance bacterial survival and growth capability in anaerobic soils, denitrification among rhizobia is rare, and only *B. japonicum* and *A. caulinodans* have been shown to be true denitrifiers, i.e. to reduce NO_3^- simultaneously to both NH_4^+ (assimilation) and N_2 (denitrification) when cultured microaerobically with nitrate as not only the terminal electron acceptor but also the sole source of N. Other soil microorganisms have been shown to possess higher rates of denitrification activity than rhizobia [6], but the vast area all over the world of cultivated legumes makes contribution of rhizobia to total denitrification highly significant.

Denitrification genes encoding either periplasmic nitrate reductase (Nap), Cu-containing nitrite reductase (Nir), nitric oxide reductase (Nor), or nitrous oxide reductase (Nos) (see Chapter 1 for description of the gene products) have been identified in several rhizobia species such as *R. sullae* (formerly *R. hedysari*) [7], *R. etli* [8], *S. meliloti* [9, 10], and *B. japonicum* [11, 12] (Table 6-1). None of the denitrification genes have been detected in *M. loti* strain MAFF 303090 (http://www.kazusa.or.jp/rhizobase).

6.2.1 Rhizobium sullae

R. sullae induces nodule formation on *Hedysarum coronarium*. *R. sullae* strain HCNT1 has been shown to contain a Cu-containing Nir encoded

Table 6-1 Denitrification genes in N_2-fixing endosymbiotic rhizobia.

Species	Genes	Reference
R. sullae	*NirK*	[7]
R. etli	*nirK, norCBQD*	[8]
S. meliloti	*napEFDABC, nirK, norCBQD, nosRZDFYLX*	[9, 10]
B. japonicum	*NapEDABC, nirK, norCBQD, nosRZDFYLX*	[11, 12]

by a *nirK* gene. Expression of *nirK* is atypical in that it does not require the presence of a nitrogen oxide, but only a decrease in O_2 concentration [13]. Reduction of NO_2^- by strain HCNT1 results in inhibition of growth due to the accumulation of NO to toxic levels, suggesting that *R. sullae* does not contain any Nor [13]. No amplicons were obtained after hybridization of strain HCNT1 genomic DNA against various *nor* probes, which confirmed previous results on the absence of Nor in this species [7]. Nodulation, plant growth, and rates of N_2 fixation are similar between wild-type and *nirK*-deficient strains. *R. sullae* strain HCNT1 has been shown to enter a viable-but-not-culturable status when O_2 is limiting, and NirK produces NO from nitrite present in the growth medium [14]. A physiological role for the truncated denitrification chain in *R. sullae* is not obvious. It is possible that Nir activity allows the bacteria to convert into a viable-but-not-culturable form, which, in turn, would survive for long periods under stress conditions without loss of the ability to recover the vegetative state [15].

6.2.2 *Rhizobium etli*

R. etli forms N_2-fixing symbiosis with *Phaseolus vulgaris*. Studies of genome structure and sequence analyses in the type strain *R. etli* CFN42 permitted the identification of genes encoding proteins closely related to denitrification enzymes; among them the *norCBQD* genes coding for the cytochrome *c*-type Nor and the *nirK* gene encoding a Cu-containing Nir. Strain CFN42 lacks genes coding for Nap and Nos. In fact, *R. etli* CFN42 is unable to grow under denitrifying conditions with NO_3^- and does not express nitrate reductase activity [8]. However, *R. etli* CFN42 is capable of growth with NO_2^- as a terminal electron acceptor. Mutational analysis has demonstrated that a *nirK* deficient mutant is unable to grow under NO_2^--respiring conditions, and that microaerobic growth of the mutant was inhibited in the presence of NO_2^-. Since Nir activity and NO_2^--uptake were very low in the *nirK* mutant, it is possible that the Cu-containing Nir may have both a respiratory and a NO_2^- detoxifying role in *R. etli* [8].

6.2.3 *Sinorhizobium meliloti*

DNA sequences showing homology with those published coding for the rhizobial *nap*, *nir*, *nor*, and *nos* denitrification genes have been found in the symbiotic plasmid pSymA complete genome sequence of *S. meliloti* strain 1021 (http://www.kazusa.or.jp/rhizobase). Intriguingly, despite possessing the complete set of denitrification genes, *S. meliloti* 1021 does not grow under O_2-limiting conditions with either NO_3^- or NO_2^- as terminal electron acceptors. None of the Tn5 insertions in the *S. meliloti* strain JJ1c10 *nos* region affected N_2-fixing ability in symbiosis with alfalfa, which demonstrated that denitrification is not essential for N_2 fixation [9].

6.2.4 *Bradyrhizobium japonicum*

B. japonicum, the N_2-fixing microsymbiont of soybeans, is a slow growing, true denitrifying rhizobium that has been shown to reduce $^{15}NO_3^-$ simultaneously to NH_4^+ and N_2 when cultured under O_2-limiting conditions. In this bacterium, denitrification depends on the *napEDABC*, *nirK*, *norCBQD*, and *nosRZDFYLX* genes encoding nitrate-, nitrite-, nitric oxide-, and nitrous oxide-reductase, respectively [12]. Mutational analyses of the *B. japonicum nap* genes have shown that Nap is the primary enzyme responsible for anaerobic growth under NO_3^--respiring conditions. Although it has been considered that membrane-bound nitrate reductase (Nar) catalyzes the first step of anaerobic denitrification, and that the Nap system was more important for aerobic denitrification or in redox balancing (see Chapter 2), the *B. japonicum* Nap enzyme can support anaerobic growth by reducing NO_3^- to NO_2^-, which can be further reduced by the reactions of denitrification. Regulatory studies using transcriptional *lacZ* fusions to the *napEDABC*, *nirK*, *norCBQD*, and *nosRZDFYLX* promoter regions indicated that microaerobic induction of denitrification genes is dependent on the *fixLJ* and *fixK*$_2$ genes whose products form the FixLJ-FixK$_2$ regulatory cascade. Another transcriptional regulator, the NnrR protein (see Chapter 1), is responsible for N-oxide regulation of the *B. japonicum nirK* and *norCBQD* genes. Thus, the FixLJ-FixK$_2$-NnrR cascade integrates both O_2 limitation and the presence of an N-oxide that are critical for maximal induction of the *B. japonicum* denitrification genes [16].

6.3 Denitrification in nodules

When a rhizobial species is used to inoculate the corresponding legume host, bacteroids within the nodules are also able to express the denitrification pathway. Thus, bacteroids isolated from nodules of N_2-dependent plants that were incubated with nitrate may produce large amounts of NO_2^-, N_2O, and N_2, depending on whether the species used for inoculation contains a complete or incomplete set of denitrification genes. Emission of NOx from legume nodules also contributes to the release of greenhouse gases into the atmosphere. Soil nitrate has restricted access to the bacteroid-infected zone within the nodule; consequently, despite the presence of an active denitrification system, the process within the nodules could be limited by substrate availability. Moreover, most rhizobia species so far analyzed contain Nap. Location of the enzyme in the periplasm may be an evolutionary adaptation to protect nitrogenase from damage produced by NOx.

6.3.1 *Alfalfa nodules*

Measurements of denitrification activity in intact, root nodules of alfalfa upon NO_3^- supply demonstrated that NO_3^- reached the infected zone and was metabolized by the bacteroids to produce NOx. However, the extent of such an activity was extremely low, representing only 2% of the potential activity of the isolated bacteroids. It was concluded that the environmental and agronomic importance of denitrification by *S. meliloti* bacteroids within the nodules is negligible [6].

6.3.2 Soybean nodules

Expression of *B. japonicum* wild-type strain USDA110 *nirK*, *norC*, and *nosZ* denitrification genes in soybean root nodules has been reported by in situ histochemical detection of β-galactosidase activity due to transcriptional fusions of the *nirK*, *norC*, and *nosZ* genes to the reporter gene *lacZ* (Figure D, see Color Plate Section) [17]. Similarly, isolated bacteroids also expressed the P_{nirK}-*lacZ*, P_{norC}-*lacZ*, and P_{nosZ}-*lacZ* fusions. Levels of β-galactosidase activity were similar in both bacteroids and nodule sections from plants that were solely N_2-dependent or grown in the presence

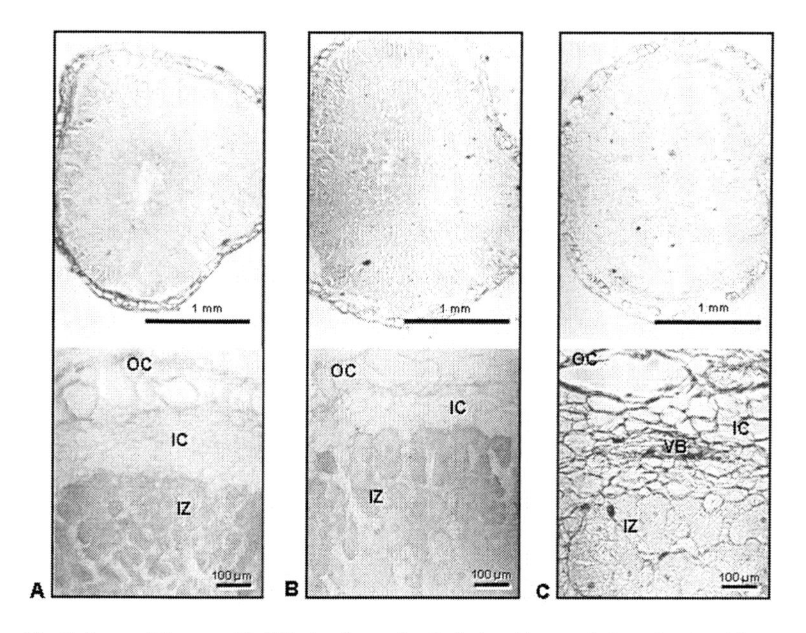

Figure D Colour Figure D Histochemical detection of β-galactosidase activity in soybean nodules isolated from plants inoculated with *B. japonicum* cells containing a *nirK-lacZ* fusion. Nodules were obtained from plants grown in the absence (panel A) or in the presence (panel B) of 4 mM KNO_3. Nodule sections from nitrate-treated soybeans inoculated with *B. japonicum* USDA110 containing a promoterless *lacZ* gene were used as a control (panel C). OC, outer cortex; IC, inner cortex; IZ, infected zone; VB, vascular bundles. (See Colour Plate Section in back of this volume)

of 4 mM KNO_3, which suggests that O_2, and not NO_3^-, is the main factor controlling expression of denitrification genes in soybean nodules.

The symbiotic phenotype of *B. japonicum* strains carrying a mutation in either the *nirK, norC*, or *nosZ* strucural genes has also been reported in comparison with that of the wild-type strain. In soybean plants not supplemented with NO_3^-, *B. japonicum* mutants showed a wild-type phenotype with regard to nodule number and nodule dry weight as well as plant dry weight and N-content. In the presence of 4 mM KNO_3, plants inoculated with either the *nirK* or the *norC* mutants showed fewer nodules, and lower plant dry weight and N-content, relative to those of wild-type and *nosZ* mutant strains. Thus, although not essential for N_2 fixation, mutation of either the *nirK* or *norC* genes encoding respiratory Nir and Nor, respectively, confers on *B. japonicum* a reduced ability for nodulation in soybean plants grown with nitrate. Because nodules formed by the parental and mutant strains exhibited similar nitrogenase activities, it is possible that denitrification enzymes play a role in nodule formation rather than in nodule function [17].

6.3.3 Role of denitrification in nodules

6.3.3.1 Rhizobial denitrification as an energy-producing mechanism

Oxygen concentration within the nodules is extremely low, ranging from 3 to 22 nM. This environment not only allows the maintenance of nitrogenase activity, but also the expression of genes involved in the denitrification process [17]. During denitrification, reduction of NOx is coupled to energy conservation. Although an excess of fixed N inhibits the formation and functioning of nodules, nitrate, in small doses, could be used for ATP generation by denitrifying bacteroids, thus improving energy availability for nitrogenase activity. In fact, *B. japonicum* bacteroids isolated from nodules were capable of N_2 fixation using NO_3^- as an electron acceptor for energy conservation [5]. The cbb_3-type oxidase encoded by the *fixNOQP* is the high-affinity oxidase responsible for ATP formation in *B. japonicum* free-living cells growing under denitrifiying conditions and in endosymbiotic bacteroids. The critical balance between O_2-input and consumption within the nodule may easily be disturbed in the natural environment of the symbionts, e.g. by flooding, drought, etc. This, in turn, would result in transient anoxic conditions close to the roots. Under these conditions, NO_3^- can function as an alternate electron acceptor to O_2, thereby providing cellular ATP needed for N_2 fixation in the nodules of legumes.

6.3.3.2 Rhizobial denitrification as a nitrite and nitric oxide detoxifying mechanism

An associated role of denitrification in nodules would also be the detoxification of cytotoxic compounds such as NO_2^- or NO produced as intermediates during denitrification reactions or emerging from the host-plant. Apart from the denitrification process, NO_2^- is also produced in the nodule from NO_3^- reduction by the plant nitrate reductase. This enzyme not only reduces NO_2^- to NH_4^+, but also produces NO from NO_2^- using NADH [18]. NO may also be formed by a nitric oxide synthase activity present in roots and nodules of *Lupinus* [19]. UV-visible [20, 21] and electron paramagnetic resonance (EPR) spectroscopy studies [22] have demonstrated the presence of leghemoglobin (Lb) bound to nitric oxide within soybean nodules. NO_2^- and NO, which have been identified as inhibitors of nitrogenase, may bind to Lb forming nitrosylleghemoglobin complexes (NO–Lb). Thus, nitrosylation of Lb may work as a protective mechanism used by the nodule to prevent nitrogenase inhibition. Alternatively, the accumulation of NO–Lb may result in the inhibition of nitrogenase activity, as the binding of NO to Lb may competitively inhibit the binding of O_2, thus diminishing the O_2 supply available to bacteroids and thereby reducing N_2 fixation [20]. NO may also interfere with heme-based sensors such as the O_2 sensor FixL [23], a protein involved in O_2 regulation of the N_2-fixation genes. A clear function for NO during symbiosis has not yet been established.

The identification of NO–Lb complexes within soybean nodules has led to the question of how *B. japonicum* bacteroids overcome toxicity due to NO. As indicated above, a candidate for nitric oxide detoxification is the respiratory Nor, whose expression in nodules has already been demonstrated [17]. The role of this enzyme in NO detoxification has been examined by analyzing the EPR-detected NO–Lb levels within nodules treated with nitrate and formed by the wild-type and a *B. japonicum norC* mutant [24]. The EPR spectrum of the *norC* nodules was similar to that of the wild-type nodules demonstrating that a mutation in *nor* genes does not affect the level of NO within nodules. These results imply that Nor is not solely responsible for NO detoxification within soybean nodules, but other systems may be involved in NO detoxification.

References

[1] H. Sawada, L.D. Kuykendall, J.M. Young, Changing concepts in the systematics of bacterial nitrogen-fixing legume symbionts, J. Gen. Appl. Microbiol. 49 (2003) 155–179.

[2] F.R. Minchin, Regulation of oxygen diffusion in legume nodules, Soil Biol. Biochem. 29 (1997) 881–888.

[3] C.A. Appleby, The origin and functions of hemoglobin in plants. Sci. Prog. 76 (1992) 365–398.

[4] M.J. Delgado, E.J. Bedmar, J.A. Downie, Genes involved in the formation and assembly of rhizobial cytochromes and their role in symbiotic nitrogen fixation, Adv. Microb. Physiol. 40 (1998) 193–222.

[5] G.M. O'Hara, R.M. Daniel, Rhizobial denitrification: a review, Soil Biol. Biochem. 17 (1985) 1–9.
[6] J.I. Garcia-Plazaola, J.M. Becerril, C. Arrese-Igor, C. Gonzalez-Murua, P.M. Aparicio-Tejo, Denitrifying ability of thirteen *Rhizobium meliloti* strains, Plant Soil 157 (1993) 207–213.
[7] A. Toffanin, Q. Wu, M. Maskus, S. Casella, H.D. Abruña, J.P. Shapleigh, Characterization of the gene encoding nitrite reductase and the physiological consequences of its expression in the nondenitrifying *Rhizobium "hedysari"* strain HCNT1, Appl. Environ. Microbiol. 62 (1996) 4019–4025.
[8] E. Bueno, N. Gómez-Hernández, L. Girard, E.J. Bedmar, M.J. Delgado, Function of the *Rhizobium etli* CFN42 *nirK* gene in nitrite metabolism, Biochem. Soc. Trans. 35 (2005) 166–167.
[9] P. Holloway, W. McCormick, R.J. Watson, Y.K. Chan, Identification and analysis of the dissimilatory nitrous oxide reduction genes, *nosRZDFY*, of *Rhizobium meliloti*, J. Bacteriol. 178 (1996) 1505–1514.
[10] F. Galibert, T.M. Finan, S.R. Long, A. Puhler, *et al.* The composite genome of the legume symbiont *Sinorhizobium meliloti*, Science 293 (2001) 668–672.
[11] T. Kaneko, Y. Nakamura, S. Sato, K. Minamisawa, T. Uchiumi, S. Sasamoto, A. Watanabe, K. Idesawa, M. Iriguchi, K. Kawashima, et al., Complete genomic sequence of nitrogen-fixing symbiotic bacterium *Bradyrhizobium japonicum*, DNA Res. 9 (2002) 189–197.
[12] E. Bedmar, E.F. Robles, M.J. Delgado, The complete denitrification pathway of the symbiotic N_2-fixing bacteria *Bradyrhizobium japonicum*, Biochem. Soc. Trans. 35 (2005) 11–16.
[13] S. Casella, J.P. Shapleigh, W.J. Payne, Nitrite reduction in *Rhizobium "hedysari"* strain HCTN1, Arch. Microbiol. 146 (1986) 233–238.
[14] A. Toffanin, M. Basaglia, C. Ciardi, P. Vian, S. Povolo, S. Casella, Energy content decrease and viable-not-culturable status induced by oxygen limitation coupled to the presence of nitrogen oxides in *Rhizobium "hedysari"*, Biol. Fertil. Soils. 31 (2000) 484–488.
[15] S. Casella, J.P. Shapleigh, A. Toffanin, M. Basaglia, Investigation into the role of the truncated denitrification chain in *R. sullae* strain HCNT1, Biochem. Soc. Trans. 34/1 (2006) 130–132.
[16] S. Mesa, E.J. Bedmar, A. Chanfon, H. Hennecke, H.-M. Fischer, *Bradyrhizobium japonicum* NnrR, a denitrification regulator, expands the FixLJ-FixK_2 regulatory cascade, J. Bacteriol. 185 (2003) 3978–3982.
[17] S. Mesa, J.D. Alché, E.J. Bedmar, M.J. Delgado, Expression of *nir*, *nor* and *nos* denitrification genes from *Bradyrhizobium japonicum* in soybean root nodules, Physiol. Plant 120 (2004) 205–211.
[18] H. Yamasaki, S. Sakihama, Simultaneous production of nitric oxide and peroxynitrite by plant nitrate reductase: in vitro evidence for the

NR-dependent formation of active nitrogen species, FEBS Lett. 68 (2000) 89–92.
[19] M. Cueto, O. Hernández-Perera, R. Martín, M.L. Bentura, J. Rodigo, S. Lamas, M.P. Golvano, Presence of nitric oxide synthase activity in roots and nodules of *Lupinus albus*, FEBS Lett. 398 (1996) 159–164.
[20] Y. Kanayama, Y.Yamamoto, Inhibition of nitrogen fixation in soybean plants supplied with nitrate II. Accumulation and properties of nitrosylleghemoglobin in nodules, Plant Cell Physiol. 31 (1990) 207–214.
[21] Y. Kanayama, I. Watanabe, Y. Yamamoto, Inhibition of nitrogen fixation in soybean plants supplied with nitrate I. Nitrite accumulation and formation of nitrosylleghemoglobin in nodules, Plant Cell Physiol. 31 (1990) 341–346.
[22] C. Mathieu, S. Moreau, P. Frendo, A. Puppo, M.J. Davies, Direct detection of radicals in intact soybean nodules: presence of nitric oxide-leghemoglobin complexes, Free Radic. Biol. Med. 24 (1998) 1242–1249.
[23] W.C. Winkler, G. Gonzalez, J.B. Wittemberg, R. Hille, N. Dakappagari, J. Jacob, L.A. Gonzalez, M.A. Gilles-Gonzalez, Nonsteric factors dominate binding of nitric oxide, azide, imidazole, cyanide, and fluoride to the rhizobial heme-based oxygen sensor FixL, Chem. Biol. 3 (1996) 841–850.
[24] G.E. Meakin, B. Jepson, D.J. Richardson, E.J. Bedmar, M.J. Delgado, The role of *Bradyrhizobium japonicum* nitric oxide reductase in NO detoxification in soybean nodules, Biochem. Soc. Trans. 1 (2006) 195–196.

Chapter 7

The Dissimilatory Reduction of Nitrate to Ammonia by Anaerobic Bacteria

Sudesh B. Mohan and Jeff A. Cole

7.1 Dissimilatory nitrate reduction to ammonia, a process distinct from denitrification and nitrate assimilation

Nitrate reduction to NO_2^- is the first step of three major biological processes, NO_3^- assimilation, denitrification and the dissimilatory reduction of NO_3^- to ammonia. This brief review is concerned with the third of these processes that, because it bypasses denitrification and N_2 fixation, has been called the short circuit of the biological N-cycle [1, 2]. Like NO_3^- assimilation, it is a two-step process involving NO_3^- reduction to NO_2^- followed by NO_2^- reduction to NH_4^+. It is strictly an anaerobic process that dominates NO_3^- and NO_2^- reduction in reductant-rich environments such as anaerobic marine sediments [3] and S^{2-}-rich thermal vents [4], the human gastrointestinal tract [5] and the bodies of warm-blooded animals. There is an obvious explanation for this. Electron acceptors are by definition scarce in an environment where reductants are abundant, so optimal use must be made of any available oxidant to regenerate NAD^+ from NADH

Biology of the Nitrogen Cycle
Edited by H. Bothe, S.J. Ferguson and W.E. Newton

and hence sustain substrate oxidation and growth. It is therefore beneficial for bacteria to exploit this process rather than denitrification because NO_2^- reduction to NH_4^+ consumes six electrons compared with just two or three electrons consumed when NO_2^- is reduced via NirS or NirK to N_2O or N_2.

Various names have been given to the process, including the ugly term, NO_3^- ammonification, respiratory reduction of NO_3^- to NH_4^+, and the title used in this article. All of them emphasize – exaggerate – one aspect of the process. Genes required for NO_2^- reduction to NH_4^+ are frequently misannotated as encoding assimilatory NO_2^- reductases with NAD(P)H designated as the electron donor, so it is important to begin by noting criteria that distinguish it from denitrification or NO_3^- assimilation (Table 7-1). First, it is an anaerobic process that, unlike denitrification in some bacteria, has no aerobic counterpart. In facultative anaerobes like *Escherichia coli* and *Staphylococcus carnosus*, expression of the NO_3^- and NO_2^- reductase genes is tightly repressed in the presence of O_2, induced during anaerobic growth and further regulated by the availability of NO_3^- and NO_2^-. A critical point is that gene expression requires the housekeeping sigma factor, σ^{70}, rather than the specialized sigma factor, σ^{54}. NH_4^+ does not repress gene expression, which, unlike NO_3^- assimilation in all bacteria so far studied in detail, is insensitive to the general N-control circuit mediated by NtrB–NtrC two-component regulatory system (Table 7-1). Unlike NO_3^- assimilation, which is always a cytoplasmic process, it can

Table 7-1 Differences between denitrification, nitrate assimilation and dissimilatory reduction of nitrate to ammonia.

Process	Enzymes involved	Regulation by				Location of the active site
		O_2	NH_4^+	σ^{70}	σ^{54}	
Nitrate assimilation	Nas; Nir	None	Repressed	No	Yes	Cytoplasm
Denitrification	Nar, Nap, NirK, NirS	Repressed	None	Yes	No	Cytoplasm/ periplasm
Dissimilation to ammonia	Nar, Nir	Repressed	None	Yes	No	Cytoplasm
Dissimilation to ammonia	Nap, Nrf	Repressed	None	Yes	No	Periplasm

Nas, assimilatory nitrate reductase; Nir, siroheme-dependent nitrite reductase; Nar, membrane-associated nitrate reductase; Nap, periplasmic nitrate reductase; NirK, copper-containing nitrite reductase; NirS, cytochrome cd_1 nitrite reductase; Nrf, cytochrome *c* nitrite reductase.

occur in the cytoplasm, or in the periplasm, or both, depending on the bacterial species and growth conditions.

7.2 The cytoplasmic pathway for nitrate dissimilation to ammonia

The cytoplasmic pathway for the respiratory reduction of NO_2^- to NH_4^+ was first characterized in *E. coli* K-12 and its close relative, *Klebsiella pneumoniae*, but it has subsequently been studied in the bacterium traditionally used to ferment German salami sausages, *S. carnosus* [6, 7]. In each case, NO_3^- is reduced to NO_2^- by a respiratory NO_3^- reductase A that is coupled to a cytoplasmic siroheme-containing NO_2^- reductase. As described in detail in Chapter 2, NO_3^- reductase A consists of three structural components: these are the catalytic molybdoprotein, NarG; a non-heme, Fe–S protein, NarH, which forms a tight complex with NarG; and a *b*-type cytochrome, NarI, which transfers electrons from the quinol pool to the NarGH complex. These components are invariably encoded in a four or more gene operon, *narGHJI*, in which the fourth component, NarJ, is a pathway-specific chaperone or assembly factor required for the post-translational assembly of the functional complex [8]. Other genes in the *narGHJI* operons of various bacteria encode two-component regulatory systems that enable the bacteria to sense and respond to NO_3^- or NO_2^-, other transcription factors, NO_3^- transport proteins, and in a few examples, NO_2^- transport and reduction genes. The *Thermus thermophilus nar* operon includes a gene encoding a *c*-type cytochrome that is essential for the synthesis of a functional complex. These energy-conserving NO_2^- reductases are predominantly synthesized only during anaerobic growth, mainly (exclusively?) in obligate or facultative anaerobes that encounter high concentrations of NO_3^-. Similar enzymes have been found in the α, β and γ proteobacteria as well as in some Archaea. Multiple copies of *narG* or even the complete operon are found in some bacteria, for example, *E. coli*, *S. typhimurium* and *Mycobacterium tuberculosis*. The different physiological roles of these duplicated systems, which are far from “redundant”, are just beginning to be resolved. The diversity of the *narG* nucleotide sequence has been used to characterize a NO_3^--reducing bacterial community [9], but in general is more conserved than sequences of the assimilatory or periplasmic NO_3^- reductases. It has been argued that this high degree of sequence similarity reflects an ancient common origin, but conversely it is also consistent with horizontal gene transfer.

The crystal structure of the *E. coli* enzyme revealed an organisation that complements that of a major electron donor pathway for NO_3^- reduction, the formate dehydrogenase encoded by the *fdnGHI* operon [10, 11]. This anaerobic electron transfer chain is so far unique in that proton motive force is generated both during electron transfer from the primary electron donor, formate, to ubiquinone, and also from ubiquinol to NO_3^-. The same is true when NADH or other electron donors are used. In terms of conservation of free energy released, it is the most efficient electron transfer chain so far characterizsed in anaerobic metabolism.

In contrast to the efficient cytoplasmic reduction of NO_3^- to NO_2^-, there is a profligate waste of energy as NO_2^- is reduced to NH_4^+ in the cytoplasm by the NADH-dependent NO_2^- reductase. Note, however, a point that is often overlooked, namely that when fermentative bacteria reduce NO_2^- to NH_4^+, acetate rather than ethanol is the final product of C-metabolism, resulting in the generation of one extra ATP by substrate level phosphorylation for each acetate generated.

The cytoplasmic NO_2^- reductase consists either of a single, large polypeptide (usually designated NirB), or a two-subunit enzyme, NirB–NirD [12]. It is a soluble protein that includes two nucleotide-binding domains, an Fe–S centre and a siroheme binding site. Siroheme is a specialized heme group that is found in many SO_3^{2-} and NO_2^- reductases, enzymes that catalyse six-electron transfer reactions by reducing NO_2^- to NH_4^+, or sulphite to sulphide [13, 14]. This soluble NO_2^- reductase is so far unique in being the only enzyme involved in NO_3^- reduction that does not receive electrons directly from the quinol pool: in this respect, the term dissimilatory rather than respiratory NO_2^- reduction is more appropriate.

7.3 The periplasmic pathway for respiratory reduction of nitrate to ammonia

In parallel with the cytoplasmic pathway for NO_3^- and NO_2^- reduction induced in *E. coli* during anaerobic growth, there is a periplasmic pathway involving the periplasmic NO_3^- and NO_2^- reductases, Nap and Nrf (Table 7-1). Similar, though far from identical, enzymes are found in many other bacteria ranging from members of the *Enterobacteraceae* to obligate anaerobes such as *Sulfospirillum delayiani* and *Desulfovibrio desulfuricans* to obligate microaerophiles such as *Campylobacter jejuni*. Relatively few sulphate-reducing bacteria so far studied are able to reduce NO_3^-, but many more express the NrfA NO_2^- reductase. For example, *D. vulgaris* strain Hildenborough, for which the complete genome sequence is known, lacks *nap* genes but expresses a periplasmic NO_2^- reductase [15]. Although there is currently no evidence that proton motive force is generated as electrons flow from the quinol pool to NapA or NrfA, these enzymes indirectly allow energy to be conserved as electrons are transferred from primary substrates such as formate, NADH and possibly also H_2 to the re-oxidized quinone. This periplasmic pathway is therefore aptly described as the respiratory reduction of NO_3^- to NH_4^+.

Only a few periplasmic NO_3^- reductases have been intensively studied in the laboratory, so much of our current information is deduced from bioinformatic analysis of the genomic databases. These databases in turn are highly biased towards the genomes of pathogenic bacteria, so it is still dangerous to draw general conclusions from the limited data available. With this caveat, the conservation of *nap* gene clusters in pathogens with small genomes, especially when coupled with the absence of the *narGHJI* operon, is strong evidence that Nap provides a selective advantage for these organisms in the bodies of warm-blooded animals. In this sense, Nap is both a pathogenicity determinant and physiologically more important than the membrane-associated NO_3^- reductase A.

The primary structures of the periplasmic NO_3^- reductases are far more varied than those of the other enzymes mentioned in this chapter. Even before sequence or structural data were available, this became obvious from attempts to design generic probes for detecting *nrf* and *nap* genes in environmental samples. Universal primers suitable for amplifying *nrf* genes were readily designed [16], but attempts to design probes for *nap* were unsuccessful. The primers designed by Flanagan et al. [17] detect *nap* in *Pseudomonas* strains and related denitrifying bacteria, but cannot detect *napA* from obligate anaerobes, and vice versa. The 90 kDa NapA from α-proteobacteria forms a tight complex with its immediate electron donor, NapB, whereas at the other extreme, the smaller, 70 kDa NapA from *D. desulfuricans* was purified as a single subunit [18], and no *napB* was found in the *nap* gene cluster from this bacterium [19]. The NapA proteins from γ-proteobacteria such as *E. coli* and other members of the *Enterobactericeae* have intermediate properties: these bacteria express both NapC and NapB, but the di-heme NapB polypeptide associates only loosely with its partner, NapA [20].

Several points immediately emerge from the organisation of some of the *nap* gene clusters from bacteria that reduce NO_3^- to NH_4^+ (Figure 7-1). First, eight different types of polypeptide are encoded by the various *nap* clusters, three of which, *napD, napA* and *napG*, have so far always been found together in bacteria that reduce NO_3^- to NH_4^+. These encode the catalytic sub-unit, NapA, the non-heme, iron–sulphur cluster protein, NapG, and the pathway-specific chaperone, NapD, that is essential for post-translational assembly of a functional Nap complex [21] (Table 7-2). However, extreme variability can be found even within a single genus, the most spectacular example to date being the various *Shewanella* species that have been sequenced. At least four different combinations occur in this genus: *napAB, DAB, DABC* and *DAGHB*, and duplicated *nap* clusters occur in *S. frigidimarina*, *S. amzonensis*, *S. putrefaciens* and Shewanella sp.PV4.

Until recently, the di-heme cytochrome *c*, NapB, was considered to be an essential component, but there is no *napB* gene in either the

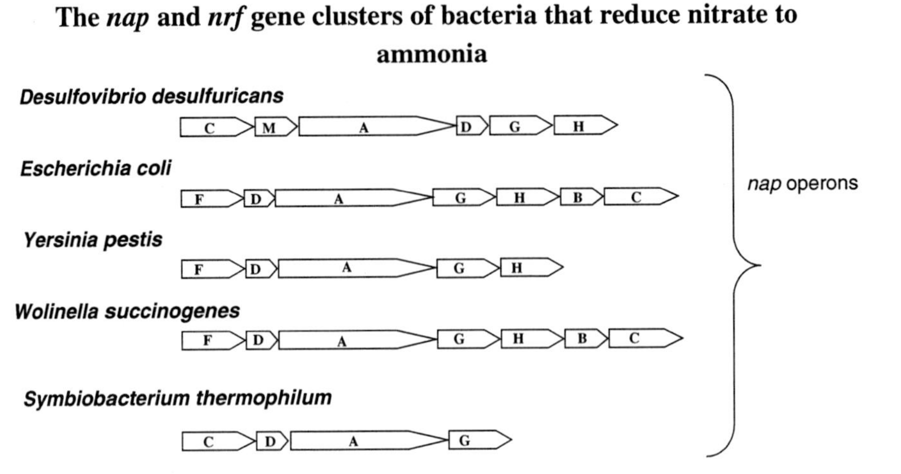

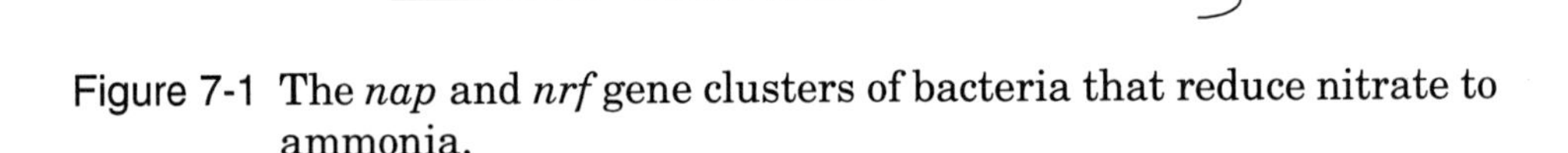
Figure 7-1 The *nap* and *nrf* gene clusters of bacteria that reduce nitrate to ammonia.

Table 7-2 Known or proposed functions of proteins encoded in *nap* gene clusters of bacteria that reduce nitrate to ammonia.

Gene	Proposed or established function
NapA	Catalytic molybdoprotein sub-unit
NapB	Di-heme *c*-type cytochrome that donates electrons to NapA
NapC	Tetra-heme *c*-type cytochrome that functions as a quinol dehydrogenase
NapD	Essential for NapA maturation
NapF	Non-heme iron–sulphur protein, possibly required for insertion of the iron–sulphur centre into NapA
NapG	Non-heme iron–sulphur protein that transfers electrons from a quinol dehydrogenase to the terminal Nap components
NapH	Non-heme iron–sulphur protein that probably functions as an alternative quinol dehydrogenase to NapC
NapM	Tetra-heme *c*-type cytochrome that might function as a quinol dehydrogenase

Symbiobacterium thermophilum or the *Desulfovibrio desulfuricans nap* gene clusters [19]. The Nap system of *Wolinella succinogenes* was the first to be described that lacks the tetra-heme cytochrome, NapC: neither of the only two tetra-heme *c*-type cytochromes encoded elsewhere on the *Wolinella* chromosome is essential for NO_3^- reduction [22].

Certain components tend to be found in bacteria of a particular physiological type (Figure 7-1). The seven-gene *nap* operons of facultatively anaerobic bacteria such as *E. coli* and the small-genome pathogen, *Haemophilus parainfluenzae*, include all four components, NapB, C, G and H. Electrons are transferred preferentially from menadiol via NapC and NapB to NapA, but there is also a slower rate of electron transfer from ubiquinol via NapH and NapG to NapB–NapA, again via NapC [23]. While this implies that NapH–NapG form a ubiquinol dehydrogenase, it is clear that the homologous proteins in *Wolinella succinogenes*, which lacks ubiquinol, function as a menaquinol dehydrogenase [22]. This again emphasizes the danger of trying to draw general conclusions from just a few examples. Although the non-heme Fe–S proteins, NapG and NapH, seem always to be present in obligate anaerobes, they do not seem to occur in non-fermentative, denitrifying bacteria of the α and β proteobacteria, which instead always include both the *c*-type cytochromes, NapB and NapC [19].

In contrast to the highly variable structures of different NapA proteins, NrfA has been relatively conserved throughout evolution [16]. Nevertheless, two distinct and well-characterized types of Nrf complex can be recognized, those encoded by the seven-gene *nrfABCDEFG* operon and those encoded by the four-gene *nrfHAIJ* operon. The former arrangement is typical of the γ-proteobacteria, whereas the latter is found in the δ- and ε-proteobacteria. Significantly, *nrf* genes have not been reported in the α- and β-proteobacteria, which include essentially aerobic bacteria, or facultatively anaerobic bacteria that are denitrifiers. Bacteria in which a *nrfA* gene has been identified include obligate anaerobes such as *D. desulfuricans*, *Sulfospirillum deleyianum, Geobacter metallireducens* and *Wolinella succinogenes*, the fermicute *Desulfitobacterium hafniense* and also the planctomycete, *Pirellula sp.*

7.4 Regulation of the cytoplasmic and periplasmic pathways for nitrate reduction to ammonia

Strong clues to the physiological role of any biochemical process can often be deduced by studying how transcription of the structural genes is regulated. For example, the key role of Nap in aerobic denitrification by *Thiosphaera pantotrophus* is consistent with both a redox-balancing role for this enzyme, and with the observation that *nap* gene expression is optimal during aerobic growth on a highly reduced C-source such as butyrate [24]. In *E. coli*, expression of both *nap* and *nrf* genes is repressed during growth on excess NO_3^-, raising the question why expression of genes encoding a NO_3^--reductase should be repressed by the pathway substrate. The explanation is simple, at least for *E. coli* (there are no comparable data for other bacteria). When NO_3^- is sufficiently abundant, enteric bacteria exploit the energy-efficient, highly active but low-affinity NarG enzyme to reduce NO_3^- in the cytoplasm. When NO_3^- is scarce, Nap provides a high affinity but low activity pathway that does not require NO_3^- transport for NO_3^- to serve as an effective electron sink. These reciprocal roles for Nap and NarG were demonstrated in chemostat competition experiments between strains expressing either only Nap, or only NarG [25]. Furthermore, reciprocal regulatory circuits described below involving two two-component regulatory systems, NarX-NarL and NarQ-NarP, consolidate this kinetic advantage by enhancing the expression of *nap* genes during NO_3^--limited growth [26]. Nap can also fulfil a redox-balancing role in *E. coli* [23], one of the few bacteria in which multiple physiological roles for Nap have been documented experimentally.

In the best-studied bacterium, *E. coli* K-12, synthesis of the NADH-dependent NO_2^--reductase is regulated almost, but not quite, coordinately with NO_3^--reductase A. Both are repressed during aerobic growth, induced during anaerobic growth by the O_2-sensing transcription factor, FNR (regulator of fumarate and NO_3^--reduction), and are further induced by the two-component regulatory system, NarX-NarL. NarX is a membrane-spanning environmental sensor that detects and responds primarily to moderately high concentrations of NO_3^-, which stimulates its autokinase activity, resulting in the transfer of the phosphate group to the response regulator, NarL. NarL-P activates expression of both the *narGHJI* and *nirBDC* operons (NirC is a NO_2^- transport protein). NO_2^- promotes dephosphorylation of NarL. The significant difference between the two operons is that the *narG* promoter is essentially insensitive to a second two-component system, NarQ–NarP, which responds to lower concentrations of NO_3^- than NarX–NarL [27], but the *nirB* promoter is strongly activated by phosphorylated NarP in response to NO_2^-. It is significant that the *fdnGHI* operon is regulated in parallel with expression of NO_3^--reductase A [28]. Conversely, expression of the *nap* and *nrf* genes for the periplasmic pathway are repressed by NO_3^--activated NarL, but induced by NarP in response to low concentrations of NO_3^- [26, 28].

Superficially, NO_3^- dissimilation to NH_4^+ by *Staphylococcus carnosus* is regulated similarly to the *E. coli* pathway: it is strongly repressed during aerobic growth, induced during anaerobic growth, especially in the presence of NO_3^- and is insensitive to the NH_4^+ concentration. However, the details of how this is achieved are significantly different. Neither an FNR homologue nor NarXL have been found in *S. carnosus* (or in *S. aureus*). Instead, there is a two-component regulatory system encoded by genes located between the *narGHJI* operon and the structural gene for a NO_3^- transport protein, NarT [29]. This is part of a cluster of three genes, *nreABC*, in which *nreB* encodes a soluble, cytoplasmic environmental sensor and NreC is a sequence-specific transcription regulator. As in FNR of *E. coli*, there are four N-terminal cysteines in NreB, at least one of which, C62, has been shown to be essential for sensing O_2 and for transcription activation [29]. The NreB–NreC system does not appear to regulate the NO_3^--response. NreA is a GAF domain-containing protein of unknown function: how *S. aureus* detects and responds to NO_3^- or NO_3^- remain to be resolved.

7.5 Distribution of enzymes for nitrate reduction to ammonia and their ecological significance

There are many examples of bacteria that are able to both assimilate NO_3^- and either denitrify or catalyse rapid anaerobic reduction of NO_3^- to NH_4^+. There are no well-established examples of bacteria that both denitrify and dissimilate NO_3^- to NH_4^+: the very few occasional statements to the contrary await confirmation.

Generalizations can also be made – at least based on the current inadequately small database – concerning the distribution of genes for the various NO_3^-- and NO_2^--reductases. In different bacteria, NO_3^--reductase A (NarGHI) can apparently couple with any of the cytoplasmic or periplasmic NO_2^--reductases involved in denitrification or NO_2^- reduction to NH_4^+.

In contrast, no bacterium has yet been shown to couple a periplasmic NO_3^--reductase solely to the cytoplasmic NO_2^--reductase, NirB. The cytoplasmic pathway for NO_3^- reduction to NH_4^+ is almost exclusively restricted to a few groups of facultative anaerobic bacteria that encounter high concentrations of environmental NO_3^-. Although the enteric bacteria have a complete fermentative metabolism when no inorganic terminal electron acceptors are available, they also have a fully competent respiratory metabolism in which O_2, NO_3^-, NO_2^- and other oxidants can be used to conserve energy. *K. pneumoniae* typically lives outside the bodies of warm-blooded animals, for example in soil. It expresses NO_3^- reductase A and Nir, but lacks *nap* and *nrf* genes, as well as the *ccm* genes that are essential for cytochrome *c* assembly in the periplasm. Does this reflect a response to evolutionary pressures on organisms living in environments in which NO_3^- is more abundant than the human body? Consistent with this suggestion is that NO_3^- reductase A and Nir are also exploited in German salami sausage production when *S. carnosus* is inoculated into meat mixed with a high concentration of NO_3^-. The same pathway also occurs in *Bacillus subtilis*, which lacks a fermentative metabolism. Conversely, pathogenic bacteria with small genomes that survive in O_2-limited environments where both NO_3^- and O_2 are scarce have retained Nap and Nrf, but lack NO_3^- reductase A and Nir. This suggests that Nap and Nrf might provide a selective advantage to bacteria other than *E. coli* where it is essential to be able to compete for trace concentrations of any terminal electron acceptor that might come available. It is significant that denitrifying bacteria are rarely isolated from normal human flora [5].

7.6 Nitrate reduction to ammonia as an evolutionary link between nitrate assimilation and denitrification

Despite current limited data, biochemical studies clearly show that the structure of NapA from different bacteria is more variable than that of Nas or NarG. The cytoplasmic assimilatory NO_3^--reductases are more similar to the periplasmic NO_3^- reductases than to the membrane-bound NO_3^- reductase A. A fourth class of NO_3^- reductase has recently been recognized [30]. Like NarG, they are membrane associated, but their catalytic sites are located outside the membrane [30] – not necessarily in the periplasm, as the first examples of this class of NO_3^--reductase were found in Gram-positive bacteria that lack a periplasm. A line of evolution of the dimeric NapAB complexes was proposed from Nas via a primordial Nap typified by the *Symbiobacterium thermophilum* NO_3^- reductase to the *D. desulfuricans* NapA. The subsequent selective pressure has been the need for NapA to evolve from an enzyme that accepts electrons from low mid-point redox potential Fe–S proteins in the reducing environment of the cytoplasm to be able to function in the increasingly oxidizing environment of the periplasm [30]. For obligate anaerobes such as *S. thermophilum*, *D. desulfuricans* or *W. succinogenes*, electrons are delivered by Fe–S proteins such as NapGH [19, 22]. Next to evolve would be the dual systems involving both NapGH and NapBC, as found in fermentative facultative anaerobes of the γ-proteobacteria, culminating in enzymes dedicated to accepting electrons only from the *c*-type cytochromes, NapC and NapB, in the α- and β-proteobacteria [30]. If the above ideas are correct, it is relevant that bacteria that rely on *c*-type cytochromes to transfer electrons from the quinol pool to NapA are all denitrifying bacteria, whereas those that use Fe–S proteins either alone, or in combination with *c*-type cytochromes, reduce the NO_2^- formed to NH_4^+. This implies that NO_3^- reduction to NH_4^+ pre-dates denitrification in the evolutionary timescale. In view of the vestigial similarity between *Desulfovibrio napMC* and *E. coli nrfAB*, did Nap evolve from Nrf, as proposed recently [19, 30]? If so, can other similar evolutionary links be recognized?

7.7 Current challenges and unanswered questions

It has been assumed that there are only four types of NO_3^- reductase, Nas, Nap, Nar and the membrane-associated proteins with active sites facing outside the cytoplasmic membrane. Already there are reports of molybdenum-independent NO_3^- reductases, the most convincing being vanadium-dependent NO_3^- reduction [31]. Will preliminary reports of iron-dependent NO_3^- reductases be confirmed?

NO_3^- reduction to NH_4^+ is now recognized to be widely distributed and ecologically significant. Mutants defective in periplasmic NO_3^- reduction were isolated by signature tagged mutagenesis of *Salmonella typhimurium*, implicating Nap as a pathogenicity determinant. However, its potential for exploitation remains unexplored.

Perhaps the most fundamental question is whether there are mechanisms for energy conservation as electrons are transferred from the quinol pool to NapA. Pertinent to this question is why the quinol dehydrogenase, NapGH, in some organisms is coupled directly to NapA, but in other organisms it is linked via NapC. Does this simply reflect progressive evolution of the Nap complex, or differences in biochemical function?

One of the few physiological roles not yet assigned to Nap is NO_3^- assimilation: does any bacterium rely on Nap and Nrf to generate NH_4^+ for NO_3^- assimilation? While the list is endless, one final intriguing question is how NO_3^--reduction in the periplasm of sulphate-reducing bacteria is regulated. Given that the SO_4^{2-}:SO_3^{2-} redox couple (or more correctly, the adenosine phospho- SO_4^{2-}:SO_3^{2-} redox couple) is strongly electronegative compared with the NO_3^-:NO_2^- couple, it would be fascinating to determine whether NO_3^- reduction by *D. desulfuricans* is repressed by sulphate. If so, this would be contrary to one of the central dogmas of anaerobic bacterial metabolism, namely that bacteria have evolved control mechanisms that enable them to use the energetically most favourable electron acceptor first.

References

[1] J.A. Cole, The rapid accumulation of large quantities of ammonia during nitrite reduction by *Escherichia coli*, FEMS Microbiol. Lett., 4 (1978) 327–329.

[2] J.A. Cole, C.M. Brown, Nitrite reduction to ammonia by fermentative bacteria: a short circuit in the biological nitrogen cycle, FEMS Microbiol. Lett. 7 (1980) 65–72.

[3] J. Sørensen, Capacity for denitrification and reduction of nitrate to ammonia in a coastal marine sediment, Appl. Environ. Microbiol. 35 (1978) 301–305.

[4] J. Zopfi, T. Kjaer, L.P. Nielsen, B.B. Jorgensen, Ecology of *Thioploca* spp.: nitrate and sulfur storage in relation to chemical microgradients and influence of *Thioploca* spp. on the sedimentary nitrogen cycle, Appl. Environ. Microbiol. 67 (2001) 5530–5537.

[5] S.J. Forsythe, J.M. Dolby, A.D.B. Webster, J.A. Cole, Nitrate and nitrite-reducing bacteria in the achlorhydric stomach, J. Med. Microbiol. 25(1988) 253−259.

[6] H. Neubauer, I. Pantel, P.-E. Lindgren, F. Götz, Molecular characterization of the nitrite-reducing system of *Staphylococcus carnosus*, J. Bacteriol. 181 (1998) 1481–1488.

[7] I. Pantel, P.-E. Lindgren, H. Neubauer, F. Götz, Identification and characterization of the *Staphylococcus carnosus* nitrate reductase operon, Mol. Gen. Genet. 259 (1998) 105–114.

[8] T. Palmer, C.L. Santini, C. Iobbi-Nivol, D.J. Eaves, D.H. Boxer, G. Giordano, Involvement of the *narJ* and *mob* gene products in distinct steps in the biosynthesis of the molybdoenzyme nitrate reductase in *Escherichia coli*, Mol. Microbiol. 20 (1996) 875–884.

[9] L.G. Gregory, P.L. Bond, D.L. Richardson, S. Spiro, Characterization of a nitrate-respiring bacterial community using the nitrate reductase gene (*narG*) as a functional marker, Microbiology 149 (2003) 229–237.

[10] D. Richardson, G. Sawers, PMF through the redox loop, Science 295 (2002) 1842–1843.

[11] M. Jormakka, D. Richardson, B. Byrne, S. Iwata, Architecture of NarGH reveals a structural classification of Mo-*bis*-MGD enzymes, Structure 12 (2004) 95–104.

[12] N.R. Harborne, L. Griffiths, S.J.W. Busby, J.A. Cole, Transcriptional control, translation and function of the products of the five open reading frames of the *Escherichia coli nir* operon, Mol. Microbiol. 6 (1992) 2805–2813.

[13] J.Y. Wu, L.M. Siegel, N.M. Kredich, High-level expression of *Escherichia coli* NADPH-sulfite reductase: requirement for a cloned *cysG* plasmid to overcome limiting siroheme cofactor, J. Bacteriol. 173 (1991) 325–333.

[14] H. Macdonald, J.A. Cole, Molecular cloning and functional analysis of the *cysG* and *nirB* genes of *Escherichia coli* K12, two closely-linked genes required for NADH-dependent nitrite reductase activity, Mol. Gen. Genet. 200 (1985) 328–334.

[15] S.A. Haveman, E.A. Greene, C.P. Stillwell, J.K. Voordouw, G. Voordouw, Physiological and gene expression analysis of inhibition of *Desulfovibrio vulgaris* Hildenborough by nitrite, J. Bacteriol. 186 (2004) 7944–7950.

[16] S.B. Mohan, M. Schmid, M. Jetten, J.A. Cole, Detection and widespread distribution of the *nrfA* gene encoding nitrite reduction to ammonia, a short circuit in the biological nitrogen cycle that competes with denitrification, FEMS Microbiol. Ecol. 49 (2004) 433–443.

[17] D.A. Flanagan, L.G. Gregory, J.P. Caret, A. Karakas-Sen, D.J. Richardson, S. Spiro, Detection of genes for periplasmic nitrate reductase in nitrate respiring bacteria and in community DNA, FEMS Microbiol. Lett. 177 (1999) 263–279.

[18] J.M. Dias, M.E. Than, A. Humm, R. Huber, G.P. Bourenkov, H.D. Bartunik, S. Bursakov, J. Calvete, J. Calderia, C. Carneiro, J.J.G. Moura, I. Moura, M.J. Ramao, Crystal structure of the first dissimilatory nitrate reductase at 1.9Å solved by MAD methods, Structure 7 (1999) 65–79.
[19] A. Marietou, D.J. Richardson, J. Cole, S. Mohan, Nitrate reduction by *Desulfovibrio desulfuricans*; a periplasmic nitrate reductase that lacks NapB, but includes a unique tetraheme *c*-type cytochrome, NapM, FEMS Microbiol. Lett. 248 (2005) 217–225.
[20] J. Grove, S. Tanapongpipat, G. Thomas, L. Griffiths, H. Crooke, J. Cole, *Escherichia coli* K12 genes essential for the synthesis of *c*-type cytochromes and a third nitrate reductase located in the periplasm, Mol. Microbiol. 19 (1996) 476–481.
[21] L.C. Potter, J.A. Cole, Essential roles for the products of the *napABCD* genes, but not *napFGH*, in periplasmic nitrate reduction by *Escherichia coli* K-12, Biochem. J. 344 (1999) 69–76.
[22] J. Simon, M. Sanger, S.C. Schuster, R. Gross, Electron transport to periplasmic nitrate reductase (NapA) of *Wolinella succinogenes* is independent of a NapC protein, Mol. Microbiol. 4 (2003) 69–79.
[23] H.J. Sears, G. Sawers, B.C. Berks, S.J. Ferguson, D.J. Richardson, Control of periplasmic nitrate reductase gene expression (*napEDABC*) from *Paracoccus pantotrophus* in response to oxygen and carbon substrates, Microbiol. 146 (2000) 2977–2985.
[24] T.H.C. Brondijk, A. Nilavongse, N. Filenko, D.J. Richardson, J.A. Cole, The NapGH components of the periplasmic nitrate reductase of *Escherichia coli* K-12: location, topology, and physiological roles in quinol oxidation and redox balancing, Biochem. J. 379 (2004) 47–55.
[25] L. Potter, P. Millington, G. Thomas, J. Cole, Competition between *Escherichia coli* strains expressing either a periplasmic or a membrane-bound nitrate reductase: does Nap confer a selective advantage during nitrate-limited growth? Biochem. J. 344 (1999) 77–84.
[26] H. Wang, C.P. Tseng, R.P. Gunsalus, The *napF* and *narG* nitrate reductase operons in *Escherichia coli* are differentially expressed in response to submicromolar concentrations of nitrate but not nitrite, J. Bacteriol. 181 (1999) 5303–5308.
[27] V. Stewart, Nitrate regulation of anaerobic gene expression in *Escherichia coli*, Mol. Microbiol. 9 (1993) 425–434.
[28] A.J. Darwin, K.L. Tyson, S.J.W. Busby, V. Stewart, Differential regulation by the homologous response regulators NarL and NarP of *Escherichia coli* K-12 depends on DNA binding site arrangement, Mol. Microbiol. 25 (1997) 583–595.
[29] I. Fedtke, A. Kamps, B. Krismer, F. Götz, The nitrate reductase and nitrite reductase operons and the *narT* gene of *Staphylococcus*

carnosus are positively controlled by the novel two-component system NreBC, J. Bacteriol. 184 (2002) 6624–6634.

[30] B.J.N. Jepson, A. Marietou, S. Mohan, J.A. Cole, C.S. Butler, D.J. Richardson, (2006) Evolution of the soluble nitrate reductase: defining the monomeric periplasmic nitrate reductase subgroup, Biochem. Soc. Trans. 34 (2006) 122–126.

[31] A.N. Antipov, D.Y. Sorokinj, N.P. L'vov, J.G. Kuenen, New enzyme belonging to the family of molybdenum-free nitrate reductases, Biochem. J. 369 (2003) 185–189.

Part II

Biological Nitrogen Fixation

Chapter 8

Physiology, Biochemistry, and Molecular Biology of Nitrogen Fixation

William E. Newton

8.1 Which organisms fix N_2?

8.1.1 Introduction

The major source of fixed-N to the biosphere is through biological N_2-fixation by micro-organisms known collectively as diazotrophs; they provide about 60% of the total annual input. No eukaryote is known to have this capability. Almost all diazotrophs use essentially the same enzyme called nitrogenase, which operates at ambient T and pressure in the soil with the sun as the ultimate energy source. Diazotrophs have little in common apart from being prokaryotes and are found in the Eubacteria and Archaea. They are generally grouped on the basis of their lifestyle, being either free-living, symbiotic, or in a loose association usually with plant roots. Free-living diazotrophs fix N_2 for their own benefit and may do so under aerobic, anaerobic, or micro-anaerobic conditions; they may be chemotrophs or phototrophs. Symbiotic diazotrophs almost

Biology of the Nitrogen Cycle
Edited by H. Bothe, S.J. Ferguson and W.E. Newton

always live and fix N_2 under micro-aerobic conditions inside a specialized structure on plant roots and provide fixed-N to their host [1].

8.1.2 Free-living microbes

Free-living chemotrophic microorganisms do not provide significant fixed-N input into agriculture, but they are scientifically important because they are easy to culture and their nitrogenase is virtually identical to that from the agriculturally important symbiotic diazotrophs called rhizobia. The obligate aerobes are best represented by some *Azotobacter* species, all of which efficiently fix N_2 in air. Most other aerobic diazotrophs are less well adapted to grow in air, but all of them possess mechanisms to prevent O_2 damaging their nitrogenase. Aerobic phototrophs are exemplified by the cyanobacteria, e.g., *Anabaena* and *Nostoc*. These supremely independent organisms not only fix their own N_2, but also use green-plant-type photosynthesis to generate their fixed-C and energy needs. However, their photosynthetic capacity presents them with a problem because it produces O_2, which would compromise nitrogenase. Their O_2-protection strategy is different from the aerobic chemotrophs. Some of them locate nitrogenase in specialized cells, called heterocysts, where a degraded photosystem-II component of photosynthesis is no longer capable of producing O_2, but where very active cyclic photophosphorylation continues to provide ATP for N_2-fixation.

The anaerobic diazotrophs are best represented by the clostridia, which have many species capable of fixing N_2. These anaerobes ferment carbon sources mainly to butyric acid, CO_2, and H_2. The anaerobic sulfate-reducing bacteria also include diazotrophs, e.g., *Desulfovibrio*, which uses SO_4^{2-} instead of O_2 as terminal respiratory electron acceptor and so produces H_2S. *Desulfovibrio* is the principal nonphotosynthetic N_2-fixing contributor to the formation of N-containing marine sediments. Other important anaerobic diazotrophs include the methanogens, which belong to the third domain of living things, the Archaea. These organisms are responsible for the CH_4 produced by ruminant animals and in stagnant water. N_2-fixation also occurs in anaerobic phototrophs, e.g., the purple nonsulfur bacterium, *Rhodospirillum rubrum*. Although photosynthetic, these diazotrophs lack a functional photosystem II, hence no O_2 is produced to impact nitrogenase adversely. Photosynthetic bacteria include both sulfur and nonsulfur bacteria and, like cyanobacteria (see above), they can utilize CO_2 as sole C-source.

A third group consists of the facultative anaerobes. They can grow with or without O_2 if provided with fixed-N, but they can only fix N_2 anaerobically. The best-studied examples are of the genus *Klebsiella*, which is related to *Escherichia coli* of the Enterobacteriaceae and contains a number of species that fix N_2. They can be isolated from soils, water, and animal intestines. There is no known natural diazotrophic strain of *E. coli*, but genetically modified diazotrophic strains have been constructed. Diazotrophic species of the related genera, *Citrobacter*, *Enterobacter*, and *Bacillus*, are also known [1].

8.1.3 Plant symbioses and associations

As described in the following chapters, there are a variety of systems, in which bacteria fix N_2 in various symbioses and associations with the roots (usually) of various plant hosts. The best understood are the *Rhizobium*-legume symbioses in which nodules result from colonization of the legume roots by rhizobia (see Chapter 10). There is a single example of rhizobia nodulating a nonlegume, *Parasponia*, which is a woody member of the elm (Ulmaceae) family. The best-studied symbioses are those involving important crops, such as peas, beans, clovers, and alfalfa. Actinomycetes, such as *Frankia*, also form symbioses but these involve the roots of nonleguminous shrubs and trees, such as alder. These plants are often the first to colonize either poor or devastated soils and, therefore, have an important ecological role. Cyanobacterial symbioses include lichens, liverworts, pteridophytes, gymnosperms, and a single angiosperm genus. Most of these cyanobacteria are also capable of independent growth and N_2-fixation. In the symbiosis, the cyanobacterium supplies fixed-N for both partners (see Chapter 11).

There are also less-formalized associations of various grasses with *Acetobacter*, *Azoarcus*, *Azotobacter*, *Azospirillum*, *Gluconacetobacter*, and *Herbaspirillum* (see Chapter 12). Although heterotrophic bacteria are often found associated with the roots of some grasses, some associations appear quite tight, even though no specialized structure is developed. In the *Digitaria-Azospirillum* association, the spirilla do invade the root tissue where they form a layer beneath the rhizodermis and stop growing but continue fixing N_2. The exact benefit to the plants is uncertain. More formalized endophytic associations involve both *Gluconacetobacter diazotrophicus* and *Herbaspirillum* spp. with sugar cane and *Azoarcus* spp. with Kallar grass and rice. Some of these associations can supply up to 60% of the fixed-N needed for the host plant's growth, indicating a significant agricultural and economical potential (see also [1]).

8.2 Nitrogenases

8.2.1 Nitrogenase types

Currently, there are four genetically distinct nitrogenases, three of which are closely related. The best studied of the three related enzymes is the conventional molybdenum-based nitrogenase (Mo-nitrogenase). The other two in this group are the alternative nitrogenases, the vanadium-based enzyme (V-nitrogenase) and an enzyme system based on iron alone

(Fe-nitrogenase). Although each enzyme has a different heterometal (Mo, V, or Fe), they are otherwise so similar that they must have arisen from a common ancestor [2]. In contrast, the fourth and so far unique nitrogenase, which was isolated from the thermophilic organism, *Streptomyces thermoautotrophicus*, is so different that it may well have been an evolutionary "independent invention" [3].

Controversy exists as to which one of the Mo-, V-, and Fe-only nitrogenases arose first. One view is that an aboriginal "nitrogenase" once existed in the form of a pyrite-forming Fe–S cluster, which produced C-bound N rather than NH_3 [4]. This view has N_2-fixation occurring prior to the existence of enzymes and suggests that Fe–nitrogenase is the oldest of these three nitrogenases. An alternative view is that nitrogenase first evolved as an assimilatory HCN/CN–reductase to detoxify local environments [5] and, because Mo-nitrogenase is most effective for catalyzing HCN reduction, it could be the most ancient nitrogenase. A third suggestion cites one primal nitrogenase from which the three related nitrogenases developed in more recent time. Support for this suggestion comes from the observation that, although separately encoded, the three sets of structural genes, which encode the polypeptide subunits of the nitrogenase proteins, were likely formed by gene iteration. Moreover, the products of certain N_2-fixation-specific genes are absolutely required for all three nitrogenases [1, 2].

Little is known about the origin and evolution of the N_2-fixation genes and the mechanisms involved in shaping the process [6]. Similarly, just how diazotrophy became distributed among the genera of Eubacteria and Archaea that fix N_2 remains unclear [7, 8]. From a classical evolutionary viewpoint, the seemingly haphazard distribution of diazotrophy among microbes could be due to a random loss of a common ancestral property during divergent evolution. Alternatively, the N_2-fixation genes could be of more recent origin and are being spread laterally, like antibiotic resistance, among diverse prokaryotic genera [9]. If the former explanation were true, there should be a match between the phylogenies (relatedness among microbes) on the basis of the sequences of both the 16S rRNA (ribosomal) genes and the genes encoding the subunits of nitrogenase. It turns out that sometimes there is a match and sometimes there is not. As more and more genomes are sequenced and the consequences of gene duplication during evolution are considered, it appears that either multiple losses or multiple transfers of the N_2-fixation genes or a combination of both could have occurred. Whatever the future outcome of this discussion, the close genetic relatedness of the three nitrogenases, including even the order in which many of the genes are found in the genome, supports a common ancient ancestry.

With the single exception of the organism that carries the unique nitrogenase, all N_2-fixing organisms have Mo-nitrogenase, but the distribution of V-nitrogenase and Fe-nitrogenase appears completely random.

Some organisms, e.g., *Klebsiella pneumoniae* and all rhizobia, have only Mo-nitrogenase, whereas others, e.g., *Azotobacter vinelandii*, have all three. The other two combinations are also found; *A. chroococcum* has both a Mo- and V-nitrogenase, whereas *Rhodobacter capsulatus* has a Mo- and Fe-nitrogenase [1, 2]. Which nitrogenase is expressed depends on the availability of Mo and V in the growth medium [10]. With Mo present, Mo-nitrogenase expression is stimulated, whereas both V- and Fe-nitrogenase expression is repressed. With V available and Mo absent, expression of only V-nitrogenase occurs. If both Mo and V are absent, then just Fe-nitrogenase is expressed. This control by metal availability is physiologically reasonable because Mo-nitrogenase is the most efficient catalyst for N_2-reduction, followed by V-nitrogenase, with the Fe-nitrogenase being the least efficient.

All three related nitrogenases consist of two component metalloproteins, which can be separately purified, but which have no activity alone. The component proteins are generally known by their trivial names, e.g., those of Mo-nitrogenase are called the MoFe protein and the Fe protein to reflect their metal contents. Sometimes, they are referred to respectively as component 1 and component 2; as Av1 and Av2, when isolated from *Azotobacter vinelandii*, for example; or as dinitrogenase and dinitrogenase reductase. For V-nitrogenase, either VFe protein and Fe protein-2 or Av1* and Av2* or $Av1^{(V)}$ and $Av2^{(V)}$ are used, whereas for Fe-nitrogenase, either FeFe protein and Fe protein-3 or $Av1^{(Fe)}$ and $Av2^{(Fe)}$ are common designations [11–13]. When either component protein from Mo-nitrogenase is mixed with the complementary component from V-nitrogenase, an active hybrid nitrogenase is produced. But, when either component protein from the Fe-nitrogenase is crossed with the complementary protein from either the Mo- or V-nitrogenase, the resulting hybrid nitrogenases are completely inactive [1, 2]. The physicochemical properties of all Mo-nitrogenases are remarkably similar, regardless of the N_2-fixing organism from which they are isolated [14]. However, there may be more differences among V- and Fe-nitrogenases, but they are much less well studied. Even so, their structures and mechanism of action appear to be very similar to those of Mo-nitrogenase.

It was not until 1960 that N_2-fixation by cell-free extracts from *Clostridium pasteurianum* was reliably demonstrated. Since then, after developing techniques for handling the extremely O_2-sensitive nitrogenase and recognizing its requirement for adenosine triphosphate (MgATP), Mo-nitrogenase has been isolated and purified from many different species. In addition, the three related nitrogenases require a reductant of sufficiently electronegative redox potential in order to function. In vitro, sodium dithionite is the reductant of choice, whereas in vivo either a flavodoxin or a ferredoxin is used, depending on the source organism. The structures of the two-component proteins of Mo-nitrogenase, both separately and as a complex, have been solved by X-ray diffraction techniques [11, 15–22].

8.2.2 Mo-nitrogenase component proteins

The Fe protein is a homodimer with a M_r of 64 kDa (see Figure 8-1). It contains one solvent-exposed 4Fe–4S cluster that bridges the two identical subunits and is bound by two cysteinyl residues from each subunit. Each subunit also possesses a MgATP/MgADP-binding site [11, 19]. The Fe protein has three major roles in the mechanism of N_2-fixation [1, 2, 12]. First, it is the obligate proximal electron donor to the MoFe protein. Attempts to drive N_2-fixation in the absence of the Fe protein have been uniformly unsuccessful. Second, the Fe protein contains the signal-transducing mechanism, whereby MgATP hydrolysis is coupled to efficient unidirectional electron transfer to the MoFe protein [1, 2, 12, 19, 23]. Third, it is involved with the biosynthesis and maturation of the MoFe protein; it is required (i) for the biosynthesis of the FeMo-cofactor, which is the N_2-binding prosthetic group (see below) that is synthesized independently from the remainder of the MoFe protein, and (ii) in the separate process of inserting the preformed FeMo-cofactor into the immature MoFe protein [1, 2, 12].

The MoFe protein is a $\alpha_2\beta_2$ heterotetramer with a molecular mass of 230 kDa (see Figure 8-1). It contains 2 Mo, 30 Fe, and 32 S^{2-} atoms per molecule that are organized into two pairs of unique metal clusters, the FeMo-cofactor and the P cluster [15–22]. The MoFe protein is often viewed as two functionally independent $\alpha\beta$ dimers; however, there is evidence of communication between the two $\alpha\beta$ dimers, specifically between the Fe protein-binding site on each dimer [24]. One FeMo-cofactor resides within each of the α-subunits, whereas one P cluster is located at each $\alpha\beta$-subunit interface near the protein's surface.

The Fe_8S_7 P cluster is bound within the protein by three cysteinyl residues from the α-subunit and three cysteinyl residues from the β-subunit. The P cluster consists of a [4Fe–4S] subcluster that shares one of its S^{2-} with a [4Fe–3S] subcluster. This very unusual situation results in a central S^{2-} being shared by the six central Fe atoms. The [4Fe–4S] subcluster is bound to both α-cysteinyl-62 and α-cysteinyl-154, whereas the [4Fe–3S] subcluster is bound to the equivalent β-subunit cysteinyls (β-cysteinyl-70 and β-cysteinyl-153). In addition, two other cysteinyl residues, α-cysteinyl-88 and β-cysteinyl-95, form thiolate bridges between the two subclusters. On oxidation by redox-active dyes, the P cluster structurally rearranges to a more open structure with two Fe atoms losing contact with the central shared S^{2-}. One of these Fe atoms then binds to β-serinyl-188 and the other to the deprotonated backbone amide-N of the already bound and bridging α-cysteinyl-88 (see Figure 8-2). Because both of these new contacts are protonated when unbound, but unprotonated when bound to the Fe atoms, these redox-induced changes suggest that oxidation of the

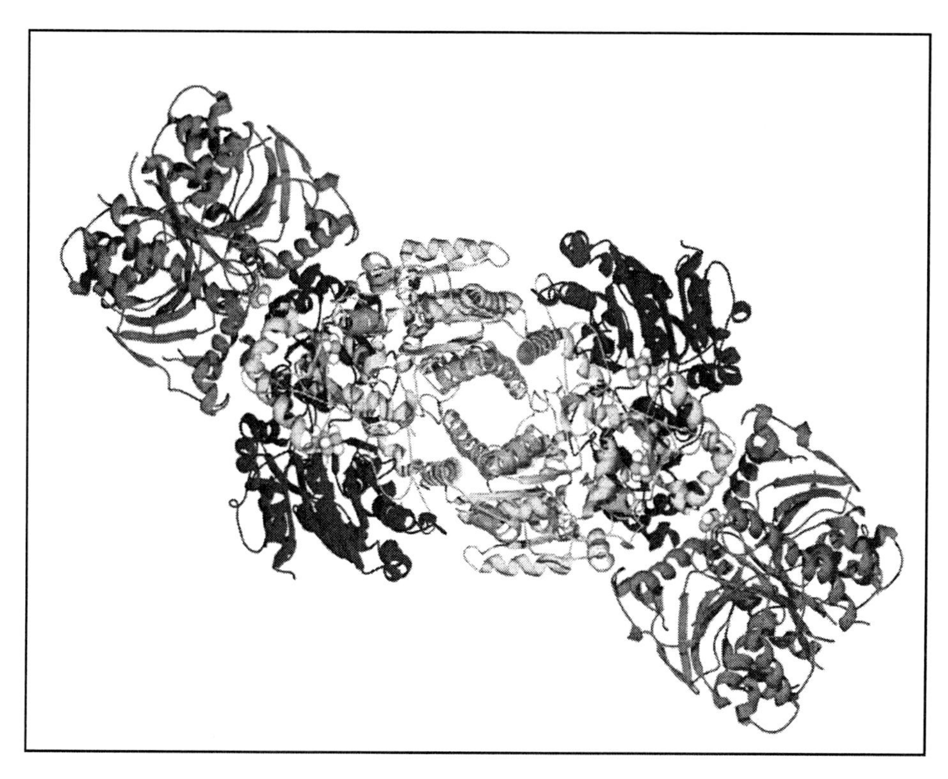

Figure 8-1 The structure of the stabilized 2:1 Fe protein-MoFe protein complex of the *Azotobacter vinelandii* nitrogenase. Each Fe-protein component (at top left and bottom right) docks at a MoFe-protein α/β-subunit interface (the α-subunit is of darker color). The [4Fe–4S] cluster of the Fe protein is just visible in the center of each interface and in direct line with the P cluster of the MoFe protein. The FeMo-cofactor is located either directly below (at left) or above (at right) of the P cluster [20]. (PDB code: 1G20).

P cluster during catalysis could cause the coupled release of two protons from the P cluster [17]. This suggestion is in line with the generally accepted role for the P cluster, which is to accept and store the electrons delivered by the Fe protein, prior to electron transfer to substrate bound at the FeMo-cofactor.

The FeMo-cofactor is the site of substrate binding and reduction. It is bound to the α-subunit polypeptide by only one cysteinyl and one histidinyl residue. Its composition is $Mo_1Fe_7S_9$ plus one homocitrate bound to the Mo [1, 2, 11]. The function of homocitrate is yet to be fully defined, but replacement by citrate clearly affects the enzyme's activity [1, 2, 21, 25]. The FeMo-cofactor can be extruded intact from the MoFe protein, most

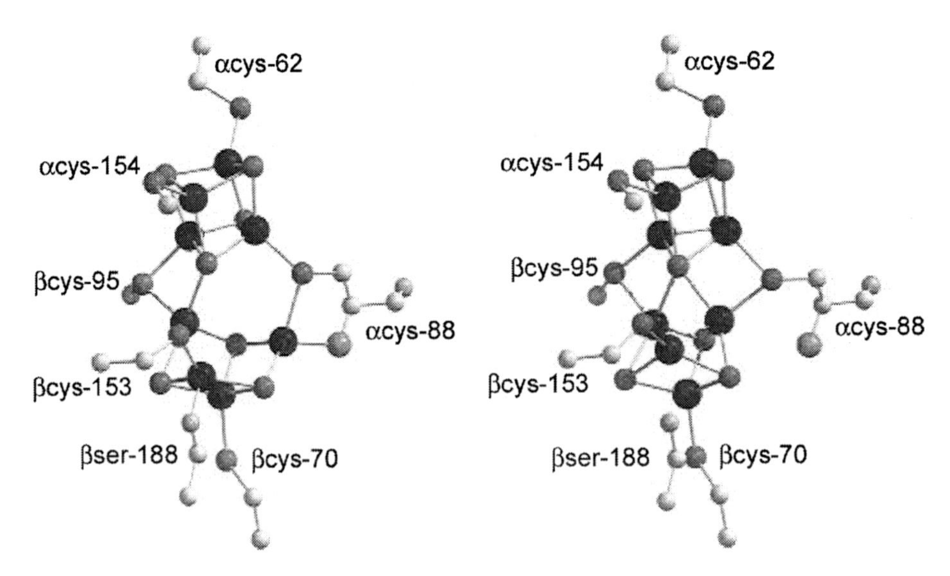

Figure 8-2 The structure of the P cluster of *Azotobacter vinelandii* nitrogenase MoFe protein in its dye-oxidized (left) and normal dithionite-reduced (right) states with amino-acid residues (α-cysteinyl62, α-cysteinyl-88, α-cysteinyl154, β-cysteinyl70, β-cysteinyl95, β-cysteinyl153, and β-serinyl188) provided by both the α- and β-subunits. The Fe atoms are the larger darker spheres and the S atoms are the medium-colored and medium-size spheres [17]. PDB code: 3MIN and 2MIN.

effectively using *N*-methylformamide (NMF) [1, 2, 12, 26]. After isolation, it no longer catalyzes N_2-reduction and cannot be crystallized. The FeMo-cofactor consists of two subclusters, one [Mo–Fe_3–S_3] and one [Fe_4–S_3], each apparently derived from the typical [4M–4S] but with one missing sulfide. The original description of the structure indicated that the two subclusters were bridged to one another by only three nonprotein-based sulfides [15], which resulted in the very unusual trigonal geometry for its six central Fe atoms. A more recent, very high-resolution structure [22] revealed a light atom (C, N, or O) in the central cavity of the FeMo-cofactor (Figure 8-3), but whether this light atom has a functional or structural role is unknown. However, it does serve to impose the much more usual tetrahedral geometry on all Fe atoms of FeMo-cofactor.

8.2.3 Interactions of the Mo-nitrogenase component proteins

Complex formation between the MoFe and Fe proteins must occur if substrate is to be reduced. Only the complex of the two proteins can hydrolyze MgATP and transfer electrons intermolecularly. X-ray crystallography [18–20] shows that a complex of two *A. vinelandii* proteins, when

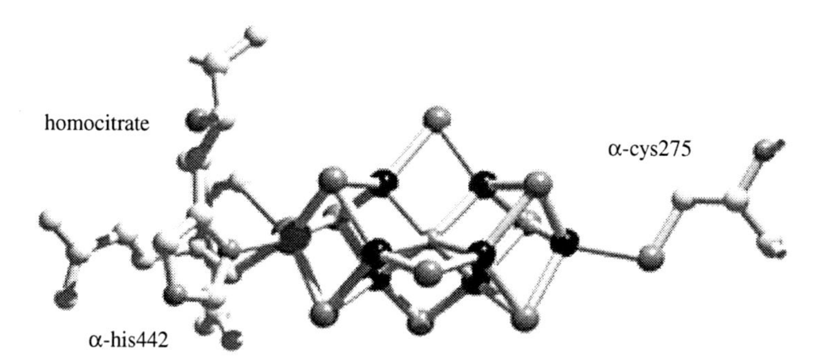

Figure 8-3 The FeMo-cofactor of *Azotobacter vinelandii* nitrogenase MoFe protein, including the α-subunit residues, α-cysteinyl275 and α-histidinyl442, and homocitrate. The Mo is the largest sphere at left of center, the Fe atoms are the darkest spheres, and the S atoms are paler in color. The identity of the central atom is unknown ([22]; PDB code: 1M1N).

trapped in a form that is unable to turn over, consists of an $(\alpha\beta\gamma_2)_2$ octamer, which has one Fe protein (a γ_2-dimer) bound to each αβ-subunit pair of the MoFe protein, i.e., two Fe proteins are bound to each MoFe protein (see Figure 8-1). When compared to crystal structures of the individual proteins, most of the structural changes within this complex occur in the Fe protein, where its structure becomes more compact. Whether a similar complex occurs during enzyme turnover is unknown, an alternative scenario is that the Fe protein alternates its servicing of the two α,β-dimers such that only one Fe protein is bound at any given time.

8.2.4 Mo-nitrogenase and substrate reduction

The minimum requirements for biological N_2-fixation are both proteins, N_2, reductant (flavodoxin, ferredoxin, or sodium dithionite), MgATP (plus an ATP-regenerating system) and an anaerobic environment. An ATP-regenerating system is needed because the MgADP formed from MgATP hydrolysis inhibits nitrogenase catalysis by competing for the MgATP-binding sites on the Fe protein. The optimal stoichiometry for N_2-reduction requires hydrolysis of four molecules of ATP for each pair of electrons incorporated into substrate, when dithionite is used as reductant.

$$N_2 + 8e^- + 8H^+ + 16MgATP \rightarrow 2NH_3 + H_2 + 16MgADP + 16P_i$$

The direction of electron flow is from reductant to Fe protein to MoFe protein and finally to bound substrate. A comprehensive kinetic model of nitrogenase action quantifies many of the individual steps of the catalytic reaction [12, 27]. The basis of this model is two interconnected cycles, the

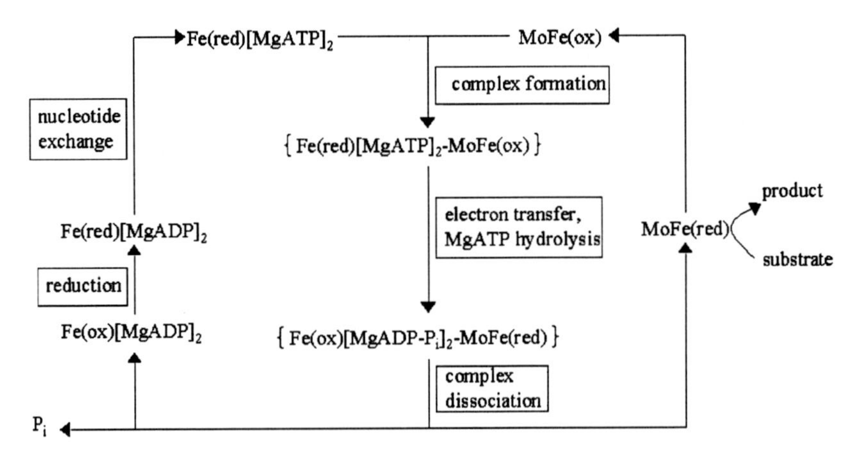

Figure 8-4 Reactions of the Fe protein (Fe) with the MoFe protein (MoFe) during catalysis. The result of the illustrated reactions is the transfer one electron from the Fe protein to the MoFe protein [27]. Therefore, several rounds must occur and a succession of electrons must be transferred from the Fe protein and accumulated on the MoFe protein before substrate reduction can proceed. (ox) is oxidized; (red) is reduced; P_i is phosphate.

Fe-protein cycle and the MoFe-protein cycle. The Fe-protein cycle describes complex formation, electron transfer from the Fe protein to the MoFe protein, MgATP hydrolysis, and the subsequent re-reduction of now oxidized Fe protein for the next round of the cycle (Figure 8-4). The MoFe-protein cycle depicts the sequence of partial reactions that occur within the MoFe protein during N_2-reduction to NH_3. For each N_2 reduced by six electrons, one H_2 (requiring two electrons) is evolved, even under a N_2 pressure of 50 atmospheres [28]. The purpose of this apparently mandatory evolution of H_2 is unknown, but it might simply be to clear the active site so that N_2 can bind. Whatever the reason, it means that each round of the MoFe-protein cycle requires eight electrons and involves eight individual partial reactions, each of which represents one turn of the Fe-protein cycle. This model also indicates that N_2 is bound to the active site only after three electrons have been accumulated within the MoFe protein. How and where these electrons are stored prior to N_2 binding is unknown.

Because N_2-reduction has such rigorous requirements, it should not be surprising that nitrogenase can catalyze the reduction of a wide variety of other more easily reduced small molecules containing N–N, N–O, N–C, and C–C multiple bonds. Most notable of these is C_2H_2, which is reduced by Mo-nitrogenase to C_2H_4 only. This reaction is very often used as an indicator of N_2-fixation activity in preparations from suspected diazotrophs (see Chapter 13). When no other reducible substrate is present,

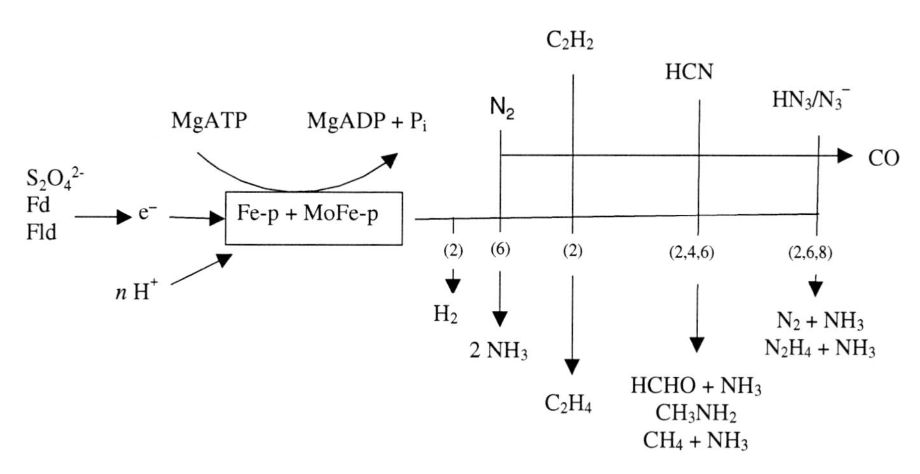

Figure 8-5 Electron donors, substrates, and products of Mo-nitrogenase catalysis. CO is a potent reversible inhibitor of all catalyzed reductions except for H^+ to H_2. Fe-p is Fe protein, MoFe-p is MoFe protein, Fld is flavodoxin, and Fd is ferredoxin. The numbers in parentheses are the electrons required for each reaction shown.

nitrogenase will reduce H^+ to H_2. Figure 8-5 shows an abbreviated selection of substrates and their products that includes CO, which is not a substrate but is a potent inhibitor of all nitrogenase-catalyzed substrate reductions except that of H^+ to H_2 [15]. Substrate reduction is not rate limiting and so nitrogenase turns over at the same rate no matter which substrate is being reduced [27]. H_2 is not only a product of Mo-nitrogenase action, it is also a specific inhibitor of N_2-reduction, but H_2 affects neither the reduction of any other substrate nor its own evolution. Furthermore, under a mixed atmosphere of N_2 and D_2, HD is formed in a reaction that has all the requirements of a nitrogenase-catalyzed reaction, including inhibition by CO [1, 2, 12, 29].

8.2.5 V-nitrogenase and Fe-nitrogenase

V-nitrogenase has been found in aerobes (*A. vinelandii* and *A. chroococcum*), phototrophs (*Anabaena variabilis*), and Archaea (*Methanosarcina barkeri*); Fe-nitrogenase has been cloned and sequenced from *A. vinelandii*, *Rhodobacter capsulatus*, and *Rhodospirillum rubrum*. Major differences with respect to Mo-nitrogenase involve only the larger component protein [1, 2, 13]. Both the VFe and FeFe proteins have δ-subunits in addition to the α- and β-subunits that result in an $\alpha_2\beta_2\delta_2$-subunit composition. This small (M_r ~15 kDa) δ-subunit is essential for activity. The δ-subunits are apparently bound during biosynthesis of the VFe (and probably FeFe) protein and remain bound thereafter. In contrast, the equivalent of the δ-subunit for the MoFe protein is lost in the late stages of its maturation. Both the VFe and FeFe proteins contain P clusters and cofactors, with either a vanadium or a Fe atom replacing Mo, respectively, in the latter. The high level of primary sequence identity recognized among the Mo-nitrogenases also extends to the V- and Fe-nitrogenases. This identity strongly suggests that all nitrogenases share common structural and mechanistic features.

8.2.6 Streptomyces thermoautotrophicus *nitrogenase*

This unique nitrogenase (St-nitrogenase) is found, so far, only in the thermophile, *Streptomyces thermoautotrophicus* [3], which grows by reducing CO_2 to organic matter, but instead of using sunlight as reductant like a plant, it uses CO. Unlike conventional Mo-nitrogenase, CO is not an inhibitor, C_2H_2 is not a substrate, and O_2 has no effect on St-nitrogenase. Even though it consists of two-component proteins, the larger of which contains Mo, Fe, and S^{2-}, this is where the similarity to Mo-nitrogenase ends. It has no Fe-protein component but rather a manganese-containing superoxide oxidoreductase, which oxidizes O_2^- to O_2 and transfers the electron to the MoFeS-containing protein. This MoFeS-containing protein is a $\alpha\beta\gamma$ heterotrimer, quite different from the $\alpha_2\beta_2$ composition of Mo-nitrogenase, and it furnishes the site at which N_2 is reduced to NH_3 accompanied by H_2-evolution. This eight-electron reaction requires less MgATP hydrolyzed per N_2 reduced than does Mo-nitrogenase, but the rate of N_2-reduction is less than 1% of that of the conventional Mo-nitrogenase. Further, the electron donor is neither a flavodoxin nor a ferredoxin but a Mo-containing CO dehydrogenase, which couples the oxidation of CO to the reduction of O_2 to produce O_2^-. Clearly, only additional research will establish the structural and mechanistic properties of this unusual enzyme.

8.2.7 Where do substrates and inhibitors bind?

The substrates and inhibitors of nitrogenase are all small multiple-bonded molecules, like N_2 and CO, which are well known to bind to metal atoms. Although assumed to do so with nitrogenase, direct evidence to support this assumption has been difficult to get. A major problem is that neither component protein alone binds substrate but, if both components (plus reductant and MgATP) are present, then enzymatic turnover occurs immediately! Circumstantial evidence for a role for the FeMo-cofactor in substrate binding came from observing the effects of disrupting various N_2-fixation-specific genes (see later). Initially, these studies showed that the FeMo-cofactor is separately biosynthesized and then inserted later on during the maturation of the MoFe protein. Furthermore, those mutant strains, which were unable to biosynthesize FeMo-cofactor, were also unable to catalyze N_2-fixation, but could be reactivated by adding isolated

FeMo-cofactor to a crude extract [1, 2, 26]. In addition, when homocitrate synthesis was interrupted (by deletion of *nifV*), the homocitrate-deficient FeMo-cofactor had a different pattern of reactivity toward various substrates [25]. Later, when amino acid substitutions were placed within the FeMo-cofactor's polypeptide environment, they were found to alter simultaneously both the FeMo-cofactor-based spectroscopic properties and the enzyme's catalytic activities [30].

Early direct spectroscopic evidence clearly indicated that substrates and inhibitors only bind to nitrogenase under turnover conditions. It was work with CO that provided the lead in locating potential binding sites. CO, as a potent noncompetitive inhibitor of all substrate reduction except that of H^+, remains bound to its site while nitrogenase is producing H_2. When Mo-nitrogenase turns over under CO, a unique spectroscopic feature (the $S = 3/2$ electron paramagnetic resonance [EPR] signal arising from the FeMo-cofactor within the MoFe protein) disappears and either of two new $S = 1/2$ EPR signals become visible on freeze-quenching the sample [31]. One signal appears at low (<0.1 atm) CO concentrations and the second at high (>0.5 atm) CO concentrations. After isotopic labeling of both the CO (with ^{13}C) and the FeMo-cofactor (by growing cells on ^{57}Fe), a technique called electron nuclear double resonance (ENDOR) was applied to these signals to determine if there was an interaction between CO and the FeMo-cofactor. The results showed that, at low CO concentrations, one CO interacts with the Fe atoms, likely forming a bridge between two Fe atoms. At high CO concentrations, two nonbridging CO molecules are present, each of which likely binds to a different Fe atom. Moreover, because the two $S = 1/2$ EPR signals interconvert by simply adding or removing CO, the single bridging CO (under low CO concentrations) likely converts to a nonbridging CO under high CO concentrations and a second CO occupies the vacated Fe site [32]. These results are entirely consistent with a sophisticated infrared spectroscopy study of Mo-nitrogenase turning over under CO [33]. Two vibrations, which were indicative of nonbridging bound CO molecules, were observed under high CO concentration, whereas only one lower frequency band was seen under low CO conditions.

Using variants of Mo-nitrogenase isolated from mutant strains, similar freeze-quench EPR/ENDOR studies showed that $HC{\equiv}C{-}R$ (where $-R{=}H$, $-CH_2OH$, or $-CH_2NH_2$) and CN^- can individually interact with the FeMo-cofactor [34]. These studies have been extended to $^{15}N_2$, $^{15}N_2H_4$, and $^{15}NH{=}N{-}CH_3$ as substrates. Using either wild-type or variant Mo-nitrogenases, these species (or derivatives of them) have been trapped and observed spectroscopically interacting with the FeMo-cofactor [35]. Detailed analysis of these and similar interactions should provide insight into exactly where and how substrates are bound to the FeMo-cofactor.

Early attempts to use theoretical calculations as a means to pinpoint binding sites were hampered by the physical size and unique chemical nature of the FeMo-cofactor. In addition, limited knowledge of its total spin state and the oxidation states of its constituent metal atoms, especially during catalysis, also led to uncertainties. Even so, most studies favored N_2 binding via a bridge between either two or four of the central Fe atoms [36, 37]. All of the putative metal–N_2 interactions resulted in a small N_2-binding energy and, therefore, a weak bond. Furthermore, direct N_2 cleavage to metal-nitrides was not favorable, suggesting that protonation of bound N_2 must occur. Then, the central light atom within the FeMo-cofactor was discovered, assumed (maybe incorrectly) to be N, and included in new calculations [38, 39]. The major consequence is that all of the previously favored two or four Fe atom combinations become less able to accommodate bridging Fe–N_2–Fe interactions, so much so that N_2 binding at a Fe_4 face was ruled out in favor of either an Fe_2 edge or a single Fe atom. An alternative view is that Mo may provide the N_2-binding site; why else is there a heterometal in the enzyme [40–42]? In fact, there is a wealth of information on the reactions of both model compounds and isolated FeMo-cofactor that is consistent with this possibility [43, 44]. Furthermore, theoretical calculations agree that, after dissociation of the carboxylate arm of homocitrate from the Mo atom, the Mo atom could be the preferred N_2-binding site [42].

8.3 MgATP and Mo-nitrogenase catalysis

The overall reduction of N_2 to yield two molecules of NH_3 is thermodynamically favorable, so why is MgATP binding and hydrolysis required for nitrogenase catalysis? If it is not a thermodynamic requirement, it must be used for kinetic purposes. Currently, a "gating" mechanism is preferred in which MgATP hydrolysis serves both to drive electron transfer toward substrate reduction and to prevent electrons flowing back into the Fe protein. This mechanism makes the system effectively irreversible [11, 23].

The binding of MgATP to the Fe protein produces a conformational change that allows the Fe protein to form a complex with the MoFe protein. MgATP binding lowers the redox potential of the Fe protein's [4Fe–4S] cluster by ca. −200mV and complex formation lowers it by an additional ca. −200mV to a final value of ca.−600mV. Simultaneously on complex formation, the redox potential of the P cluster is also lowered by ca. −100mV to ca. −400 mV. So complex formation increases the driving force (by increasing the difference in redox potential of the donor and acceptor) for electron transfer from the Fe protein's [4Fe–4S] cluster to the P cluster. In contrast, complex formation has no effect on the redox potential of the FeMo-cofactor [45]. In addition, complex formation triggers MgATP hydrolysis, although whether hydrolysis occurs shortly before, concomitantly with, or shortly after electron transfer, or varies depending on other factors is unknown. It is clear, however, that phosphate release (which is usually the work step) from the complex occurs after electron transfer and does not drive the dissociation of the complex. These several reactions cause the conformational change of the Fe protein to relax and the complex dissociates, thus preventing any backflow of electrons to the Fe protein. Complex dissociation is the rate-limiting step in Mo-nitrogenase catalysis [27]. In support of this model, primary amino acid sequence and structural comparisons align the Fe protein within a large class of signal-transduction proteins that undergo conformational changes on nucleotide binding and hydrolysis.

8.4 Genetics of N_2-fixation

By the early 1980's, the suite of genes (called *nif* genes) required for N_2-fixation by Mo-nitrogenase was established for the facultative anaerobe, *K. pneumoniae.* It encompassed genes whose products encoded not only the polypeptides of the two component proteins, but also those required for maturation of these proteins, for biosynthesis of the FeMo-cofactor, for components of a specific electron–donor system, and for the regulation of the expression of these genes. This organism was found to have 20 *nif* genes arranged in eight contiguous transcriptional units spanning ca. 23kb on the chromosome. These transcriptional units are not all read in the same direction and so either single or groups of genes can be read independently of each other. The specific designations used for individual *K. pneumoniae nif* genes are also used with the genes of other diazotrophs whose products have homologous functions [46]. Thus, *nifH*, *nifD*, and *nifK* are used to designate the *nif* structural genes of all diazotrophs; they encode the Fe protein polypeptide and the MoFe protein α- and β-subunits, respectively. However, some organisms do not have homologues of all the *K. pneumoniae nif* genes and other organisms have N_2-fixation-related genes that are not present in *K. pneumoniae.* Under these circumstances, N_2-fixation-related genes present in other organisms without functional counterparts in *K. pneumoniae* are given other designations, e.g., "*fix*" is commonly used with rhizobia. Significant differences in organization also occur. The putative amino acid sequences of all *nif*-gene products are known and the majority of the products identified (see Figure 8-6). A number of genes associated with either V- or Fe-nitrogenase have also been identified. The genes unique to V-nitrogenase are designated by *vnf* and those for Fe-nitrogenase by *anf.* These include the structural genes for both the V-nitrogenase (*vnfHDGK*) [47, 48] and the Fe-nitrogenase (*anfHDGK*) [49]. In all cases, the same following capital letter indicates a similar sequence for the gene and function for the gene product; *vnfG* and *anfG* encode the δ-subunit.

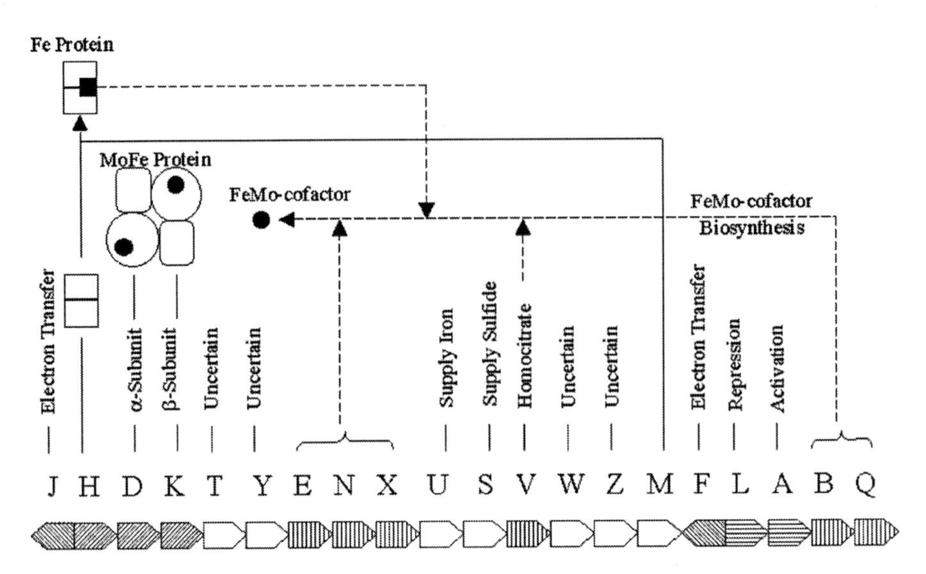

Figure 8-6 The nitrogen-fixation (*nif*) genes of *Klebsiella pneumoniae* together with the putative functions of their products. Genes with the same shading have related functions.

The primary translation products of the structural genes are inactive and the products of other *nif*-specific genes are required to activate these immature structural components. Several of these products catalyze the formation and insertion of the metalloclusters into immature forms of the Fe protein and MoFe protein. Maturation of the Fe protein requires only the *nifM* gene product. In contrast, maturation of the MoFe protein is much more complicated. The formation and insertion of the FeMo-cofactor into immature "MoFe protein" involves the products of, at least, the *nifH*, *nifE*, *nifN*, *nifB*, *nifV*, and *nifQ* genes. The *nifH* requirement is expected because, as mentioned above, the Fe protein is required for both formation and insertion of the FeMo-cofactor. Two other *nif*-gene products, those of *nifS* and *nifU*, catalyze reactions involved in the general mobilization of Fe and S for metallocluster assembly [1, 5].

When the V- and Fe-nitrogenases are considered, the complications increase because the products of the *nifMBVUS* genes are essential for their functional activity as well as for that of the Mo-nitrogenase [1, 2]. The *nifM* requirement indicates that all three Fe proteins are so similar that a single NifM protein can process them all. The common requirement for the *nifBV*-gene products, both of which are involved with FeMo-cofactor biosynthesis, suggests similar cofactors in all three systems. Moreover, the *nifV* requirement shows that homocitrate is a constituent of all three cofactors as are S^{2-} and Fe, which would be provided through the requirement for the products of the *nifUS* genes.

8.5 Regulation of N_2-fixation

The high-energy demand for N_2-fixation, the sensitivity to O_2 by the three related nitrogenases, and the heterometal-ion requirement necessitate rigorous control of *nif*-gene expression. Regulation is exerted at several different levels, but varies from organism to organism. The first environmental regulatory factor is the fixed-N status. This factor is very important and bacteria will totally deplete any fixed-N source before expressing nitrogenase to fix their own N_2. This is a particularly stringent control mechanism for free-living N_2 fixers, like *A. vinelandii*, but less so for symbiotic rhizobia, which are adapted to export fixed-N to their host. Photosynthetic bacteria use a reversible posttranslational modification to inactivate the Fe protein in response to NH_3. The second factor, O_2 tension, is important because nitrogenases are very susceptible to oxidative damage. Some aerobes respire at very high rates to maintain a low internal O_2 concentration, others induce their nitrogenase to form a transient protein aggregate, when the O_2 tension increases, to provide "conformational protection" against O_2 damage. A completely different protection mechanism occurs in some filamentous cyanobacteria, which developmentally regulate *nif*-gene expression so that nitrogenase is only synthesized in the specialized heterocysts that do not produce O_2 photosynthetically. The third regulatory factor is the availability of certain metal ions. As mentioned earlier, the presence of Mo in the growth medium represses the expression of both the V- and Fe-nitrogenases.

The effects, in many of the above examples, are mediated through a common mechanism of transcriptional regulation of the *nif* genes by the products of *nif*-specific regulatory genes, *nifLA* [50]; however, the *nifLA* system is not present in all N_2-fixing organisms. The *nifA*-gene product (NifA) activates transcription from all *nif* promoters, whereas NifL is a flavin-containing sensor protein that binds to and inactivates NifA in response to either high fixed-N or high O_2 concentrations. The expression of the *nifLA* genes is, in turn, controlled by the *ntr* (*n*itrogen *r*egulation) system (see Chapter 9).

8.6 Perspectives and future research

Many mysteries concerning biological N_2-fixation remain, especially those related to the intimate details of N_2 binding and reduction. The requirements for N_2-reduction may be so stringent that only a single site on the MoFe protein can meet them. In contrast, it is now clear that many of the alternative more easily reduced substrates, like C_2H_2, have more than one binding and reduction site. Sorting out these different sites and determining their physiological relevance may prove very useful. We now have Mo-nitrogenase variants that will bind but not reduce N_2. Here, N_2 is acting as a reversible inhibitor of electron flow through the enzyme resulting in a decreased H_2-evolution rate. Will it be possible to construct a variant that binds N_2 as an irreversible inhibitor and so determine exactly where and how N_2 is bound to the FeMo-cofactor? Another challenge is to determine the function of the P clusters. Is electron transfer and storage really their role in N_2-fixation? If so, what are the logistics of handling the eight electrons involved in N_2-reduction and concomitant H_2-evolution? Finally, how will we apply what we have learned about the biochemistry and genetics of nitrogenases to enhance the N_2-fixation capabilities of microorganisms, to endow new organisms with this ability, and to develop new commercial N-fertilizer production systems? Only time and creative research will tell. The reader is referred to a series of seven specialized volumes (Nitrogen Fixation: Origins, Applications, and Research Progress; William E. Newton, chief editor; published by Springer, 2004–2006) for additional insights and information.

References

[1] W.E. Newton, Nitrogen fixation, in: Kirk-Othmer Encyclopedia of Chemical Technology, John Wiley & Sons, New York, NY, 2005. DOI:10.1002/0471238961.1409201814052320.a01.pub2

[2] K. Fisher, W.E. Newton, Nitrogen fixation – a general overview, in: G.J. Leigh (Ed.) Nitrogen Fixation at the Millennium, Elsevier, Amsterdam, 2002, pp. 1–34.

[3] M. Ribbe, D. Gadkari, O. Meyer, N_2 fixation by *Streptomyces thermoautotrophicus* involves a molybdenum-dinitrogenase and a manganese-superoxide oxidoreductase that couple N_2 reduction to the oxidation of superoxide produced from O_2 by a molybdenum-CO dehydrogenase, J. Biol. Chem. 272 (1997) 26627–26633.

[4] G. Wächsterhäuser, Before enzymes and templates: theory of surface metabolism, Microbiol. Rev. 52 (1988) 452–488.

[5] J.R. Postgate, R.R. Eady, The evolution of biological nitrogen fixation, in: H. Bothe, F.J. DeBruijn, W.E. Newton (Eds.) Nitrogen Fixation: Hundred Years After, Gustav Fischer, Stuttgart, 1988, pp. 31–40.

[6] R. Fani, R. Gallo, P. Lio, Molecular evolution of nitrogen fixation: the evolutionary history of the *nifD*, *nifK*, *nifE*, and *nifN* genes, J. Mol. Evol. 51 (2000) 1–11.

[7] J.P.W. Young, Phylogenetic classification of nitrogen-fixing organisms, in: G. Stacey, R.H. Burris, H.J. Evans (Eds.) Biological Nitrogen Fixation, Chapman and Hall, New York, 1992, pp. 43–86.

[8] J.P.W. Young, The phylogeny and evolution of nitrogenases, in: R. Palacios, W.E. Newton (Eds.) Genomes and Genomics of

Nitrogen-Fixing Organisms, Springer, Dordrecht, The Netherlands, 2005, pp. 221–241.
[9] J.R. Postgate, Evolution within nitrogen-fixing systems, Symp. Soc. Gen. Microbiol. 24 (1974) 263–292.
[10] R.N. Pau, Metals and nitrogenase, in: G.L. Eichhorn, L.G. Marzilli (Eds.) Advances in Inorganic Biochemistry, Vol. 10, PTR Prentice Hall, New Jersey, 1994, pp. 49–70.
[11] J.B. Howard, D.C. Rees, Structural basis of biological nitrogen fixation, Chem. Rev. 96 (1996) 2965–2982.
[12] B.K. Burgess, D.J. Lowe, Mechanism of molybdenum nitrogenase, Chem. Rev. 96 (1996) 2983–3011.
[13] R.R. Eady, Structure-function relationships of alternative nitrogenases, Chem. Rev. 96 (1996) 3013–3030.
[14] D.W. Emerich, R.H. Burris, Complementary functioning of the component proteins of nitrogenase from several bacteria, J. Bacteriol. 134 (1978) 936–943.
[15] J. Kim, D.C. Rees, Structural models for the metal centers in the nitrogenase molybdenum-iron protein, Science 257 (1992) 1677–1682.
[16] M.K. Chan, J. Kim, D.C. Rees, The nitrogenase FeMo-cofactor and P-cluster pair: 2.2 Å resolution structures, Science 260 (1993) 792–794.
[17] J.W. Peters, M.B.H. Stowell, S.M. Soltis, M.G. Finnegan, M.K. Johnson, D.C. Rees, Redox-dependent structural changes in the nitrogenase P-cluster, Biochemistry 36 (1997) 1181–1187.
[18] H. Schindelin, C. Kisker, J.L. Schlessman, J.B. Howard, D.C. Rees, Structure of ADP.AIF$_4^-$-stabillized nitrogenase complex and its implications for signal transduction, Nature 387 (1997) 370–376.
[19] S.B. Jang, L.C. Seefeldt, J.W. Peters, Insights into nucleotide signal transduction in nitrogenase: structure of an iron protein with MgADP bound, Biochemistry 39 (2000) 14745–14752.
[20] H.-J. Chiu, J.W. Peters, W.N. Lanzilotta, M.J. Ryle, L.C. Seefeldt, J.B. Howard, D.C. Rees, MgATP-bound and nucleotide-free structures of a nitrogenase protein complex between the Leu127Δ-Fe protein and the MoFe protein, Biochemistry 40 (2001) 641–650.
[21] S.M. Mayer, C.A. Gormal, B.E. Smith, D.M. Lawson, Crystallographic analysis of the MoFe protein of nitrogenase from a *nifV* mutant of *Klebsiella pneumoniae* identifies citrate as a ligand to the molybdenum of iron molybdenum cofactor (FeMoco), J. Biol. Chem. 277 (2002) 35263–35266.
[22] O. Einsle, A. Tezcan, S.L.A. Andrade, B. Schmid, M. Yoshida, J.C. Howard, D.C. Rees, Nitrogenase MoFe-protein at 1.16 Å resolution: a central ligand in the FeMo-cofactor, Science 297 (2002) 1696–1700.
[23] L.C. Seefeldt, D.R. Dean, Role of nucleotides in nitrogenase catalysis, Acc. Chem. Res. 30 (1997) 260–266.

[24] S. Maritano, S.A. Fairhurst, R.R. Eady, Long-range interactions between Fe protein binding sites of the MoFe protein of nitrogenase, J. Biol. Inorg. Chem. 6 (2001) 590–600.

[25] P.A. McLean, B.E. Smith, R.A. Dixon, Nitrogenase of *Klebsiella pneumoniae nifV* mutants, Biochem. J. 211 (1983) 589–597.

[26] V.K. Shah, W.J. Brill, Isolation of an iron-molybdenum cofactor from nitrogenase, Proc. Natl. Acad. Sci. USA 74 (1977) 3249–3253.

[27] D.J. Lowe, R.N.F. Thorneley, The mechanism of *Klebsiella pneumoniae* nitrogenase action, Biochem. J. 224 (1984) 877–909.

[28] F.B. Simpson, R.H. Burris, A nitrogen pressure of 50 atmospheres does not prevent evolution of hydrogen by nitrogenase, Science 224 (1984) 1095–1097.

[29] K. Fisher, M.J. Dilworth, W.E. Newton, Differential effects on N_2 binding and reduction, HD formation, and azide reduction with α-195His- and α-191Gln-substituted MoFe proteins of *Azotobacter vinelandii* nitrogenase, Biochemistry 39 (2000) 15570–15577.

[30] D.J. Scott, H.D. May, W.E. Newton, K.E. Brigle, D.R. Dean, Role for the nitrogenase MoFe protein α-subunit in FeMo-cofactor and catalysis, Nature 343 (1988) 188–190.

[31] L.M. Cameron, B.J. Hales, Investigation of CO binding and release from Mo-nitrogenase during catalytic turnover, Biochemistry 37 (1998) 9449–9456.

[32] H.-I. Lee, L.M. Cameron, B.J. Hales, B.M. Hoffman, CO binding to the FeMo cofactor of CO-inhibited nitrogenase: ^{13}CO and ^{1}H Q-band ENDOR investigation, J. Am. Chem. Soc. 119 (1997) 10121–10126.

[33] S.J. George, G.A. Ashby, C.W. Wharton, R.N.F. Thorneley, Time-resolved binding of carbon monoxide to nitrogenase monitored by stopped-flow infrared spectroscopy, J. Am. Chem. Soc. 119 (1997) 6450–6451.

[34] P.C. Dos Santos, R.Y. Igarashi, H.-I. Lee, B.M. Hoffman, L.C. Seefeldt, D.R. Dean, Substrate interactions with the nitrogenase active site, Acc. Chem. Res. 38 (2005) 208–214.

[35] B.M. Barney, T.-C. Yang, R.Y. Igarashi, P.C. Dos Santos, M. Laryukhin, H.-I. Lee, B.M. Hoffman, D.R. Dean, L.C. Seefeldt, Intermediates trapped during nitrogenase reduction of N≡N, CH_3–N=NH, and H_2N–NH_2, J. Am. Chem. Soc. 127 (2005) 14960–14961.

[36] H. Deng, R. Hoffmann, How N_2 might be activated by the FeMo-cofactor in nitrogenase, Angew. Chem., Intl. Ed. Engl. 32 (1993) 1062–1065.

[37] I. Dance, The binding and reduction of dinitrogen at an Fe_4 face of the FeMo cluster of nitrogenase, Aust. J. Chem. 47 (1994) 979–990.

[38] I. Dance, The consequences of an interstitial N atom in the FeMo cofactor of nitrogenase, Chem. Commun. (2003) 324–325.

[39] T. Lovell, T. Liu, D.A. Case, L. Noodleman, Structural, spectroscopic, and redox consequences of a central ligand in the FeMoco of nitrogenase: a density functional theoretical study, J. Am. Chem. Soc. 125 (2003) 8377–8383.

[40] K.L.C. Grönberg, C.A. Gormal, M.C. Durrant, B.E. Smith, R.L. Henderson, Why R-homocitrate is essential to the reactivity of FeMo cofactor of nitrogenase: studies on the NifV$^-$-extracted FeMo cofactor, J. Am. Chem. Soc. 120 (1998) 10613–10621.

[41] C.J. Pickett, The Chatt cycle and the mechanism of enzymic reduction of molecular nitrogen, J. Biol. Inorg. Chem. 1 (1996) 601–606.

[42] M. Durrant, An atomic level mechanism for molybdenum nitrogenase. Part 1. Reduction of dinitrogen, Biochemistry 41 (2002) 13934–13945.

[43] R.L. Richards, Reactions of small molecules at transition metal sites: studies relevant to nitrogenase an organometallic enzyme, Coord. Chem. Rev. 154 (1996) 83–97.

[44] D.V. Yandulov, R.R. Shrock, Catalytic reduction of dinitrogen to ammonia at a single molybdenum cluster, Science 301 (2003) 76–78.

[45] W.N. Lanzilotta, L.C. Seefeldt, Changes in midpoint potentials of the nitrogenase metal centers as a result of iron protein-MoFe protein complex formation, Biochemistry 36 (1997) 12976–12983.

[46] M.R. Jacobson, K.E. Brigle, L.T. Bennett, R.A. Setterquist, M.S. Wilson, V.L. Cash, J. Beynon, W.E. Newton, D.R. Dean, Physical and genetic map of the major *nif* gene cluster from *Azotobacter vinelandii*, J. Bacteriol. 171 (1989) 1017–1027.

[47] R.R. Eady, Current status of structure function relationships of vanadium nitrogenase, Coord. Chem. Rev. 237 (2003) 23–30.

[48] R.D. Joerger, T.M. Loveless, R.N. Pau, L.A. Mitchenall, B.H. Simon, P.E. Bishop, Nucleotide sequences and mutational analysis of the structural genes for nitrogenase-2 of *Azotobacter vinelandii*, J. Bacteriol. 172 (1990) 3400–3408.

[49] R.D. Joerger, M.R. Jacobson, R. Premakumar, E.D. Wolfinger, P.E. Bishop, Nucleotide sequences and mutational analysis of the structural genes (*anfHDGK*) for the second alternative nitrogenase from *Azotobacter vinelandii*, J. Bacteriol. 171 (1989) 1075–1086.

[50] R. Dixon, The oxygen-responsive NIFL-NIFA complex: a novel two-component regulatory system controlling nitrogenase synthesis in γ-proteobacteria, Arch. Microbiol. 169 (1998) 371–380.

Chapter 9

Regulatory Cascades To Express Nitrogenases

Bernd Masepohl and Karl Forchhammer

9.1. Introduction

Dinitrogen accounts for about 80% of the atmosphere, but it must be reduced to NH_3 before it can be assimilated. Reduction of N_2 to NH_3 is catalyzed by an enzyme called nitrogenase, which is highly sensitive toward O_2 damage. To date, nitrogenases have been isolated from more than 30 different diazotrophic bacteria, and DNA sequences of N_2-fixation (*nif*) genes are available for more than 100 species. The organization of *nif* genes and the structure of their products suggest that N_2 fixation is based on similar modules in all N_2-fixing bacteria and archaea. However, despite the high degree of conservation of nitrogenases and of the central module for sensing the cellular (fixed) N status (the so-called P_{II} protein; see below), the regulatory mechanisms to express the *nif* genes in response to environmental signals differs remarkably among different phylogenetic groups. Many diazotrophic bacteria do not "rely" on a single mechanism to control the N_2-fixation process but instead have developed regulatory cascades (Figure 9-1).

Biology of the Nitrogen Cycle
Edited by H. Bothe, S. Ferguson and W.E. Newton

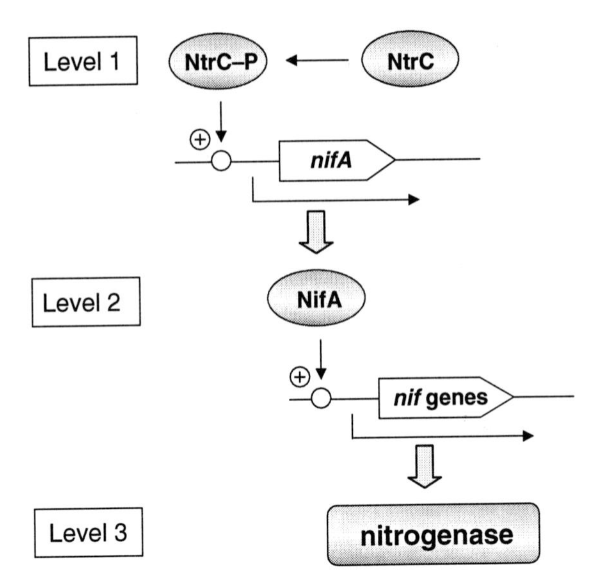

Figure 9-1 The levels of regulation of N_2-fixation in proteobacteria. The scheme depicts three levels of a regulatory cascade controlling synthesis (levels 1 and 2) and activity of nitrogenase (level 3). It is worth to note, however, that all diazotrophic proteobacteria analyzed so far control NifA activity (level 2) and most of these species regulate, in addition, transcription of the *nifA* gene (level 1). In contrast, regulation of nitrogenase activity (level 3) is confined to a subset of proteobacteria. Regulation at all three levels involves P_{II}-like signal transduction proteins. For further details, see text.

Since by far most work concerning regulation of N_2-fixation has been carried out in proteobacteria, this chapter will focus on this group of bacteria, before the situation for selected nonproteobacterial species will briefly be described. The description within the proteobacteria will be confined to free-living diazotrophic bacteria, since species that fix N_2 either in symbiosis or association with plants will be covered elsewhere.

9.2 Environmental signals regulating expression of nitrogenases

Nitrogenase-dependent reduction of N_2 to NH_3 is a highly energy-demanding process. Logically, expression of nitrogenase is inhibited by NH_4^+ in all free-living diazotrophs analyzed so far. Therefore, in this review, NH_4^+-regulation of N_2 fixation will be the center of attention.

The most commonly occurring nitrogenases have Mo in their active center, the iron–molybdenum cofactor. In addition to Mo-nitrogenase, some species possess homologous alternative nitrogenases, namely, a vanadium-containing nitrogenase and/or the so-called iron-only nitrogenase [1]. At least at moderate temperatures, Mo-nitrogenases exhibit higher specific activities than the alternative nitrogenases with respect to N_2 reduction rates. Consequently, expression of alternative nitrogenases is repressed by Mo.

Since all nitrogenases are extremely sensitive toward O_2, diazotrophic bacteria have developed different strategies to protect nitrogenase. One strategy is to express nitrogenase only under anaerobic conditions as brought about by many bacteria including *Klebsiella pneumoniae* and *Rhodobacter capsulatus*. A second strategy is the protection of nitrogenase by increased respiration as in *Azotobacter vinelandii*. Yet another strategy has been developed by some filamentous cyanobacteria like *Anabaena variabilis*, which form specialized cells (called heterocysts), where N_2-fixation takes place.

Compared to the amount of work done on NH_4^+-, Mo-, and O_2-regulation of N_2-fixation much less effort has been spent on the role of other environmental factors that influence synthesis and/or activity of nitrogenase. Among these are the C/N ratio at which C- and N-sources are consumed, Fe-availability, temperature, and in the case of photosynthetic bacteria, light intensity.

9.3 Levels of regulation of N_2-fixation in proteobacteria and the central role of P_{II}

Ammonium is the preferred source of fixed N for most bacteria, and consequently, the N_2-fixation process is highly regulated by NH_4^+. In proteobacteria, regulation of N_2-fixation operates at (up to) three levels (Figure 9-1). At the first level of the regulatory cascade to express nitrogenase, the nitrogen regulation system (Ntr system, see below) senses the fixed N status. In case of low levels of fixed N (absence of NH_4^+), the NtrC protein becomes activated, leading to the expression of the *nifA* gene, which codes for the central activator protein of all the other *nif* genes [2]. The second level of regulation affects the activity of NifA. At this level, the transcriptional activity of the NifA protein is modulated by the N status as well as by the redox status. Different mechanisms of NifA regulation have been evolved in various phylogenetic lineages. Finally, a third level of regulation affects the enzymatic activity of nitrogenase in response to environmental cues. Some bacteria, like *Rb. capsulatus*, control N_2-fixation at all three levels [2, 3]. Other species, like *K. pneumoniae*, regulate N_2 fixation only at the first two levels.

The response toward NH_4^+ at all three levels of regulation involves the highly conserved P_{II}-signaling system. However, it is not NH_4^+ per se, which is sensed, but rather its assimilation through the glutamine synthetase (GS)–glutamate synthase (GOGAT) pathway. The key metabolites of this pathway, 2-oxoglutarate (2-OG) and glutamine (Gln), are the molecules, to which the P_{II}-signaling systems responds [4]. In most proteobacteria, two paralogous P_{II}-signaling proteins, encoded by *glnB* and *glnK* genes, are present [4, 5]. The P_{II} proteins are trimers, which bind in a synergistic manner 2-OG and ATP. In addition to binding these effector molecules, the P_{II} proteins may be covalently modified through uridylylation at a surface-exposed tyrosyl residue in each subunit. The bifunctional uridylyltransferase/UMP-removing enzyme (briefly termed UTase/UR or GlnD) is the Gln sensor in this system, and a high degree of P_{II} uridylylation (corresponding to low internal Gln levels) is the major signal for N-deficiency (Figure 9-2). Depending on the state of uridylylation and 2-OG/ATP binding, the P_{II} proteins are able to interact with various targets at all three levels of the regulatory cascade that controls N_2-fixation. In addition to control of N_2-fixation, the P_{II} proteins coordinate also other aspects of N-assimilation, like control of GS. The P_{II} paralogues GlnB and GlnK can substitute for each other in various (but not all) aspects. Generally, the

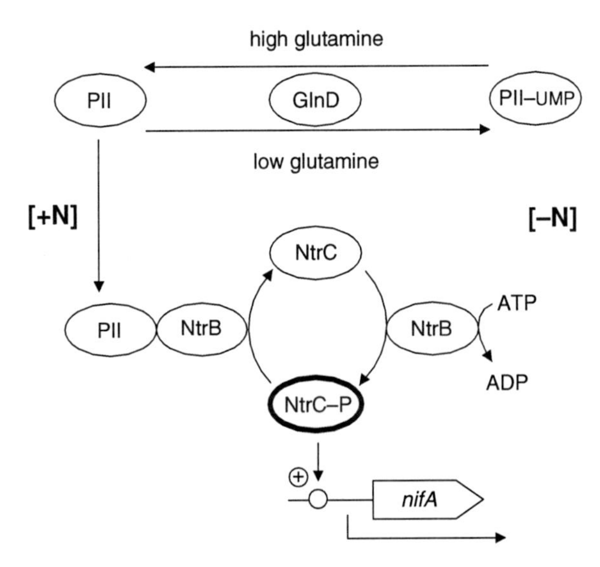

Figure 9-2 The enteric model of transcriptional regulation of *nifA* in response to the fixed N status. GlnD functions essentially as a Gln-sensing apparatus. Low levels of fixed N (low Gln, [−N]) lead to uridylylation of P_{II} thus preventing interaction with NtrB. Under these conditions NtrB phosphorylates NtrC, which in turn activates transcription of *nifA*. For further details, see text.

presence of two (or more) P_{II} paralogues in one species is believed to allow fine-tuning of the diverse processes.

9.4 Transcriptional control of the *nifA* gene represents the first level of the regulatory cascade in diazotrophic proteobacteria

The Ntr system, which senses the fixed N status, seems to be widespread in but limited to proteobacteria. It has been analyzed in great detail mainly in enteric bacteria, namely, *Escherichia coli* and *K. pneumoniae* (Figure 9-2). The Ntr system consists of the UTase/UR enzyme (GlnD), the P_{II} signal transduction protein, and the two-component regulatory system NtrB/NtrC, in which NtrB acts as a sensor kinase and NtrC functions as a response regulator.

In the absence of NH_4^+ (depletion of fixed N, [−N] conditions), P_{II} becomes uridylylated, with P_{II}-UMP being unable to interact with NtrB. Under these conditions NtrB promotes the phosphorylation of NtrC. In turn, NtrC-P activates transcription of the *nifA* gene, whose product mediates expression of all the other *nif* genes. Members of the NtrC family typically activate transcription of their target genes in concert with RNA polymerase containing the alternative sigma factor RpoN (also called NtrA or σ^{54}). The only known exception is *Rb. capsulatus* NtrC, which does not require RpoN but instead activates transcription together with RNA polymerase containing the σ^{70}-like house-keeping sigma factor RpoD.

Upon addition of NH_4^+ (fixed N-surplus, [+N]) to a N-depleted culture the intracellular Gln levels increase, leading to P_{II} deuridylylation by UTase/UR. In contrast to P_{II}-UMP, the unmodified form of P_{II} binds to NtrB, as long as the 2-OG levels are not in excess. In the presence of high 2-OG concentrations, P_{II} is unable to bind to NtrB. Thus, the C-signal 2-OG is antagonistic to the N-signal Gln. Depending on the proteobacterial species, one of the two (or both) P_{II} paralogs transduces the Gln/2-OG signal to NtrB. In complex with P_{II}, NtrB no longer phosphorylates NtrC, but instead, NtrB promotes the dephosphorylation of activated NtrC-P [5]. As a result, the activator properties of NtrC are tuned down and expression of NtrC-dependent promoters is switched off.

In some proteobacteria including *Ab. vinelandii* and *Azospirillum brasilense*, this first level of *nif* regulation is absent. In these species, expression of *nifA* is independent of NtrC, and *nif*-specific regulation is exerted at the level of NifA activity (see below).

It should be emphasized that proteobacteria (in contrast to diazotrophic archaea) do not contain a N-responsive repressor, which prevents transcription of *nifA* in the presence of NH_4^+. Therefore, one should avoid talking about "repression" of *nif* genes by NH_4^+.

9.5 NifA activates *nif* gene expression in concert with a specific sigma factor, RpoN

The NifA protein is the central regulator of N_2-fixation, which (under [−N] conditions) activates transcription of all the other *nif* genes in proteobacteria [2]. Common to all NifA proteins is a modular structure consisting of an N-terminal domain, which is involved in control of NifA activity, a highly conserved central domain, which interacts with RNA polymerase, and a C-terminal DNA-binding domain containing a helix–turn–helix motif.

NifA recognizes and binds to enhancer elements located about 100–200 base pairs upstream of the transcriptional start sites of its target genes (Figure 9-3). The consensus sequence for the NifA-binding site (upstream activator-binding site, UAS) is TGT–N_{10}–ACA. NifA-dependent target genes are the structural genes of nitrogenase, *nifHDK*, and all the other *nif* genes (except *nifA* itself). NifA proteins activate transcription in concert with RNA polymerase containing the alternative sigma factor

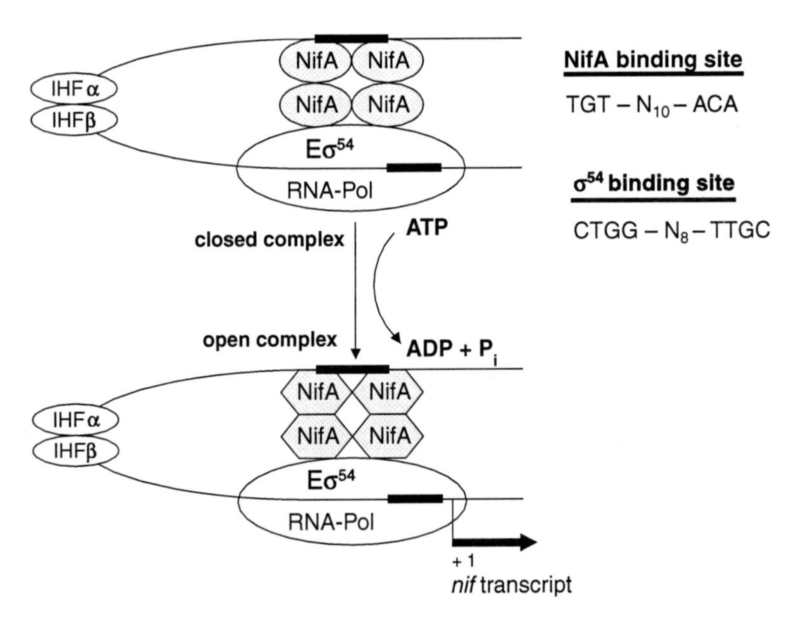

Figure 9-3 NifA-dependent transcriptional activation of N_2-fixation genes. Binding of NifA to a specific DNA site (UAS) promotes assembly of a functionally active oligomeric complex. Contact between NifA and RNA polymerase-σ^{54} holoenzyme is facilitated through IHF-mediated DNA looping. ATP hydrolysis is thought to involve conformational changes in NifA leading to open complex formation accompanied by initiation of *nif* gene transcription.

RpoN (σ^{54}, NtrA). RpoN binds to a characteristic sequence motif, CTGG–N_8–TTGC, that is typically located at position −24/−12 upstream of the transcription start site. RpoN-dependent activators are unique among transcriptional activators in that they must hydrolyze ATP (or other nucleoside triphosphates) to activate transcription. There is no evidence that NifA uses ATP to phosphorylate itself, but instead the coupling of ATP hydrolysis to open complex formation likely involves conformational changes in NifA that are communicated to RNA polymerase [6].

The interaction between NifA and RNAP-σ^{54} can be stimulated by IHF (integration host factor). Originally, IHF was described as a factor required for integration of phage λ DNA into the chromosome of *E. coli*. IHF is a small DNA-binding protein belonging to the family of prokaryotic histone-like proteins. It binds preferentially to specific sites in DNA – in case of *nif* promoters between the binding sites of NifA (UAS) and RpoN (−24/−12) – thereby inducing a sharp bend in the DNA.

9.6 Control of NifA activity represents the second level of the regulatory cascade

While transcriptional control of the *nifA* gene constitutes the first level of the regulatory cascade to express nitrogenase, control of the activity of the NifA protein itself forms the second level. NifA activity is inhibited in response to the addition of NH_4^+ and also in response to O_2.

With regard to O_2-sensitivity, NifA proteins may be divided into two classes. O_2-sensitive NifA proteins (like *Bradyrhizobium japonicum* NifA) are characterized by a cysteine-rich interdomain linker between the central and C-terminal domains, which is believed to confer O_2 reactivity on this class of NifA proteins. In contrast, O_2-tolerant NifA proteins (like *K. pneumoniae* NifA) lack such an interdomain linker. O_2-dependent regulation of this group of NifA proteins is mediated by another regulatory protein, NifL, which controls the activity of NifA by protein–protein interaction that is modulated by redox changes, ligand binding, and interaction with GlnK [6]. The response toward O_2 is directly achieved by a redox-sensitive domain in NifL proteins (namely, the N-terminal PAS domain containing a FAD cofactor). In its oxidized state (sensing O_2), NifL complexes and thereby inactivates NifA.

NH_4^+-dependent modulation of NifA activity is mediated by P_{II}-like proteins as shown for many diazotrophic proteobacteria. However, three different mechanisms have evolved in the various bacteria, to link N signaling of P_{II} proteins to NH_4^+ control of NifA activity. In the first case, found in *Ab. vinelandii* and *K. pneumoniae*, GlnK is required for regulation of NifL-mediated inhibition of NifA activity. In *Ab. vinelandii*, GlnK favors NifL–NifA interaction under [+N] conditions, whereas in

K. pneumoniae, GlnK inhibits NifL–NifA interaction under [−N] conditions. Second, in bacteria like *As. brasilense, Herbaspirillum seropedicae*, and *Rhodospirillum rubrum*, NifA activity (under [−N] conditions) depends on the presence of GlnB. Third, in *Azorhizobium caulinodans* and *Rb. capsulatus*, neither GlnB nor GlnK is required for NifA activity, but both GlnB and GlnK can inhibit NifA activity in the presence of NH_4^+.

9.7 Control of nitrogenase activity represents the third level of the regulatory cascade

Several diazotrophic bacteria harbor a regulatory system that rapidly and reversibly controls nitrogenase activity in response to sudden changes in the environment [7]. This so-called switch-off/switch-on effect has been described for eubacteria (including the proteobacteria *Azoarcus* BH72, *As. brasilense*, *Rb. capsulatus* and *Rs. rubrum*, the cyanobacterium *Gloeothece*, and the green sulfur bacterium *Chlorobium tepidum*) as well as for archaea (like *Methanococcus maripaludis*). Many other species (like *K. pneumoniae* and *Ab. vinelandii*) are devoid of such a system. Switch-off may be caused by addition of NH_4^+ to a N_2-fixing culture, increase of the O_2-partial pressure, or (in case of phototrophic bacteria) by darkness.

The molecular basis for switch-off/switch-on has been best studied in the phototrophic purple bacterium *Rs. rubrum*. In this bacterium, post-translational regulation of nitrogenase involves the reversible ADP-ribosylation of dinitrogenase reductase mediated by the DraT/DraG system (Figure 9-4). After a negative stimulus (e.g., upon addition of NH_4^+ to a nitrogenase-derepressed culture), DraT (dinitrogenase reductase ADP-ribosyl transferase) becomes transiently active and carries out the transfer of the ADP-ribose from NAD^+ to the arginine residue 101 of one subunit of the dinitrogenase reductase homodimer, leading to inactivation of that enzyme. After NH_4^+ has been used up, DraG (dinitrogenase reductase activating glycohydrolase) becomes active and removes the ADP-ribose from the covalently modified enzyme, thereby restoring its activity. In contrast to *Rs. rubrum*, other bacteria, like *Rb. capsulatus*, control nitrogenase activity not only by a DraT/DraG system but, in addition, by a second DraT/DraG-independent mechanism [1].

DraT and DraG themselves are subject to post-translational regulation, and there is increasing evidence for many species including *Azoarcus*, *As. brasilense*, *Rb. capsulatus*, and *Rs. rubrum* that P_{II}-like proteins are involved in control of the DraT/DraG system. Interestingly, the switch-off mechanism in the archaeon *M. maripaludis* also involves P_{II}-like proteins, namely, NifI1 and NifI2 [8]. However, post-translational regulation of nitrogenase activity in *M. maripaludis* does not involve ADP-ribosylation or any other detectable covalent modification of dinitrogenase reductase.

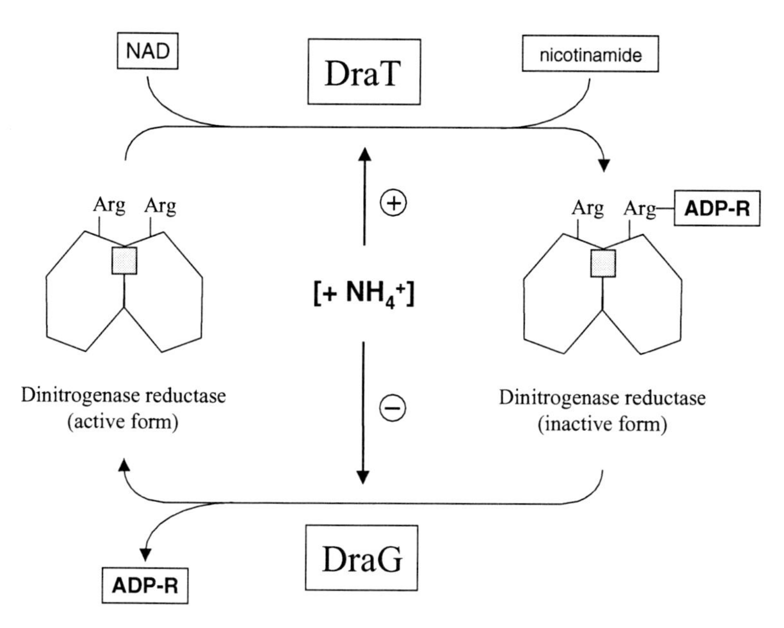

Figure 9-4 Reversible ADP-ribosylation of nitrogenase reductase. Upon addition (+) of NH_4^+, DraT catalyzes the transfer of ADP-ribose to an arginine residue of one subunit of dinitrogenase reductase. When NH_4^+ is exhausted (−), DraG removes the ADP-ribose.

On the other hand, nitrogenase reductase in the unicellular cyanobacterium *Gloeothece* is subjected to modification by palmitoylation [9]. Therefore, despite some similarities concerning post-translational control of nitrogenase throughout the eubacterial and archaebacterial kingdoms, the underlying mechanisms may differ from species to species.

9.8 Regulation of alternative nitrogenases

As outlined above for Mo-nitrogenases, most work on regulation of alternative nitrogenases has been done in proteobacteria. Both alternative nitrogenases are encoded by distinct sets of N_2-fixation genes with the *vnf* genes coding for V-nitrogenase and the *anf* genes coding for the Fe-only nitrogenase. Similar to *nif* genes, expression of *vnf* and *anf* genes in proteobacteria requires central NifA-like activator proteins, VnfA and AnfA, respectively, which mediate transcriptional activation of their respective target genes in the absence of NH_4^+. Like NifA, both VnfA and AnfA activate transcription of their target genes in concert with RNA polymerase containing sigma factor RpoN. In *Rb. capsulatus*, transcription of the *anfA* gene itself (like *nifA*) depends on activation by NtrC. In

Ab. vinelandii, however, expression of *vnfA* and *anfA* (like *nifA*) is independent of NtrC.

In contrast to *nifA*, transcription of *vnfA* and *anfA* is inhibited by Mo [1]. This inhibition involves Mo-dependent repressor proteins similar to *E. coli* ModE, which bind to dyad symmetric promoter elements that overlap the transcription start [10].

Alternative nitrogenases are advantageous under conditions of Mo limitation. Such a limitation can be due either to natural low availability of Mo or to formation of Mo-depleted microenvironments by competition between different organisms. In addition, temperature affects the expression of *vnf* and *anf* genes in *Ab. vinelandii*. In contrast to growth at 30°C, Mo-repression is overcome at temperatures below 20°C since Mo is poorly transported into cells grown at the lower temperature [1]. Therefore, Mo-independent nitrogenases may be advantageous for diazotrophic growth at lower temperatures even in the presence of Mo.

9.9 Regulation of N_2-fixation in nonproteobacterial species

Nonproteobacterial prokaryotes are devoid of a Ntr system, and they do not use a NifA homologue to control transcription of N_2-fixation genes. Instead, the various phylogenetic groups have evolved quite different regulatory systems, both at transcriptional and post-translational levels. The only module of the N-control system conserved in both prokaryotic kingdoms is the P_{II} signal transduction protein, which is engaged in remarkably diverse processes related to N-assimilation [5, 8]. A common feature of P_{II}-signaling proteins is the cooperative binding of 2-OG and ATP, whereas the response toward the N-signal Gln (operated by reversible uridylylation by UTase/UR in proteobacteria) is not universally conserved.

In cyanobacteria, 2-OG is only synthesized as a carbon skeleton for assimilation of NH_4^+ (since the TCA cycle is incomplete) and thus, its cellular level depends on the state of NH_4^+-assimilation. It seems to be the primary signal of the cellular N status [11]. The transcription factor of N-controlled genes, NtcA, is a member of the CRP family of bacterial transcription factors [12]. Its activity also responds directly to 2-OG, but in addition, signal transduction by P_{II} was demonstrated in *Synechococcus elongatus* (Figure 9-5). The mechanism of P_{II}-NtcA interaction is still not clearly established. A group of filamentous cyanobacteria, the order of *Nostocales* encompassing *An. variabilis* and *Anabaena sp.* PCC 7120, carries out N_2-fixation in specialized cells, termed heterocysts [13]. These cells protect nitrogenase from O_2. Differentiation of heterocysts is initiated following the activation of NtcA and HetR, and is programmed by a complex cascade of transcriptional events. Again, 2-OG is the crucial effector molecule that triggers the initiation of this process [14, 15].

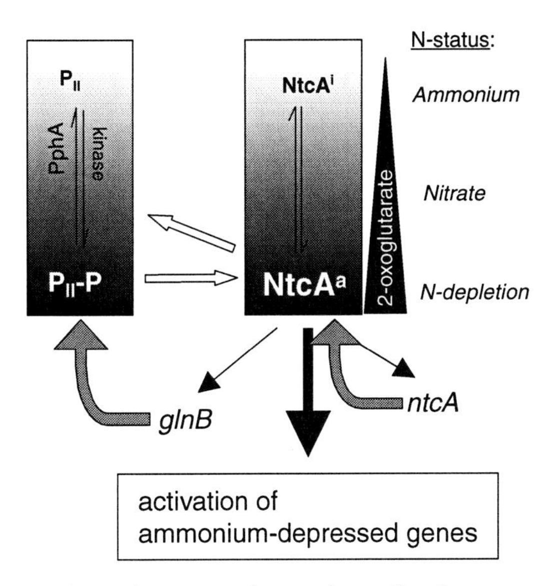

Figure 9-5 N-control in the cyanobacterium *S. elongatus*. The transcription factor of N-controlled genes, NtcA, may switch from an inactive state ($NtcA^i$) under N-excess conditions (low intracellular 2-OG levels) to an activated state ($NtcA^a$). Activation involves increased cellular 2-OG levels (indicated by the black arrow) and P_{II} signaling. The homotrimeric P_{II} protein may switch between the completely nonphosphorylated protein, and phosphorylated forms containing one, two, or three phosphorylated subunits (for clarity, only the fully phosphorylated form is shown), depending on the cellular 2-OG level. High 2-OG levels enhance phosphorylation and inhibit dephosphorylation of P_{II} [11]. The mutual interaction between P_{II} signaling and NtcA is indicated by vertical arrows. Phosphorylated P_{II} stimulates NtcA activity, whereas activated NtcA stimulates P_{II} phopshorylation. Activated NtcA enhances transcription of *ntcA* and *glnB* (leading to a positive feedback signal amplification) as well as other N-regulated genes [19].

In contrast to most proteobacteria, which contain two (or more) P_{II}-like proteins, cyanobacterial species analyzed so far harbor only a single P_{II}-like protein, GlnB. While activity of proteobacterial GlnB proteins is modulated by uridylylation of a conserved tyrosine residue, Tyr-51, cyanobacterial GlnB proteins are modified by phosphorylation at a conserved serine residue, Ser-49, in response to the intracellular 2-OG levels [11]. In *Anabaena* 7120, during the adaptation to diazotrophic growth conditions, P_{II} is phosphorylated in vegetative cells while unphosphorylated in heterocysts [16].

Most remarkably, in archaea like *M. maripaludis*, transcriptional regulation of *nif* gene expression is mediated by a repressor, NrpR [17], rather than an activator (as in proteobacteria and cyanobacteria). NrpR contains a N-terminal helix-turn-helix DNA-binding motif and two mutually homologous domains, which are thought either to sense the N-state of the cell or to interact with other proteins that do so. Only in the presence of NH_4^+, NrpR binds to specific operator sequences, and thus, prevents transcription of *nif* genes.

The *M. maripaludis* genome encodes five P_{II}-like proteins, but single null mutations in any of these genes do not affect transcriptional regulation [18]. At present it remains speculative, whether the respective P_{II} homologs can substitute for each other in regulation of NprR, or NprR itself senses N, or if NprR interacts with an unknown protein that mediates the N-signal to NprR.

In addition to transcriptional control of N_2-fixation genes by NprR, *M. maripaludis* regulates nitrogenase activity by a switch-off/switch-on-like mechanism. Again, as is the case for transcriptional regulation, the phenomenon is similar (rapid and reversible inhibition of nitrogenase activity) but the mechanism is different compared to switch-off in proteobacteria. The major difference is that (unlike proteobacteria), detectable covalent modification of nitrogenase reductase does not occur. Interestingly enough, two of the five P_{II}-like proteins in *M. maripaludis*, namely NifI1 and NifI2, are required for switch-off. The corresponding genes form part of the *nifH-nifI1-nifI2-nifD-nifK-nifE-nifN-nifX* operon, and consequently, expression of *nifI1* and *nifI2* is controlled by NprR.

9.10 Concluding remarks

Although the ability to fix N_2 is widespread among representatives of many prokaryotic genera, our knowledge about N_2-fixation is mainly based on work with proteobacterial species. This clear imbalance is explained by a remarkable resistance to genetic analysis of many species outside the proteobacteria. As one would expect from the phylogenetic relationship, there are many similarities between the proteobacterial species analyzed so far, which allow to propose a proteobacterial model of regulation of N_2-fixation (Figure 9-1). However, it should be emphasized that such a proteobacterial model cannot simply be extrapolated to all diazotrophs.

On the one hand, there are clear similarities between all diazotrophic species concerning the use of similar modules to build up nitrogenases and strategies to regulate the N_2-fixation process. First, the organization of N_2-fixation genes and the structure of their products are highly conserved among different species. In fact, the structural genes of nitrogenase, *nifHDK*, belong to the most conserved genes within prokaryotes. Second, since N_2-fixation is highly energy demanding, all diazotrophic bacteria have adopted regulatory strategies to inhibit expression of nitrogenase when fixed N is available. Furthermore, most diazotrophic species do not rely on a single regulatory mechanism but instead possess regulatory cascades controlling synthesis and/or activity of nitrogenase.

On the other hand, the mechanisms underlying N-regulation of the N_2-fixation process differ remarkably among different phylogenetic groups. First, NifA-like activator proteins are restricted to proteobacteria, while nonproteobacterial species either possess activator proteins belonging to other regulator families (as in cyanobacteria), or control *nif* gene expression by a specific repressor protein (as in methanogenic archaea). Second, although all diazotrophic species have one regulatory module in common, the P_{II} module, both the number of P_{II}-encoding genes and the function of their products may differ from one species to another. In addition, control of P_{II} by reversible modification may differ (uridylylation in proteobacteria vs. phosphorylation in cyanobacteria). Third, although several bacteria are able to switch off nitrogenase very rapidly (and reversible) in response to changes in the environment unfavorable for N_2-fixation, the underlying mechanism again seems to differ remarkably. In some proteobacteria (like *Rs. rubrum*) nitrogenase switch-off involves reversible ADP-ribosylation, whereas in other species (like the cyanobacterium *Gloeothece*) nitrogenase is subjected to modification by palmytoylation. In yet other species (like the archaeon *M. maripaludis*), no modifying group was detectable at all.

Summing up, the high functional diversity of mechanisms to regulate nitrogenase in response to environmental cues is striking. The only conserved modules are the N-signaling proteins of the P_{II} family, indicating that the regulatory pathways have evolved more rapidly than the highly conserved structural genes for P_{II} and nitrogenases and represent specific adaptations to the physiology of the various organisms.

References

[1] B. Masepohl, K. Schneider, T. Drepper, A. Müller, W. Klipp, Alternative nitrogenases, in: G.J. Leigh (Ed.) Nitrogen Fixation at the Millenium, Elsevier, Amsterdam, Boston, London, New York, Oxford, Paris, San Diego, San Francisco, Singapore, Sydney, Tokyo, 2002, pp. 191–222.

[2] R. Dixon, D. Kahn, Genetic regulation of biological nitrogen fixation, Nat. Rev. Microbiol. 2 (2004) 621–631.

[3] B. Masepohl, T. Drepper, W. Klipp, Nitrogen fixation in the photosynthetic purple bacterium *Rhodobacter capsulatus*, in: W. Klipp, B. Masepohl, J.R. Gallon, W.E. Newton (Eds.) Genetics and Regulation

of Nitrogen Fixation in Free-Living Bacteria, Kluver Academic Publishers, Dordrecht, NL, 2004, pp. 141–173.

[4] A.J. Ninfa, M.R. Atkinson, P_{II} signal transduction proteins, Trends Microbiol. 8 (2000) 172–179.

[5] T. Arcondeguy, R. Jack, M. Merrick, P_{II} signal transduction proteins, pivotal players in microbial nitrogen control, Microbiol. Mol. Biol. Rev. 65 (2001) 80–105.

[6] I. Martinez-Argudo, R. Little, N. Shearer, P. Johnson, R. Dixon, The NifL–NifA system: a multidomain transcriptional regulatory complex that integrates environmental signals, J. Bacteriol. 186 (2004) 601–610.

[7] C.M. Halbleib, P.W. Ludden, Regulation of biological nitrogen fixation, J. Nutr. 130 (2000) 1081–1084.

[8] J.A. Dodsworth, N.C. Cady, J.A. Leigh, 2-Oxoglutarate and the P_{II} homologues $NifI_1$ and $NifI_2$ regulate nitrogenase activity in cell extracts of *Methanococcus maripaludis*, Mol. Microbiol. 56 (2005) 1527–1538.

[9] J.R. Gallon, J. Cheng, L.J. Dougherty, V.A. Gallon, H. Hilz, D.M. Pederson, H.M. Richards, S. Ruggeberg, C.J. Smith, A novel covalent modification of nitrogenase in a cyanobacterium, FEBS Lett. 468 (2000) 231–233.

[10] W.T. Self, A.M. Grunden, A. Hasona, K.T. Shanmugam, Molybdate transport, Res. Microbiol. 152 (2001) 311–321.

[11] K. Forchhammer, Global carbon/nitrogen control by P_{II} signal transduction in cyanobacteria: from signals to targets, FEMS Microbiol. Rev. 28 (2004) 319–333.

[12] A. Herrero, A.M. Muro-Pastor, E. Flores, Nitrogen control in cyanobacteria, J. Bacteriol. 183 (2001) 411–425.

[13] J.C. Meeks, J. Elhai, Regulation of cellular differentiation in filamentous cyanobacteria in free-living and plant-associated symbiotic growth states, Microbiol. Mol. Biol. Rev. 66 (2002) 94–121.

[14] A. Herrero, A. Muro-Pastor, A. Valladares, E. Flores, Cellular differentiation and the NtcA transcription factor in filamentous cyanobacteria, FEMS Microbiol. Rev. 28 (2004) 469–487.

[15] S. Laurent, H. Chen, S. Bedu, F. Ziarelli, L. Peng, C.-C. Zhang, Nonmetabolizable analogue of 2-oxoglutarate elicits heterocyst differentiation under repressive conditions in *Anabaena* sp. PCC 7120, Proc. Natl. Acad. Sci. U S A 102 (2005) 9907–9912.

[16] S. Laurent, K. Forchhammer, L. Gonzalez, T. Heulin, C.-C. Zhang, S. Bedu, Cell-type specific modification of P_{II} is involved in the regulation of nitrogen metabolism in the cyanobacterium *Anabaena* PCC 7120, FEBS Lett. 576 (2004) 261–265.

[17] T.J. Lie, G.E. Wood, J.A. Leigh, Regulation of *nif* expression in *Methanococcus maripaludis*: roles of the euryarchael repressor

NrpR, 2-oxoglutarate, and two operators, J. Biol. Chem. 280 (2005) 5236–5241.
[18] J.A. Leigh, Regulation of nitrogen fixation in methanogenic archaea, in: W. Klipp, B. Masepohl, J.R. Gallon, W.E. Newton (Eds.) Genetics and Regulation of Nitrogen Fixation in Free-Living Bacteria, Kluver Academic Publishers, Dordrecht, NL, 2004, pp. 65–71.
[19] R. Schwarz, K. Forchhammer, Acclimation of unicellular cyanobacteria to macronutrient deficiency: emergence of a complex network of cellular responses, Microbiol. 151 (2005) 2503–2514.

Chapter 10

The *Rhizobium*-Legume Nitrogen-Fixing Symbiosis

Gary Stacey

10.1 Introduction

Specific Gram-negative soil bacteria, which are members of the α-proteobacteria and lie within the order Rhizobiales, possess the unique ability to infect the roots of leguminous plants and so elicit formation of a specialized organ (nodule), which they colonize. As intracellular symbionts, the bacteria fix N_2 using a conventional Mo-based nitrogenase and provide this fixed N to the plant. Thus, this symbiosis assumes agricultural and ecological importance by providing a ready source of limiting fixed N for plant growth. The last few decades have brought an explosion in our understanding of the molecular basis of this symbiosis, from the standpoint of both the symbiont and the legume host. The establishment of the symbiosis requires close coordination between the partners and is mediated by the exchange of diffusible signal molecules. Most recently, bacterial and plant genome-sequencing projects have added immensely to the resources available to study the symbiosis. A major event was the

Biology of the Nitrogen Cycle
Edited by H. Bothe, S.J. Ferguson and W.E. Newton

adoption of two genetic model legumes, *Lotus japonicus* and *Medicago truncatula*, and the genomes of both plants are currently being sequenced. Research with these model plants has now revealed the basic outlines of the plant-signaling pathways that lead to nodule formation.

10.2 An overview of nodule formation

Nodulation is a host-specific process with each rhizobium having a defined host-plant range (Table 10-1). Rhizobia, normally found in the soil, respond to the plant-root environment (rhizosphere) by increasing their population levels and attaching to the root surface. When in a compatible interaction (i.e., an appropriate symbiont with an appropriate host), the bacteria attach to root hairs and recognize flavonoid (secondary plant metabolites) signals excreted by the plant. Each rhizobium has co-evolved to recognize the specific flavonoid mixture excreted by its compatible host. Recognition of this signal leads to de novo transcription of nodulation (*nod*) genes in the symbiont. These *nod* genes, in turn, encode enzymes that synthesize a unique signal molecule, the Nod signal, which is excreted from the bacterium and recognized by the plant host. The Nod signal is a modified lipo-chitin molecule and rhizobia appear to be the only prokaryotes with the ability to make such a molecule. Indeed, possession of the *nod* genes defines, in large part, whether a rhizobium is competent for forming an N_2-fixing symbiosis.

The specific chemistry of the Nod signal is determined by the rhizobial symbiont and has co-evolved to be recognized only by the compatible legume host. Therefore, it is this chemistry and the recognition systems

Table 10-1 Selected rhizobial species and examples of host range.

Species	Host range
Rhizobium leguminosarum bv. *phaseoli*	Common bean
Rhizobium leguminosarum bv. *trifolii*	Clover
Rhizobium leguminosarum bv. *viceae*[a]	Pea, vetch
Rhizobium tropici	Common bean
Rhizobium etli[a]	Common bean
Mesorhizobium loti[a]	*Lotus japonicus*
Azorhizobium caulinodans	Sesbania
Sinorhizobium meliloti[a]	Alfalfa, *Medicago truncatula*
Sinorhizobium fredii	Soybean
Bradyrhizobium japonicum[a]	Soybean, cowpea, mungbean
Bradyrhizobium elkanii	Soybean, cowpea, mungbean

[a] Species for which the genomic sequence is known or for which sequencing projects are underway.

involving diffusible signals that determine host specificity. Recognition of the lipo-chitin Nod signal by the host initiates a chain of events leading ultimately to the formation of the nodule structure. As described in detail below, Nod-signal recognition leads to rapid events in the root hair of the compatible host and a modification of normal, polar root-hair growth, such that the hair cell curls at its tip. This often results in a "cork-screw" like appearance and the rhizobia colonize the cavity created by the curl. Subsequently, the bacteria gain entry into the root hair through a mechanism that is still not well understood. Entry appears to involve a form of endocytosis, such that the bacteria are never free in the cytoplasm but are confined within a membrane and cell-wall-enclosed tube (infection thread) that initiates at the curl and then progresses intracellularly down the root-hair cell. The infection thread then continues its growth into the cortex, where it ramifies into the newly forming nodule primordium. The cell-wall biosynthetic machinery that is normally used to produce the division between two dividing daughter cells is adapted by the invading rhizobia to form the infection thread.

The lipo-chitin molecules produced by rhizobia are biologically active at <1 nM. Addition of either purified or chemically synthesized Nod signal can induce root-hair deformation but not root-hair curling, which seems to require the presence of the bacteria. The Nod signal alone can trigger cell division in the root cortex and lead to the formation of a nodule primordium. This event involves activation of the cell cycle in quiescent cortical cells. Indeed, cells in the zone of the nodule, which will ultimately be infected, undergo rounds of endoreduplication leading to polyploidy. The mechanisms by which the Nod signal causes these changes are still unknown.

At some point, the rhizobia within the infection thread reach a cell that they will infect. Release of the bacteria from the infection thread again resembles endocytosis by resulting in the formation of a membrane-bound compartment in which the bacteria exist as intracellular symbionts. This membrane-bound compartment has been termed the "symbiosome" to draw attention to both its quasi-organellar nature and its similarity to structures found in a variety of intracellular bacterial symbionts both in plants and in animals. The symbiosome is the unit of biological N_2 fixation because its membrane mediates interaction with the host cell (e.g., for nutrient uptake and ammonia excretion), whereas the bacteria induce all of the machinery for N_2 fixation. Rhizobia within the symbiosomes differentiate and induce a variety of new enzyme systems and often take on a larger, more extended (sometimes branched) shape. For these reasons, the special term "bacteroid" is used to define the intracellular symbiont.

The nodule is a highly specialized organ [1]. Two morphological types of nodules are known and they are determined by the plant host (Figure 10-1). The first type is called indeterminate and these nodules occur on clover

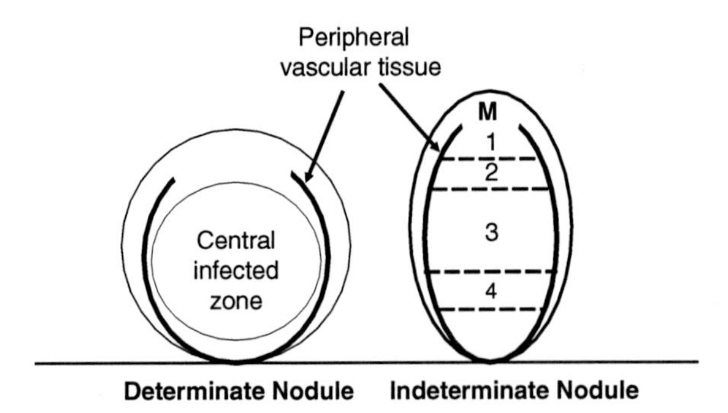

Figure 10-1 Graphical representation of the anatomical structure of determinate (left) and indeterminate (right) nodules. The peripheral vascular tissue divides the outer and inner cortex. Clear developmental zones are difficult to discern in determinate nodules. Indeterminate nodules have a persistent apical meristem (M) and show clear zones marking stages in nodule development: 1, preinfection zone; 2, infection zone; 3, fixation zone; 4, senescent zone. Baseline represents the root surface.

and alfalfa, for example. They appear as modified lateral roots with a terminal apical meristem, but with lateral vascular tissue. Because these nodules grow from their tip by means of new cell divisions, the full developmental series of the nodule can be seen in a cross-section. In order from the tip, such nodules exhibit a preinfection zone, infection zone, fixation zone, and, finally, a zone of senescent tissue. In contrast, the second type, called determinate nodules, which form on plants such as soybean and common bean, is initiated from cortical cell divisions but largely grows through cell expansion and results in a globular nodule structure. Determinate nodules also possess peripheral vascular tissue but, because development is in a radial pattern, distinct zones are more difficult to distinguish.

The interior of both determinate and indeterminate nodules contains very low O_2 levels. A physical barrier to O_2 permeation exists within the outer cortex of the nodule and leghemoglobin, which is specifically produced in nodules, binds available O_2 within the nodule. The low O_2 concentration is essential because nitrogenase is quickly denatured (inactivated) by O_2. Paradoxically, rhizobia are obligate aerobes and so require O_2 for respiration. The bacteroids circumvent this seeming paradox by expressing a high-affinity respiration system that functions using the O_2 transferred directly from leghemoglobin. Hence, the bacteroids are able to carry out aerobic metabolism, even though the free O_2 levels are very low.

The nodule is a true organ because it exhibits cellular specialization. The infected cells, which contain the symbiosomes, carry out biological N_2

fixation. Each infected cell is also in contact with at least one uninfected cell. N fixed in the infected cell is transferred to the uninfected cell where it is incorporated. Temperate-climate legumes, which mainly form indeterminate nodules, largely incorporate fixed N into amides (e.g., asparagine), which are used to transport the fixed N to the upper part of the plant. Tropical legumes, which mainly form determinate nodules, incorporate the fixed N into ureides (e.g., allantoin) that also act as fixed-N carriers. The transcription of enzymes for N assimilation, carbon utilization, nodule development, etc., is often induced to high levels during nodulation. Originally, some of these genes were thought to be nodule specific and, therefore, their products were given the special name, nodulins. However, more research has shown that these genes were likely recruited from pre-existing systems and it is only their regulation that is nodulation responsive. Nodulins produced within the first 48 h or so after rhizobial inoculation are termed "early nodulins" and thought to largely function in infection and nodule development. The "late nodulins" include gene products thought to function in nodule metabolism, N_2 fixation, and nodule maintenance. However, although these distinctions were historically useful, recent data has blurred them and nodulation is more correctly viewed as a continuum.

10.3 Molecular mechanisms of signal exchange

The interaction between a rhizobium and its compatible host is an intimate one that requires close coordination to orchestrate the highly complex developmental events leading to nodule formation [2–7]. It is now clear that a variety of signal molecules are likely recognized by both symbiont and host to mediate the communication required. These include polysaccharide moieties on both bacterial and plant surfaces, plus excreted proteins and secondary metabolites. Although research in each of these areas has produced fascinating results, here only the signaling required to induce expression of the nodulation genes in the symbiont and the subsequent recognition of the lipo-chitin Nod signal by the compatible host will be outlined (Figure 10-2).

10.3.1 Induction of nod-gene expression

All plants possess the general phenylpropanoid pathway involved in producing a variety of secondary metabolites, which include precursors for lignin biosynthesis as well as a number of metabolites involved in mediating plant–microbe interactions, such as phytoalexins that act to resist pathogen invasion. The general phenylpropanoid pathway also produces a variety of flavonoids (e.g., flavones and isoflavones) that are recognized by rhizobia. Each plant species produces a unique pattern of

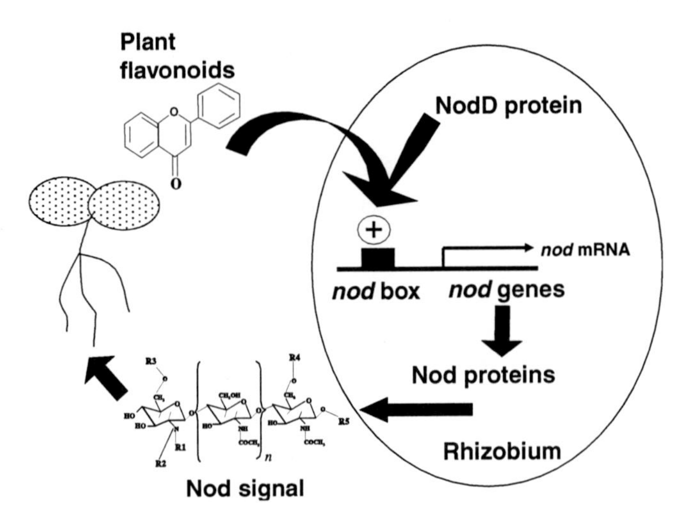

Figure 10-2 Exchange of diffusible signals between symbiont and host is crucial for nodule formation. The rhizobia are adapted to recognize the specific flavonoid inducer secreted by the compatible host. This is recognized BY the bacterial NodD protein that activates transcription of the nodulation genes by binding to the Nod box promoter element. The nodulation genes encode enzymes that synthesize the substituted lipo-chitin Nod signal. The compatible host is adapted to recognize the specific chemistry of the Nod signal produced by its compatible rhizobial partner. The Nod signal is the major factor that elicits nodule formation.

secondary metabolites. Indeed, this fact has been used historically as a tool for plant phylogenetic analysis. Rhizobia have co-evolved to recognize the specific flavonoid patterns of their respective host plant. For example, *Sinorhizobium meliloti* recognizes the flavone, luteolin, released from the roots of its host, alfalfa. Likewise, *Bradyrhizobium japonicum* recognizes the isoflavones, daidzein and genistein, produced by its host, soybean [8].

Recognition of the flavonoid inducer of *nod*-gene expression involves a LysR-like [9] transcriptional regulatory protein called NodD. The primary structure of the NodD protein determines specificity for the *nod*-gene-inducer molecule. NodD recognizes a specific promoter sequence that is localized 5′ of other nodulation genes. In the presence of the appropriate flavonoid inducer, binding of NodD to this promoter element (the Nod box) results in transcriptional activation of the downstream genes. This is a simple model.

Much more complex models exist in many rhizobia, where multiple NodD proteins can be found, including some that appear to repress, not activate, *nod*-gene expression. For example, in *B. japonicum*, mutations that disrupt NodD expression do not block nodulation by this bacterium. An additional, two-component regulatory system, NodVW, also appears to recognize the flavonoid inducer and, through a typical histidine kinase cascade, activates *nod*-gene expression. Only mutants blocked both in NodD and NodVW are unable to nodulate. Expression of *nod* genes is also under negative control. For example, NolR in some *S. meliloti* strains represses *nod*-gene expression depending on the fixed-N supply. In response to increasing bacterial population density, the *B. japonicum* NolA regulator represses *nod*-gene expression by activating the repressor NodD2.

Over-expression of the *nod* genes, which results in elevated levels of the Nod signal, is detrimental to nodulation. Moreover, expression of the *nod* genes *in planta* is repressed through mediation by NolA/NodD2 in *B. japonicum*. Ectopic expression of the *nod* genes *in planta* results in rapid senescence of the nodules. Therefore, it is clear that fine regulation of *nod*-gene expression is required to mediate the levels of Nod signal. In addition, although nodulation exhibits host specificity, many rhizobia can infect multiple host species. Thus, some of this regulatory complexity may be necessary to allow nodulation of multiple hosts, e.g., to recognize a variety of flavonoid signals.

10.3.2 Nod signal structure and host specificity

The known lipo-chitin Nod signals produced by rhizobia are substituted chitin oligomers, usually of 4–5 $\beta 1 \rightarrow 4$ linked *N*-acetylglucosamine (GlcNAc) residues, mono-*N*-acylated at the non-reducing end and carrying a variety of substitutions at both the reducing and non-reducing terminal GlcNAc residues. The core lipo-chitin structure is produced by the protein products of the most conserved rhizobial nodulation genes, *nodABC*. NodC is the chitin synthetase that produces the $\beta 1 \rightarrow 4$ *N*-acetylglucosamine backbone of 3–6, depending on the species. The NodB enzyme deacetylates the N-acetyl on the non-reducing end sugar, whereas NodA N-acylates this residue (see Figure 10-2).

The structure of the Nod signals can be used to classify rhizobia into two general groups. One group comprises those rhizobia that nodulate tropical and temperate legumes of the Genistea and Loteae tribes (e.g., *B. japonicum, Rhizobium* sp. NGR234, *Sinorhizobium fredii*, *Rhizobium loti*, and *Azorhizobium caulinodans*). These rhizobia produce Nod signals that are *N*-acylated with fatty acids of general lipid metabolism (e.g., vaccenic acid, $C_{18:1}\Delta 11$). In contrast, rhizobia that nodulate temperate herbaceous legumes, Cicereae, Trifolieae, and Vicieae, produce Nod signals that are *N*-acylated by polyunsaturated fatty acids (e.g., $C_{16:2}\Delta 2E,9Z$ for *S. meliloti*; $C_{18:4}\Delta 2E,4E,6E,11Z$ for *Rhizobium leguminosarum* bv. *viciae*; $C_{20:3}$ and $C_{20:4}$ for *Rhizobium leguminosarum* bv. *trifolii*; and $C_{18:2}$ and $C_{18:3}$ for *Rhizobium galegae*). In this latter group, the chemistry of the fatty acid clearly influences host range but in what way is not clear. Possibly, the fatty acid mediates movement of the Nod signal into the plant membrane or, alternatively, the polysaturated fatty acids may actually trigger oxylipin-like responses in the plant. Oxylipins, such as jasmonic acid, are well known to be involved in plant-signal transduction.

In addition to acylation, the Nod signals can be further modified by the addition of either a glycosyl residue or another modifying group, such as sulfate, carbamyl, acetyl, and/or glycerol residue. It is clear that many of these modifications are essential in determining the host range of Nod-signal action. For example, sulfation of the terminal-reducing N-acetyglucosamine by the *S. meliloti nodH*-gene product is required for recognition of this molecule by the alfalfa host. Likewise, fucosylation of the same residue through the action of the *B. japonicum nodZ*-gene product is required for recognition of the Nod signal by soybean. Some of the chemical modifications to the basic Nod-signal structure may act to protect the molecule from degradation by plant chitinases. However, given the high chemical specificity of Nod-signal action and their bioactivity at very low concentration, it is widely accepted that a plant-receptor protein must be involved in their recognition and subsequent action.

10.3.3 Nod-signal recognition and signal transduction; model legumes lead the way

Following elucidation of the Nod-signal structure, research attention immediately focused on identifying the plant receptor(s) involved in recognition of these potent signal molecules. Biochemical approaches were tried and identified high-affinity Nod signal-binding sites in legume-membrane preparations. However, clear evidence for the role of such binding sites in Nod-signal reception remains to be shown.

Real advances in our understanding of Nod-signal recognition came about through genetic studies using two-model legume species, *L. japonicus* (representing determinate nodule-forming plants) and *M. truncatula* (representing indeterminate nodule formers) [2–7, 10, 11]. These plants were chosen because they were out-crossing diploids (facilitating genetic analysis), were of small size (useful in greenhouse studies), produced large seed sets (again, an aid to genetics), were amenable to genetic transformation by *Agrobacterium tumefaciens*, and had relatively small genome sizes (~450 Mbp, which ultimately aided the on-going genome-sequencing projects).

A variety of mutants were isolated in both of these legume species that were defective in nodule formation or function. Over the past several years, several genes underlying these mutant phenotypes have been identified (Table 10-2). In the case of those mutants defective in nodule formation, knowledge of the underlying genetics has provided the basis for our understanding of Nod-signal recognition and subsequent signal transduction leading to a nodule structure (Figure 10-3).

Table 10-2 Examples of legume mutants (and corresponding genes) defective in the early events of nodulation.

Stage at which nodulation is blocked or defective	Legume host	Mutant designation	Identified gene
At Nod-signal perception	*L. japonicus*	*nfr1 nfr5*	LysM RLK
	M. truncatula	*nfp*	LysM RLK
	P. sativum	*sym10*	LysM RLK
After calcium flux, root-hair deformation	*M. truncatula*	*dmi2*	LRR RLK
	L. japonicus	*symrk*	LRR RLK
	M. sativa	*nork*	LRR RLK
	P. sativum	*sym19*	LRR RLK
	M. truncatula	*dmi1*	Ion channel protein
	L. japonicus	*pollux*	Ion channel protein
	L. japonicus	*castor*	Ion channel protein
After calcium spiking	*M. truncatula*	*dmi3*	CCaMK
	M. truncatula	*nsp1*	GRAS-family TF
	M. truncatula	*nsp2*	GRAS-family TF
	L. japonicus	*nin*	Putative TF
	P. sativum	*sym35*	Putative TF
	M. truncatula	*hcl*	Unknown
After root-hair curling, infection thread, and nodule-like formation	*M. truncatula*	*lin*	Unknown
	L. japonicus	*crinkle*	Unknown
	L. japonicus	*alb1*	Unknown
At control of nodule number	*L. japonicus*	*har1*	CLAVATA1-like RLK
	P. sativum	*sym29*	CLAVATA1-like RLK
	G. max	*nark*	CLAVATA1-like RLK
	M. truncatula	*sunn*	CLAVATA1-like RLK
	M. truncatula	*sickle*	Unknown
	L. japonicus	*klavier*	Unknown
	L. japonicus	*astray*	bZIP-family TF

Abbreviations: LysM-RLK, LysM domain receptor-like kinase; LRR-RLK, leucine-rich repeat domain receptor-like kinase; TF, transcription factor; RLK, receptor-like kinase.

The plant-nodulation-defective mutants cannot form nodules. A few of these mutants appeared to be blocked in all of the measurable events associated with Nod-signal action [e.g., root-hair calcium oscillations (spiking)]. As can be seen in Figure 10-3, calcium is a key intracellular messenger and Nod-signal-induced changes in intracellular calcium levels are likely a key element of nodule signaling. Positional cloning of two genes essential for calcium spiking [i.e., *L. japonicus* mutants *nfr1* (Nod factor recognition 1) and *nfr5*] revealed that they encoded receptor-like

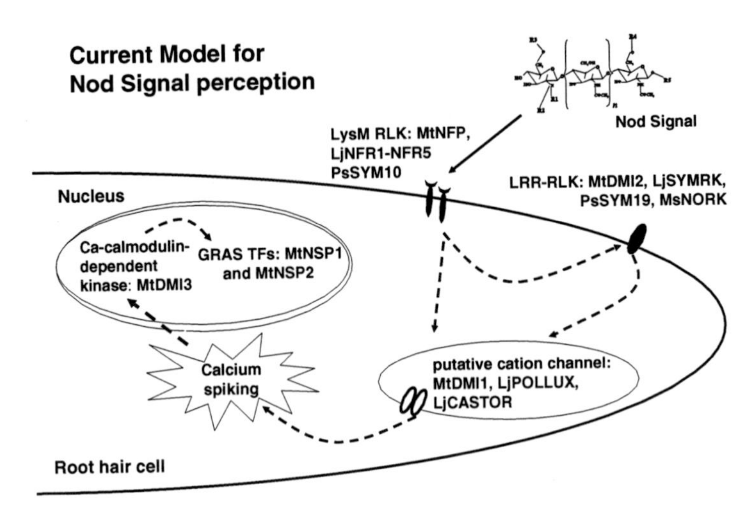

Figure 10-3 Current model for Nod-signal recognition derived primarily from studies of plant mutants and their phenotypes. The model predicts that the Nod signal is recognized by the LysM domain receptor-like kinases. The phosphorylation targets for the LysM RLKs are unknown. The LysM RLKs may directly activate the LRR-RLKs and/or may signal directly to other downstream components. DMI1 and related proteins have been localized to plastids suggesting a role in releasing calcium from plastid pools. The DMI3 protein could recognize changes in intracellular calcium flux and phosphorylate downstream elements (e.g., NSP1 and NSP2), so triggering gene transcription. Biochemical connections between the various gene products are undefined and, therefore, are shown by dotted lines. See Table10-2 for gene information. Gene prefixes referred to the plant species in which the gene was identified: Mt, *M. truncatula;* Lj, *Lotus japonicus;* Ms, *Medicago sativa;* Ps, *Pisum sativum.*

protein kinases (RLKs), which consist of an intracellular serine/threonine kinase domain, a single transmembrane domain, and an extracellular domain with 2 or 3 LysM-like domains. These LysM-like domains are found in many bacterial peptidoglycan-binding proteins. Peptidoglycan is a polymer of $\beta1\rightarrow4$ *N*-acetylglucosamine-*N*-acetylmuramic acid and, therefore, is structurally similar to chitin. It is now widely assumed that the NFRI and NFR5 proteins directly bind the Nod signal via the LysM domains and trigger an intracellular cascade by activation of the intracellular kinase domain. However, biochemical proof of a direct interaction between the Nod signal and LysM-like RLKs is still lacking.

Similar mutants were identified in *M. truncatula* (*nfp*) and pea (*sym10*) and also shown to be LysM-like RLKs. The *sym2* mutant allele of

M. truncatula is also a LysM-like RLK; however, this mutant shows some Nod-signal-induced responses (e.g., root-hair calcium oscillation). Therefore, although LysM RLKs may act during initial Nod-signal recognition, they may also have other subsequent functions. Indeed, such a model fits structure-function relationships in Nod-signal action. Studies show that the chemical specificity, which mediates initial signaling (e.g., root-hair curling, induction of some early nodulins), is less stringent than that required to mediate rhizobial entry or induction of other early nodulins. These data were interpreted as indicating two reception events, possibly mediated by two different receptors. Analysis of the *sym2* mutant is consistent with these ideas, but much work remains to be done before a clear understanding can be claimed.

Analysis of additional mutants blocked in nodulation revealed downstream signaling elements involved in Nod-signal action. For example, the does n't make infection (*dmi1,2,3*) mutants of *M. truncatula* define a signaling pathway that is essential, not only for nodulation, but also for establishment of the symbiosis with mycorrhizal fungi. Given that mycorrhizal fungi infect a much broader range of angiosperms and likely represent a more ancient plant-microbe association, it appears that rhizobia adapted this pathway for their use in nodulation. The pathway begins with both a mycorrhizal- and rhizobial-specific (e.g., represented by LysM RLKs) branch, which converges on the common *dmi1,2,3* pathway, and then diverges toward establishment of specific structures required for each symbiosis. The *dmi1,2,3* genes were identified using positional cloning and shown to encode a putative cation (calcium?) channel, a leucine-rich repeat receptor-like kinase, and a calcium calmodulin-dependent protein kinase (CCaMK), respectively. Genetic analyses, as well as studies of phenotype, suggest that DMI1 acts after NFP (the *M. truncatula* LysM RLK). DMI1 and orthologous genes in *L. japonicus* (i.e., *POLLUX* and *CASTOR*) were localized in plastids and likely mediate a calcium channel that functions to help mediate Nod-signal-induced calcium oscillations that are known to be critical for rhizobial infection. DMI2 was localized to the plasma membrane and infection-thread membrane. Mutations that block DMI2 function also block normal development of symbiosomes, again suggesting a role for the Nod-signal pathway beyond initial infection events in the root hair. Orthologous (to DMI2) receptor-like kinases were also identified in *L. japonicus* (SYMRK, symbiosis receptor kinase), *Medicago sativa* (NORK, nodulation receptor kinase), and *Pisum sativum* (SYM19). The exact function of DMI2 and its orthologues in Nod signaling is not known.

Given the importance of intracellular Nod-signal-triggered calcium oscillations in rhizobial infection, it is perhaps not surprising to find a calcium calmodulin-dependent protein kinase (DMI3) in the signaling pathway. Unlike *dmi1* and *dmi2* mutants, *dmi3* mutants do exhibit the nuclear-associated calcium spiking seen in response to Nod-signal addition. An obvious role for DMI3 would be to sense the intracellular calcium signal and then transduce this signal, presumably by phosphorylating downstream regulators (e.g., NSP1 and NSP2; see below). Exciting recent data support a critical role for DMI3 (J. Stougaard, personal communication). An *L. japonicus* mutant (*snf1*, spontaneous nodulation factor 1) was found to spontaneously (i.e., in the absence of rhizobia) form nodule-like structures on its roots. When the responsible gene was cloned, it was found to be an allele of the *L. japonicus* DMI3 orthologue. Transgenic expression of this gene (but lacking the calmodulin-like domain) replicated the mutant phenotype and resulted in both the formation of nodule-like structures and the induction of nodulin genes in the absence of either rhizobia or Nod signal.

Ultimately, Nod signaling results in the activation of gene transcription and, therefore, it came as no surprise when plant mutations that blocked nodulation were found to be in specific transcription factors. *M. truncatula* NSP1 (nodulation signaling pathway 1) and NSP2 identify two GRAS-family [12] transcription factors that are predicted to act after DMI3. Like *dmi3* mutants, *nsp1* and *nsp2* mutants exhibit calcium spiking but are blocked in nodulation. NSP1 and NSP2 were localized to the nucleus of plant epidermal and cortical cells, whereas NSP2 was found on the nuclear envelope and endoplasmic reticulum. NSP1 and/or NSP2 may be phosphorylation targets of DMI3, but this remains to be tested. The *L. japonicus nin* (nodule inception) gene encodes a putative transcription factor that is not required for initial Nod-signal perception but is essential for formation of the infection thread and nodule primordium.

The genes discussed above define the basic Nod-signaling pathway. However, a number of additional mutants, which exhibit a variety of nodulation and N_2 fixation phenotypes, remain to be characterized. Of these, one interesting group has revealed the molecular mechanism of autoregulation of nodule number in legumes. Some years ago, soybean mutants were found that would continue to nodulate in the presence of nitrate. Fixed N, either NO_3^- or NH_4^+, is well known to repress nodulation. Further analysis of these nitrate tolerant soybean (*nts*) mutants revealed a hypernodulation (elevated nodule numbers) phenotype and, thus, a breakdown of normal autoregulation. Legumes clearly regulate the number of nodules as shown by initial nodule formation suppressing any subsequent nodulation. The *nts* mutants lack this autoregulatory mechanism. Grafting experiments between the *nts* mutants and wild-type plants revealed that hypernodulation was controlled by the shoot; grafting *nts* shoots onto wild-type roots conferred hypernodulation, whereas the reverse graft (wild-type shoots onto *nts* roots) gave normal nodule numbers. This result suggests that a diffusible signal, which arises in the shoot of the plant in response to a nodulation signal, acts to suppress further nodulation. Thus, the *nts* mutants are blocked either in recognition of a root signal by the shoot or in production of the shoot-derived signal that ultimately represses further nodulation.

L. japonicus and *M. truncatula* mutants that exhibit a similar phenotype have been isolated; e.g., *L. japonicus sym78* (*har1*, hypernodulation abberant root 1) and *M. truncatula sunn* (super numeric nodules). The genes underlying the *nts*, *har1*, and *sunn* mutations were identified as leucine-rich repeat receptor-like kinases (LRR-RLK) with high homology to the *Arabidopsis thaliana CLAVATA1* gene, which is a key determinate in controlling meristem identity. CLAVATA1 protein is thought to act with another receptor-like kinase, CLAVATA2, to mediate apical meristem identity by interacting with a peptide-like signal molecule. Therefore, the current hypothesis is that the CLAVATA1-like LRR-RLKs, which mediate nodule numbers in legumes, may recognize a peptide-like signal produced during nodulation that travels to the shoot to trigger the autoregulatory response. This peptide-like signal is yet to be identified.

Other factors, besides autoregulation, also control nodule numbers. For example, ethylene, a plant hormone perhaps best known for its role in senescence and plant defense, is also a potent inhibitor of nodulation. The *M. truncatula SICKLE* mutant is ethylene insensitive and, therefore, shows increased nodule numbers. Ethylene may also play a more subtle role in nodulation by controlling the radial distribution of the nodule primordium. It is well known that not all rhizobial infections of a legume root lead to nodule-primordium formation. Those infections that occur over xylem poles, as opposed to phloem, appear to be favored. Ethylene may play an important role in this localization because 1-aminocyclopropane-1-carboxylic acid (ACC) synthase, a key enzyme in ethylene synthesis, is localized exclusively above the phloem poles. Legume plants engineered to be ethylene insensitive form nodule primordia above both the xylem and phloem poles, supporting a role for ethylene in determining the radial development of nodules.

10.4 Symbiotic nitrogen fixation and assimilation

Rhizobia possess a conventional molybdenum-based nitrogenase that is discussed in detail elsewhere in this volume. In order to use this enzyme effectively for biological N_2 fixation, the plant must provide the bacterium with both a suitable environment (e.g., a low O_2 level) and a supply of energy. Sucrose, derived from photosynthesis, is thought to be the primary carbon source to the nodule. Indeed, sucrose synthase (identified as nodulin 100) is strongly induced in nodules. Sucrose is subsequently converted into dicarboxylic acids (e.g., malate and succinate), which are the likely primary carbon source utilized by the bacteroids. Recently, the symbiosome membrane WAS shown to contain a variety of transporters [13]. Among these are transporters of NH_4^+, which leaves the symbiosomes, as well as those for dicarboxylic acids, iron and sulfate, which enter the symbiosomes. These transporters are clearly important because, if any of the latter is mutated, the resulting nodules are unable to fix N_2.

Given that nitrogenase is very sensitive to O_2, it is not surprising that O_2 is a key regulator of gene expression in bacteroids. Indeed, it seems that the majority of bacterial genes that are specifically expressed in bacteroids are under the control of regulatory cascades, which sense the O_2 level. As discussed above, a cortical O_2-diffusion barrier exists in nodules and O_2 levels are further decreased by the presence of leghemoglobin. Therefore, although rhizobia carry out aerobic respiration, they do so under very low free O_2 (<50 nM) conditions. Unlike free-living N_2-fixing bacteria, fixed-N-levels appear to play little-to-no role in regulating N_2 fixation in bacteroids.

The key transcriptional regulator controlling the expression of bacterial N_2 fixation genes (i.e., *nif*, *fix*) is NifA. Together with the RNA polymerase sigma factor, σ^{54}, which is encoded by *rpoN* [14], NifA recognizes *nif*-gene promoters and activates transcription, e.g., of the *nifHDK* genes that encode the polypeptides of nitrogenase. However, in rhizobia, *nifA* transcription is itself controlled by either of the two systems: (i) by the O_2-responsive two-component FixL-FixJ system, together with FixK, which is a member of the Crp-Fnr transcription factor superfamily [14]; or (ii) by the redox-sensing RegS-RegR system (Figure 10-4). Thus, in rhizobia, there are three proteins that act to sense O_2 levels; these are FixL and RegS, both histidine kinases, and NifA. FixL contains a heme group that binds O_2 and so regulates the phosphorylation of FixJ, which is the response regulator. In contrast, RegS has a conserved cysteinyl residue that participates in the formation of an intermolecular disulphide bond, which converts the kinase from an active dimer to an inactive tetramer in response to a redox change. RegS acts by phosphorylation of RegR in a classic histidine kinase cascade. Acting together, these regulatory proteins control transcription of a wide range of genes required for formation of nitrogenase, a high-affinity respiratory apparatus, and other functions. This basic regulatory scheme can be much more complicated in certain rhizobia due to, for example, the presence of multiple FixK or RpoN proteins.

Once fixed, ammonia (NH_3) diffuses out of the bacterium into the symbiosome space, where it is protonated (to give NH_4^+), and exits the symbiosome via an ion channel, which is specific for monovalent cations, into the infected cell cytoplasm [13]. The NH_4^+ is then assimilated via the combined action of glutamine synthetase and glutamate-oxoglutarate aminotransferase (GOGAT). Although this sequence of events is widely accepted, an alternative fixed-N-carrier has been suggested; here, alanine

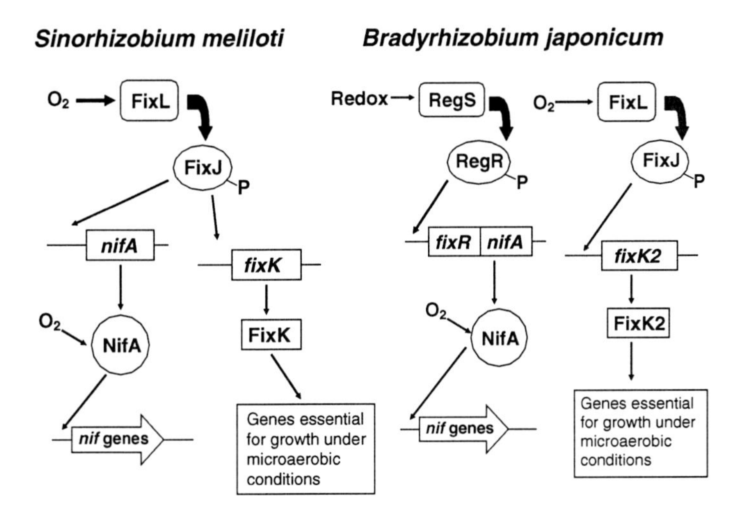

Figure 10-4 Simplified scheme of the regulatory cascades controlling *nif* gene expression in two rhizobia, *S. meliloti and B. japonicum*. Oxygen (O_2) controls NifA and FixK expression in *S. meliloti* by way of the two-component regulators, FixL and FixJ. Under conditions of low O_2, FixJ is phosphorylated, leading to expression of both NifA and FixK, which in turn induce transcription of the *nif* genes as well as other genes required for growth under microaerobic conditions. In *B. japonicum*, FixLJ also controls expression of FixK2 and its downstream targets. Another two-component regulatory system, RegSR, regulates NifA expression in response to redox state. NifA itself can sense low O_2 levels and activate *nif*-genes expression.

is thought to be the agent by which fixed N exits the symbiosome, but this situation is still being debated. Transport of the fixed N to the shoot of the plant requires further metabolism into either amides (glutamine and asparagine in temperate legumes) or ureides (e.g., allantoin and allantoic acid in tropical legumes). The ureides are formed from pathways involved in de novo purine-nucleotide biosynthesis. These reactions take place in the organelles of both the infected and uninfected cells demonstrating metabolic cooperation between the two cell types. Ultimately, the fixed N exits the nodules through a well-developed symplastic transport pathway to reach the vascular tissue.

10.5 Conclusions

The application of genetics and modern molecular methods to the study of symbiotic N_2 fixation has led to great advances in our understanding of nodule formation and function. We now have the full genome sequence of multiple rhizobial strains and sequencing of the genomes of *M. truncatula, L. japonicus*, and, most recently, soybean is well underway. The information and tools developed from knowledge of these genomes will maintain the pace of discovery in symbiotic N_2-fixation research. This work has both practical relevance, due to the agricultural and ecological importance of symbiotic N_2 fixation, and it also continues to explore new ground in basic biology. The Nod-signal-response pathway in plants is becoming one of the best characterized pathways. More and more connections are being found between nodulation and other pathways involved in normal plant development (e.g., CLAVATA1), which promises to reveal how nodulation is fully integrated into plant development and metabolism. As we now face challenges brought on by the rising cost of fossil fuels, which are needed for fertilizer production, and further degradation of our natural resources, advances in our understanding of symbiotic N_2 fixation may present opportunities for finding solutions to these problems.

References

[1] J.K. Vessey, K. Pawlowski, B. Bergman, Root-based N_2 fixing symbioses: legumes, actinorhizal plants, *Parasponia* sp., and cycads, Plant Soil 274 (2005) 51–78.

[2] J.J. Esseling, A.M.C. Emons, Dissection of Nod factor signaling in legumes: cell biology, mutants and pharmacological approaches, J. Microscopy 214 (2004) 104–113.

[3] R. Geurts, E. Fedorova, T. Bisseling, Nod factor signaling genes and their function in the early stages of *Rhizobium* infection, Curr. Op. Plant Biol. 8 (2005) 346–352.

[4] G.E.D. Oldroyd, M.J. Harrison, M. Udvardi, Peace talks and trade deals. Keys to long-term harmony in legume-microbe symbiosis, Plant Physiol. 137 (2005) 1205–1210.

[5] B.K. Riely, J.-M. An, R.V. Penmetsa, D.R. Cook, Genetic and genomic analysis in model legumes bring Nod-factor signaling to center stage, Curr. Op. Plant Biol. 7 (2004) 408–413.

[6] L. Mulder, B. Hogg, A. Bersoult, J.V. Cullimore, Integration of signalling pathways in the establishment of the legume-rhizobia symbiosis, Physiol. Plant 123 (2005) 207–218.

[7] G. Stacey, M. Libault, L. Brechenmacher, J. Wan, G.D. May, Genetics and functional genomics of legume nodulation, Curr. Opin. Plant Biol. 9(2) (2006) 110–121.

[8] J. Loh, J.G. Stacey, Nodulation gene regulation in *Bradyrhizobium japonicum*: a unique integration of global regulatory circuits, Appl. Environ. Microbiol. 69 (2003) 10–17.

[9] M.A. Schell, Molecular biology of the LysR family of transcriptional regulators, Ann. Rev. Microbiol. 47 (1993) 597–609.
[10] M.K. Udvardi, S. Tabata, M. Parniske, J. Stougaard, *Lotus japonicus*: legume research in the fast lane, Trends Plant Sci. 10 (2005) 222–228.
[11] N.D. Young, J. Mudge, T.H.N. Ellis, Legume genomes: more than peas in a pod, Curr. Op. Plant Biol. 6 (2003) 199–204.
[12] L.D. Pysh, J.W. Wysocka-Diller, C. Camilleri, D. Bouchez, P.N. Benfey, The GRAS gene family in Arabidopsis: sequence characterization and basic expression analysis of the SCARECROW-LIKE genes, Plant J. 18 (1999) 111–119.
[13] D.A. Day, B.N. Kaiser, R. Thomson, M.K. Udvardi, S. Moreau, A. Puppo, Nutrient transport across symbiotic membranes from legume nodules, Aust. J. Plant Physiol. 28 (2001) 667–674.
[14] R.A. Dixon, D. Kahn, Genetic regulation of biological nitrogen fixation. Nature Rev. Microbiol. 2 (2004) 621–631.

Chapter 11

Plant Symbioses with *Frankia* and Cyanobacteria

Katharina Pawlowski and Birgitta Bergman

11.1 Introduction

In addition to rhizobia, two other groups of prokaryotes enter N_2-fixing symbioses with plants. Gram-positive soil bacteria of the genus *Frankia* induce so-called actinorhizal nodules similar to legume nodules [1], whereas Gram-negative filamentous cyanobacteria from the genus *Nostoc* colonize a variety of plant organs [2]. As in the *Rhizobium*-legume symbioses, the microsymbiont has the capacity to fix N_2 while within the plant and exports the products of N_2-fixation to the host plant, rendering them independent of combined-N sources in the soil. For more recent comprehensive reviews, the reader is referred to [2–4].

11.2 Actinorhizal symbioses

11.2.1 Ecological role of this symbiosis

Actinorhizal nodules are formed in symbioses between members of the actinomycetous eubacterial genus *Frankia* and roots of more than 200

Biology of the Nitrogen Cycle
Edited by H. Bothe, S.J. Ferguson and W.E. Newton

dicotyledonous plant species belonging to 24 genera from eight plant families (Figure 11-1). With one exception (*Datisca glomerata*), actinorhizal plants are trees or woody shrubs. These symbioses allow actinorhizal plants to grow on marginal soils and so they are often used in soil reclamation, erosion control, agroforestry, or dune stabilization [5, 6]. *Hippophae rhamnoides* is being domesticated for its fruit [7].

11.2.2 Actinorhizal microsymbionts

Frankia comprises Gram-positive and Gram-variable actinomycetes that grow in hyphal form. Like cyanobacteria and in contrast to most rhizobia, *Frankia* strains can also fix N_2 in the free-living state. Under fixed-N limitation and aerobic conditions, *Frankia* strains form special organs for N_2-fixation; these spherical vesicles are formed at the ends of hyphae or short side hyphae. Vesicles are surrounded by envelopes of multiple layers of hopanoids [8]. The number of layers correlates with the O_2 tension, indicating a role as a gas-diffusion barrier [9]. *In planta*, the shape, septation, and subcellular localization of *Frankia* vesicles are determined by the host plant [10], which therefore can direct bacterial differentiation. *Frankia* can also form multilocular sporangia [11].

11.2.3 Phylogeny

Molecular phylogenetic analysis reveals that all plants, which form a N_2-fixing root-nodule symbiosis, belong to the eurosid I clade (Figure 11-1) [12]. These data imply that the ability to enter into symbiosis depends on a predisposition that arose only once during angiosperm evolution. Nodulated plants within the rosid I clade can be grouped into four major lineages (Figure 11-1) [13, 14]. Similar molecular phylogenetic studies on *Frankia*, using 16S rRNA, *nifD, recA*, and *glnII* gene sequences [15–18], result in three major groups, which are consistent with morphological, physiological, and host-specificity features of the *Frankia* strains [4]. One of these groups (clade III) contains no isolates, but is solely based on *Frankia* gene sequences amplified from nodules (Figure 11-1).

11.2.4 Nodule formation

Mature actinorhizal nodules are perennial organs consisting of multiple lobes, each of which represents a modified lateral root without root cap and with colonized cells in the expanded cortex (Figure E, see Colour Plate Section). Like in legumes, actinorhizal nodules are only formed when the plant cannot obtain enough combined-N from the soil [19, 20]. Two modes of colonization (intra- and intercellular) are known with the host plant determining which mode is used [21].

Members of the Fabales are colonized intracellularly, whereas members of the other plant groups are colonized intercellularly (with the possible exception of *Datisca* and *Coriaria*, where the colonization mode is

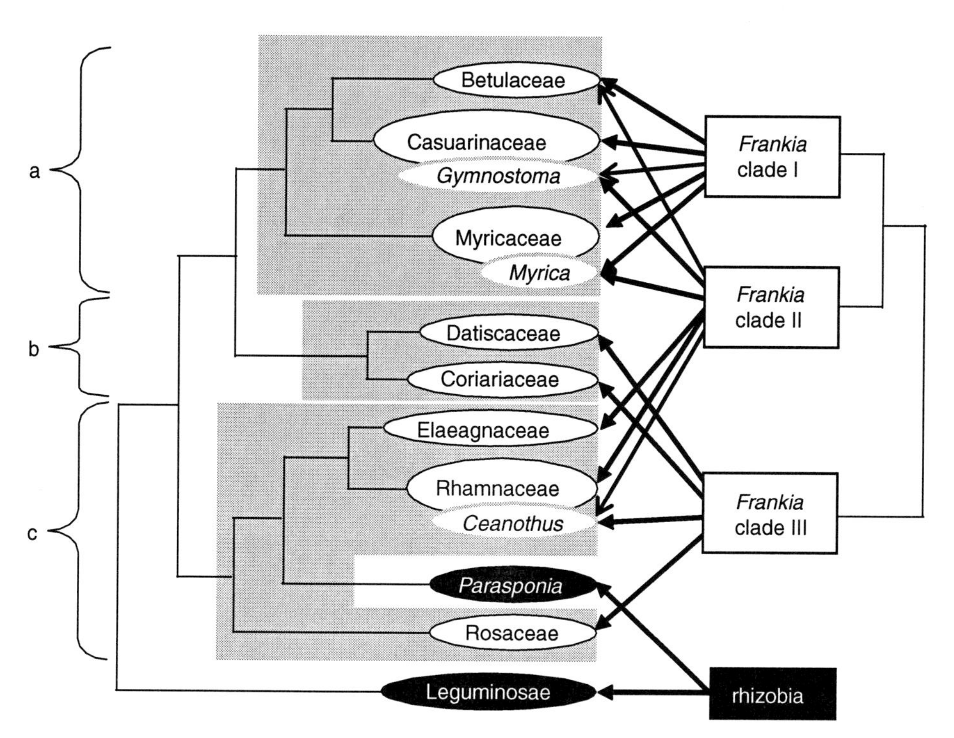

Figure 11-1 Simplified scheme of the phylogenetic relationship between legumes and actinorhizal plants based on Benson and Clawson [15]. Groups of plants colonized by rhizobia are labeled by inverse print. The grey boxes indicate the three lineages of actinorhizal plants: (a) Fabales (Betulaceae: *Alnus*; Myricaceae: *Myrica*, *Comptonia*; Casuarinaceae: *Casuarina, Allocasuarina, Ceuthostoma, Gymnostoma*), (b) Cucurbitales (Datiscaceae: *Datisca*; Coriariaceae: *Coriaria*), and (c) Rosales (Elaeagnaceae: *Elaeagnus, Hippophae, Shepherdia*; Rhamnaceae: *Ceanothus, Colletia, Discaria, Kentrothamnus, Retanilla, Talguenea, Trevoa*; and Rosaceae: *Cercocarpus, Chamaebatia, Cowania, Dryas, Purshia*). The phylogenetic relationship between the three clades of symbiotic *Frankia* is included. Actinorhizal genera (*Gymnostoma, Myrica, Ceanothus*) that differ in microsymbiont specificity from the rest of the family are indicated. Thick arrows connect *Frankia* clades with the plant groups their members are commonly associated with, while thin arrows indicate that members of that clades have been isolated from, or detected in an effective or ineffective nodule of a member of the plant group at least once. Note that not all members of a *Frankia* clade can nodulate all plants associated with that clade.

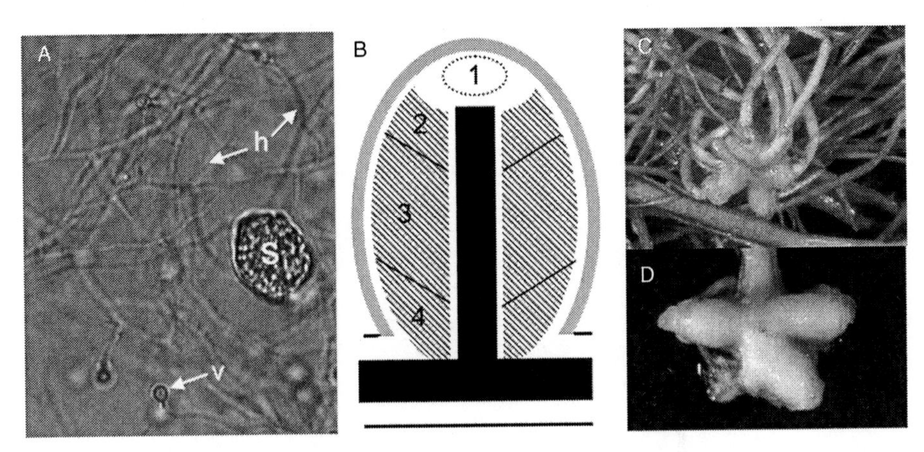

Figure E (A) *Frankia* culture grown in N-free medium under 21% O_2. h, hyphae; v, vesicle; s, sporangium. (B) Scheme of an actinorhizal nodule lobe. Due to the activity of the apical meristem (1) cortical cells are arranged in a developmental gradient. In the infection zone, (2) cortical cells become gradually filled with *Frankia* hyphae, and afterward, vesicles differentiate in the fixation zone, (3) *Frankia* fixes N_2 in the infected cells and in the senescence zone, and (4) *Frankia* hyphae and vesicles are degraded. Nodule lobes are surrounded by a periderm, which is shown in dark gray. (C) Photograph of a ca. 4—week-old nodule from *Casuarina glauca*. Aeration is provided by agraviotropically growing nodule roots that emerge from the tips of nodule lobes. (D) Photograph of a ca. 8-week-old nodule from *Datisca glomerata*. (See Colour Plate Section in back of this volume)

still unknown). During intracellular colonization, a *Frankia* hypha enters the root by means of an infection thread-like structure formed in a deformed root hair [22, 23]. Cell division is induced in the root cortex close to the colonized root hair to give rise to the formation of a so-called prenodule. The infection thread-like structures grow toward the prenodule by cell-to-cell passage and colonize some prenodule cells [24]. The primordium of the nodule lobe is initiated in the root pericycle and infection thread-like structures grow from the prenodule to the nodule primordium and colonize primordium cells. During intercellular colonization, *Frankia* hyphae penetrate the middle lamella between adjacent cells of the root epidermis and colonize the intercellular spaces of the root cortex [21, 25]. Concomitantly, the formation of the nodule lobe primordium is initiated in the root pericycle, and infection thread-like structures containing *Frankia* hyphae are formed in, then branch and fill primordium cells.

Actinorhizal nodules may contain multiple bacterial strains, not only different *Frankia* strains [26] but also, in the outer cortex, non-N_2-fixing related actinomycetes that cannot colonize the host plant on their own [27].

11.2.5 Nodule physiology

Because a dicarboxylate transporter is present in the invaginated plasma membrane that surrounds the intracellular *Frankia* in colonized cells of *Alnus glutinosa* nodules [28], it is likely that, as in legumes, actinorhizal plants supply their microsymbionts with photosynthate in the form of dicarboxylates. The fixed-N source exported by the microsymbionts to the plant is probably NH_4^+/NH_3, which then is assimilated by the plant GS/GOGAT cycle [10, 29], but *Datisca glomerata* may be an exception [30]. Furthermore, because N_2ase forms H_2 concurrently with N_2 reduction, *Frankia* forms H_2ase in hyphae as well as in vesicles under N_2-fixing conditions to recover the reductant and ATP used for H_2 production [31, 32].

11.2.6 O_2 regulation in nodules

Although N_2ase is irreversibly denatured by O_2, N_2-fixation has a high-energy demand that requires O_2 for respiratory processes in the nodules. Therefore, the supply of O_2 to the microsymbiont has to be tightly controlled. Actinorhizal plants that grow in wet or waterlogged soils have developed mechanisms for gas transport to the nodules; these include either agraviotropically growing so-called nodule roots with large air spaces in the cortex (Figure E) or thermo-osmotically mediated gas transport from the aerial parts to the roots [33]. In well-drained soils, the periderm can be interrupted by lenticels [34]. Different actinorhizal plant species use diverse strategies to protect N_2ase from O_2; these involve changes in nodule morphology, vesicle structure, plant and bacterial metabolism, and the formation of O_2-binding proteins (symbiotic hemoglobins) in colonized cells. *Casuarina* sp., e.g., resembles legumes in that the plant appears solely responsible for O_2 protection of N_2ase, whereas in *Alnus* sp. nodules, *Frankia* contributes by forming vesicles [34].

11.3 Cyanobacterial symbioses

11.3.1 General distribution of cyanobacterial symbioses

Cyanobacteria establish symbiosis with plant groups widely spread within the plant kingdom, including fungi (lichenized fungi and one non-lichenized fungus, *Geosiphon*), bryophytes, a water-fern, one gymnosperm group, the cycads, and one flowering plant (the angiosperm, *Gunnera*) [2, 35, 36]. In nature, these plants are always colonized confirming the close relationship with and a dependence on the cyanobacterial microsymbiont (cyanobiont). Indeed, a recent study implies that the organ colonized in *Gunnera* (stem glands) only develops under combined-N deprivation [37].

11.3.2 The plant organs

Which plant organs are colonized by the cyanobacteria varies considerably among hosts. In lichens composed of a fungus and a cyanobacterium, the latter forms a layer between the upper cortex and the lower parts of the thallus (e.g., *Peltigera canina*). In lichens that contain a green alga as the main microsymbiont, the cyanobacterium is restricted to cephalodia, the wart-like structures located on the thallus surface (e.g., *Peltigera aphthosa*). The fungal hyphae tightly surround the cyanobacterial filaments and may occassionally invaginate the cyanobacterial cell walls using hyphal haustoria, although these never pass the cyanobacterial cell wall (see [38]). Extracellular symbioses are also formed with the small temperate bryophyte genera, *Blasia* and *Anthoceros*. In these, the cyanobacterial filaments invade and fill cavities regularly spread over the thallus. In mosses, however, the association is looser and more sporadic [39].

In the floating water-fern genus *Azolla* spp., the cyanobacteria also remain extracellular in cavities located in the chlorophyllous dorsal leaves [40]. The only plant division in which the cyanobacteria, like rhizobia and *Frankia*, invade roots is that of the ancient cycads. In these palm-like trees, the cyanobacteria occupy a zone in special apogeotrophic roots, the coralloid roots, but remain extracellular [41]. However, intracellular host-cell penetration does occur in three separate cases. First, the terrestrial fungus *Geosiphon* engulfs and entraps cyanobacterial filaments in specific bladders. Second, cyanobacteria reside inside the frustules of some planktonic marine diatoms, e.g., *Rhizosolenia* [2, 3, 42]. Third, cyanobacterial filaments invade and fill cell clusters located in stems of the rhubarb-like angiosperm *Gunnera* [3, 43].

11.3.3 The cyanobionts

The cyanobacteria that establish symbioses with plants and fungi most often belong to the common terrestrial genus, *Nostoc*. This filamentous morphotype forms N_2-fixing thick-walled heterocysts under combined-N deprivation (Figure 11-2). *Nostoc* spp. may also develop into motile hormogonia, which act as colonization units in symbiosis (see below); although the genus may also develop resting spores (akinetes), these are rarely encountered in symbioses [39].

The largest variation in cyanobionts is found in lichens, ranging in morphology from unicellular to more complex filamentous forms. Additionally, *Calothrix* spp. occurs in the marine lichen *Lichina* and in cycads (see [2, 3]). Although the cyanobionts of the marine diatoms are filamentous and heterocystous, like *Nostoc*, they constitute a separate genus, *Richelia* [42]. The genetic affiliation of the cyanobiont(s) in the *Azolla* plants is still under debate, with both *Anabaena* and *Nostoc* being suggested [40]. *Gunnera* plants are colonized by *Nostoc* strains; however, more than one strain may occur in the same plant [44, 46], which is also possible for *Azolla* [40], cycads [41], and bryophytes [45].

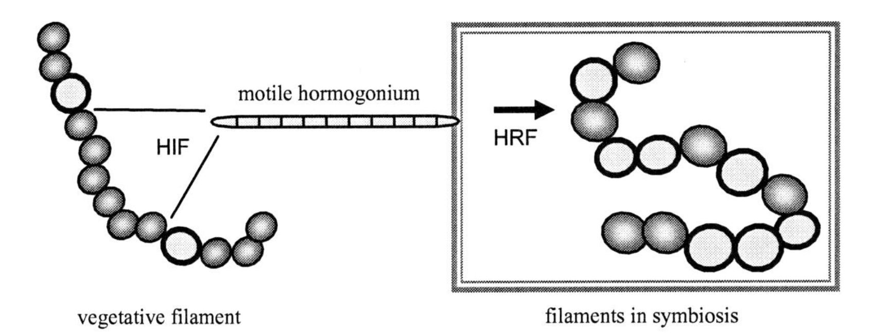

Figure 11-2 Schematic illustration of the plant colonization process. In free-living cyanobacteria (left), the filaments are composed of photosynthetic vegetative cells and thick-walled nitrogen-fixing heterocysts. These lack O_2 evolution to protect the nitrogen-fixing enzyme complex nitrogenase, located exclusively in the latter cell type. Hormogonia inducing factors (HIFs) are released by host plants and induce the development of motile small-celled hormogonia. These "colonization units" are attracted to and enter the host plant in which they may reside either extra- (e.g. in *Azolla* spp.) or intra-cellularly (e.g., in *Gunnera* spp.). At this stage, hormogonium repressing factors (HRFs) are produced by the plant to prevent hormogonium differentiation and stimulate heterocyst differentiation. Simultaneously, the heterocyst frequencies are enhanced severalfold and the heterocyst distribution pattern is disturbed, leading to a multiheterocyst phenotype (right).

The sequencing of the genome of a symbiotically competent cyanobacterium, *Nostoc* ATCC 29133 (http://www.jgi.doe.gov), has opened new avenues to cyanobacterial symbiosis research. The genome is exceptionally large for a prokaryotic organism, approaching 10 Mb with 7,432 predicted open reading frames (ORFs), which makes it comparable in size to budding yeast [47]. Among the ORFs identified, 29% are unique. This fact likely reflects the multiple phenotypic life modes of the genus and may hold keys to the symbiotic competence of *Nostoc* strains.

11.3.4 Evolutionary aspects

The eukaryotic plant hosts that form symbioses with cyanobacteria are all ancient; for instance, fossils suggest that cyanolichens may be ca. 400 million years old. Moreover, bryophytes were among the earliest land plants, cycads were major constituents of the Earth flora 300–150 million years ago, and *Gunnera*, being the youngest host, is among the oldest angiosperms known with fossilized pollen dating back about 90 million years (see [3, 48]). Extant symbioses may therefore represent ancient symbioses. The potential antiqueness of the capacity of cyanobacteria to form symbioses is also manifested in the endosymbiotic event that led to the evolution of chloroplasts. This evolutionary and highly successful step took place some 2.1 billion years ago [48].

It is also possible that the cyanobionts of some extant plant symbioses may be in the process of evolving toward becoming N_2-fixing "organelles." For instance, in the *Gunnera* symbiosis, the cyanobiont is taken up into the host cells in an endocytotic process, whereas in the other plant symbioses, the cyanobionts reside extracellularly. The cyanobacterial *Azolla* symbiosis is perpetually maintained [40] with the cyanobacteria being transferred vertically between *Azolla* generations via the sporocarps. In the other plant symbioses, *de novo* colonization of each generation is required.

11.3.5 The colonization units – hormogonia

Free-living *Nostoc* spp. filaments may undergo differentiation leading to a motile life stage, the hormogonia. These serve as escape units and promote survival in free-living cyanobacteria. The involvement of hormogonia-like filaments in the colonization of host plants is obvious in lichens, bryophytes, *Azolla*, cycads, and *Gunnera*. The vital role of hormogonia in the colonization process has also been demonstrated in reconstitution experiments [45, 46].

A hormogonium is a transient life stage lasting for 1–2 days and is characterized by small vegetative cells and a lack of heterocysts (Figure 11-2). Several rounds of rapid cell divisions, without DNA replication or cell enlargement and accompanied by filament fragmentation, lead to the formation of the hormogonium filaments [3, 49]. Their development may be induced by either abiotic environmental signals (e.g., specific spectral light qualities) or biotic plant signals [49]. The latter include hormogonium-inducing factors (HIFs) released by potential host plants, such as *Blasia*, *Anthoceros* and *Gunnera* [3, 39, 50], although their chemical identity is still unknown. *Anthoceros* may also produce hormogonium-repressing factors (HRFs) that are thought to prevent hormogonium development inside the plant. HRFs regulate hormogonia differentiation via induction of *hrmUA*, a hormogonium-regulating locus [51]. Repression of nonheterocystous hormogonia formation inside host cells/tissues will favor heterocyst formation and N_2-fixation (Figure 11-2).

11.3.6 Signals

The importance of hormogonia in symbiosis is also manifested in their chemotactic behavior. The mucilage secreted by the stem glands of *Gunnera* contains a potent chemoattractant for *Nostoc* spp. [43, 50] and hormogonia show positive chemotaxis toward extracts of bryophytes. Chemotaxis operons are present in cyanobacteria [47].

As hormogonial surface structures represent the area of first contact with potential host plants, these are likely to play pivotal roles in processes such as communication, recognition, adhesion, and colonization. There may also be cyanobacterial compounds, such as phytohormones, involved in the initiation and establishment of symbioses. Auxins, cytokinins, abscisic acid, jasmonates, ethylene, and gibberellins are all synthesized by cyanobacteria (see [52]) and may influence plant performance on contact, e.g., development of the symbiotic tissues of the hosts. For instance, the mitotic activities induced in *Gunnera* gland cells on contact with compatible hormogonia have been related to cyanobacterial IAA secretion [53].

To gain a deeper understanding of the process leading to the establishment of plant symbioses, subtractive cDNA libraries were constructed from *Nostoc* sp. PCC 9229, which were induced to form hormogonia in response to mucilage released from *Gunnera* glands [54]. Three cyanobacterial cDNA clones, termed *hieABC*, were identified as upregulated, and two novel proteins induced by *Gunnera* mucilage were also identified in *Nostoc* PCC 9229, using ^{35}S-labeling [50].

11.3.7 The colonization process

The colonization of plants by cyanobacteria has been intensively studied in the *Gunnera* symbiosis. The plant organs colonized in this host are stem glands [3, 43]. These glands secrete viscous mucilage of low pH that attracts *Nostoc* and induces differentiation of motile hormogonia. Subsequently, the hormogonia move through channels toward the dark interior of the glands and colonize certain *Gunnera* cells before reverting to nonmotile vegetative filaments.

Once the cyanobionts reach the symbiotic plant organs (whether roots, thalli, or stems) and enter the symbiotic plant tissues, they revert to the vegetative cell stage and develop high numbers of N_2-fixing heterocysts [35, 39, 43]. However, in two-component lichens, the heterocyst frequencies are only moderately enhanced (ca. 20%, compared to ca. 5% in the free-living stage), whereas in *Gunnera*, ca. 60% are formed. This increase is preceded by a severalfold increase in *ntcA* and *hetR* expression [55], the products of which are required for heterocyst development [39]. Simultaneously, the heterocyst pattern changes from a regularly spaced distribution of single heterocysts along the filaments to the occurrence of multiple heterocysts, possibly suggesting decreased expression of *patS*, which encodes a small peptide known to maintain heterocyst-pattern formation [39].

11.3.8 Adaptations and exchange of metabolites

The physiology of the cyanobionts also changes dramatically during the development of the symbiosis. For instance, the increased heterocyst frequency is accompanied by an increased N_2-fixation activity in all symbioses. The N fixed, to a large extent, by the cyanobiont is then transferred to the host plant, most often as NH_4^+. This is achieved specifically in the heterocysts by either decreased transcription of the cyanobacterial *glnA* gene, which encodes GS, or decreased translation [2, 3, 43]. In cycads, the fixed-N may be released as glutamine and/or citrulline depending on the cycad genus [56].

Most symbiotic cyanobacteria are obliged to switch from a photoautotrophic to a heterotrophic mode of C-nutrition and to use host-derived carbohydrates as the energy source [57, 58]. Little is known about the C-source that the host plant delivers to the cyanobionts living in darkness, such as in either cycad roots or *Gunnera* stems, but fructose and glucose are known to stimulate both N_2ase activity and *hetR* expression in dark grown *Nostoc* PCC 9229 [59].

Other observed modifications during symbiosis include the enlargement of the *Nostoc* cell volume, which has been interpreted as a block in cell division, possibly induced by the host plant. The host's control of cyanbiont proliferation is supported by the observation that the relative proportion of cyanobacterial biomass compared to plant biomass in symbiotic organs always remains relatively small.

References

[1] K. Pawlowski, T. Bisseling, Rhizobial and actinorhizal symbioses: what are the shared features? Plant Cell 8 (1996) 1899–1913.

[2] A.N. Rai, B. Bergman, U. Rasmussen (Eds.) Cyanobacterial-Plant Symbiosis. Kluwer Academic Publishers, Dordrecht, The Netherlands, 2002.

[3] B. Bergman, U. Rasmussen, A.N. Rai, Cyanobacterial associations, in: C. Elmerich, W.E. Newton (Eds.) Associative Nitrogen-Fixing Bacteria and Cyanobacterial Associations, Springer, Dordrecht, The Netherlands, 2006, in press.

[4] L.G. Wall, The actinorhizal symbiosis, J. Plant Growth. Regul. 19 (2000) 167–182.

[5] H.G. Diem, Y.R. Dommergues, Current and potential uses and management of Casuarinaceae in the tropics and subtropics, in: C.R. Schwintzer, J.D. Tjepkema (Eds.) The Biology of *Frankia* and Actinorhizal Plants, Academic Press, San Diego, 1990, pp. 317–342.

[6] L.F. Benoit, A.M. Berry, Methods for production and use of actinorhizal plants in forestry, low maintenance landscapes, and revegetation, in: C.R. Schwintzer, J.D. Tjepkema (Eds.) The Biology of *Frankia* and Actinorhizal Plants, Academic Press, San Diego, 1990, pp. 281-297.

[7] T. Beveridge, T.S.C. Li, B.D. Oomah, A. Smith, Sea buckthorn products: manufacture and composition, J. Agr. Food Chem. 47 (1999) 3480–3488.

[8] A.M. Berry, O.T. Harriot, R.A. Moreau, S.F. Osman, D.R. Benson, A.D. Jones, Hopanoid lipids compose the *Frankia* vesicle envelope, presumptive barrier of oxygen diffusion to nitrogenase, Proc. Natl. Acad. Sci. USA 90 (1993) 6091–6094.
[9] R. Parsons, W.B. Silvester, S. Harris, W.T.M. Gruitjers, S. Bullivant, *Frankia* vesicles provide inducible and absolute oxygen protection for nitrogenase, Plant Physiol. 83 (1987) 728–731.
[10] K. Huss-Danell, Actinorhizal symbioses and their N_2 fixation, New Phytol. 136 (1997) 375–405.
[11] C.R. Schwintzer, Spore-positive and spore-negative nodules, in: C.R. Schwintzer, J.D. Tjepkema (Eds.) The Biology of *Frankia* and Actinorhizal Plants, Academic Press, San Diego, 1990, pp. 178–195.
[12] D.E. Soltis, P.S. Soltis, D.R. Morgan, S.M. Swensen, B.C. Mullin, J.M. Dowd, P.G. Martin, Chloroplast gene sequence data suggest a single origin of the predisposition for symbiotic nitrogen fixation in angiosperms, Proc. Natl. Acad. Sci. USA 92 (1995) 2647–2651.
[13] S.M. Swensen, B.C. Mullin, Phylogenetic relationships among actinorhizal plants – the impact of molecular systematics and implications for the evolution of actinorhizal symbioses, Physiol. Plant. 99 (1997) 565–573.
[14] D.R. Benson, M.L. Clawson, Evolution of the actinorhizal plant symbiosis, in: E.W. Triplett (Ed.), Prokaryotic Nitrogen Fixation: A Model System for Analysis of a Biological Process, Horizon Scientific Press, Symondham, 2000, pp. 207-224.
[15] D.R. Benson, D.W. Stephen, M.L. Clawson, W.B. Silvester, Amplification of 16S rRNA genes from *Frankia* strains in root nodules of *Ceanothus griseus, Coriaria arborea, Coriaria plumosa, Discaria toumatou*, and *Purshia tridentata*, Appl. Env. Microbiol. 62 (1996) 2904–2909.
[16] B. Cournoyer, C. Lavire, Analysis of *Frankia* evolutionary radiation using *glnII* sequences, FEMS Microbiol. Lett. 177 (1999) 29–34.
[17] S.C. Jeong, N.J. Ritchie, D.D. Myrold, Molecular phylogenies of plant and *Frankia* support multiple origins of actinorhizal symbioses, Mol. Phylogenet. Evol. 13 (1999) 493–503.
[18] P. Normand, S. Orso, B. Cournoyer, P. Jeannin, C. Chapelon, J. Dawson, L. Evtushenko, A. Misra, Molecular phylogeny of the genus *Frankia* and related genera and emendation of the family Frankiaceae, Int. J. Syst. Bacteriol. 46 (1996) 1–9.
[19] S.J. Kohls, D.D. Baker, Effects of substrate nitrate concentration on symbiotic nodule formation in actinorhizal plants, Plant Soil 118 (1989) 171–179.
[20] K.A. Thomas, A.M. Berry, Effects of continuous nitrogen application and nitrogen preconditioning on nodulation and growth of *Ceanothus griseus* var. *horizontalis*. Plant Soil 118 (1989) 181–187.

[21] S. Racette, J.G. Torrey, Root nodule initiation in *Gymnostoma* (Casuarinaceae) and *Shephardia* (Elaeagnaceae) induced by *Frankia* strain HFPGpI1, Can. J. Bot. 67 (1989) 2873–2879.
[22] M. Lalonde, R. Knowles, Ultrastructure, composition and biogenesis of the encapsulation material surrounding the endophyte in *Alnus crispa* var. *mollis* root nodules, Can. J. Bot. 53 (1975) 1951–1971.
[23] D. Callaham, W. Newcomb, J.G. Torrey, R.L. Peterson, Root hair infection in actinomycete-induced root nodule initiation in *Casuarina, Myrica*, and *Comptonia*, Bot. Gaz. (Suppl.) 140 (1979) S1–S9.
[24] C.R. Schwintzer, A.M. Berry, L.D. Disney, Seasonal patterns of root nodule growth, endophyte morphology, nitrogenase activity and shoot development in *Myrica gale*, Can. J. Bot. 60 (1982) 746–757.
[25] I.M. Miller, D.D. Baker, The initiation development and structure of root nodules in *Elaeagnus angustifolia* (Elaeagnaceae), Protoplasma 128 (1985) 107–119.
[26] D.D. Baker, B.C. Mullin, Diversity of *Frankia* nodule endophytes of the actinorhizal shrub *Ceanothus* as assessed by RFLP patterns from single nodule lobes, Soil Biol. Biochem. 26 (1994) 547–552.
[27] H. Ramirez-Saad, J.D. Janse, A.D.L. Akkermans, Root nodules of *Ceanothus caeruleus* contain both the N_2-fixing *Frankia* endophyte and a phylogenetically related Nod(-)/Fix(-) actinomycete, Can. J. Microbiol. 44 (1998) 140–148.
[28] J.Y. Jeong, S. Suh, C.H. Guan, Y.F. Tsay, N. Moran, C.J. Oh, C.S. An, K.N. Demchenko, Pawlowski K, Y. Lee, A nodule-specific dicarboxylate transporter from alder is a member of the peptide transporter family, Plant Physiol. 134 (2004) 969–978.
[29] P. Lundberg, P.-O. Lundquist, Primary metabolism in N_2-fixing *Alnus incana-Frankia* symbiotic root nodules studied with ^{15}N and ^{31}P nuclear magnetic resonance spectroscopy, Planta 219 (2004) 661–672.
[30] A.M. Berry, T.M. Murphy, P.A. Okubara, K.R. Jacobsen, S.M. Swensen, K. Pawlowski, Novel expression pattern of cytosolic gln synthetase in nitrogen-fixing root nodules of the actinorhizal host, *Datisca glomerata*, Plant Physiol. 135 (2004) 1849–1862.
[31] A. Sellstedt, P. Reddell, P. Rosbrook, The occurrence of hemoglobin and hydrogenase in nodules of 12 *Casuarina-Frankia* symbiotic associations, Physiol. Plant. 82 (1991) 458–464.
[32] M. Leul, A. Mohapatra, A. Sellstedt, Biodiversity of hydrogenases in *Frankia*, Curr. Microbiol. 50 (2005) 17–23.
[33] P. Schrder, Aeration of the root system in *Alnus glutinosa* L. Gaertn., Ann. Sci. Forest. 46 (1989) 310-314.
[34] W.B. Silvester, S.L. Harris, J.D. Tjepkema, Oxygen regulation and hemoglobin, in: C.R. Schwintzer, J.D. Tjepkema (Eds.), The Biology of *Frankia* and Actinorhizal Plants, Academic Press, San Diego, 1990, pp. 157–176.

[35] A.N. Rai, E. Söderbäck, B. Bergman, Cyanobacterium-plant symbiosis. New Phytol. 147 (2000) 449–481.
[36] B. Bergman, The *Nostoc-Gunnera* symbiosis, in: A.N. Rai, B. Bergman, U. Rasmussen (Eds.) Cyanobacteria in Symbiosis, Kluwer Academic Publishers, Dordrecht, The Netherlands, 2002, pp. 207–232.
[37] W.L. Chiu, G. Peters, G. Levieille, P.C. Still, S. Cousins, Nitrogen deprivation stimulates symbiotic gland development in *Gunnera manicata*, Plant Physiol. 139 (2005) 224–230.
[38] J. Rikkinen, Cyanolichens: an evolutionary overview, in: A.N. Rai, B. Bergman, U. Rasmussen (Eds.) Cyanobacteria in Symbiosis, Kluwer Academic Publishers, Dordrecht, The Netherlands, 2002, pp. 31–72.
[39] J.C. Meeks, J. Elhai, Regulation of cellular differentiation in filamentous cyanobacteria in free-living and plant-associated symbiotic growth states, Microbiol. Mol. Biol. Rev. 66 (2000) 94–121.
[40] S. Lechno-Yossef, S.A. Nierzwicki-Bauer, *Azolla-Anabaena* symbiosis, in: A.N. Rai, B. Bergman, U. Rasmussen (Eds.) Cyanobacteria in Symbiosis, Kluwer Academic Publishers, Dordrecht, The Netherlands, 2002, pp. 153–178.
[41] J.K. Vessey, K. Pawlowski, B. Bergman, Root-based N_2-fixing symbioses: legumes, actinorhizal plants, *Parasponia* sp. and cycads, Plant Soil 266 (2005) 205–230.
[42] S. Janson, J. Wouters, B. Bergman, E.J. Carpenter, Host specificity in the *Richelia*-diatom symbioses revealed by *hetR* gene sequence analyses, Environ. Microbiol. 1 (1999) 431–438.
[43] B. Bergman, C. Johansson, E. Söderbäck, The *Nostoc-Gunnera* symbiosis, New Phytol. 122 (1992) 379–400.
[44] M. Nilsson, B. Bergman, U. Rasmussen, Cyanobacterial diversity in geographically related and distant host plants of the genus *Gunnera*, Arch. Microbiol. 173 (2000) 97–102.
[45] C.S. Enderlin, J.C. Meeks, Pure culture and reconstitution of the *Anthoceros-Nostoc* symbiotic association, Planta 158 (1983) 157–165.
[46] C. Johansson, B. Bergman, Reconstitution of the symbiosis of *Gunnera manicata* Linden – cyanobacterial specificity, New Phytol. 126 (1994) 643–652.
[47] J.C. Meeks, J. Elhai, T. Thiel, M. Potts, F. Larimer, J. Lamerdin, P. Predki, R. Atlas, An overview of the genome of *Nostoc punctiforme*, a multicellular, symbiotic cyanobacterium, Photosynth. Res. 70 (2001) 85–106.
[48] J.A. Raven, The evolution of cyanobacterial symbioses, in: B. Osborne (Ed.) Commentaries on cyanobacterial symbioses, Royal Irish Academy, Dublin, Ireland, 2002, pp. 3–6.
[49] N. Tandeau de Marsac, Differentiation of hormogonia and relationships with other biological processes, in: D.A. Bryant (Ed.) The

Molecular Biology of Cyanobacteria, Kluwer Academic Publishers, Dordrecht, The Netherlands, 1994, pp. 825–842.
[50] U. Rasmussen, C. Johansson, B. Bergman, Early communication in the *Gunnera-Nostoc* symbiosis: plant-induced cell differentiation and protein synthesis in the cyanobacterium, Mol. Plant-Microbe Interact. 7 (1994) 696–702.
[51] M.F. Cohen, J.C. Meeks, A hormogonia regulating locus, *hrmUA*, of the cyanobacterium *Nostoc punctiforme* strain ATCC 29133 and its response to an extract of a symbiotic plant partner *Anthoceros punctatus*, Mol. Plant Microbe Interact. 10 (1997) 280–289.
[52] A. Liaimer, B. Bergman, Phytohormones in cyanobacteria: occurrence and perspectives, in: I. Tikhonovich, B. Lugtenberg, N. Provorov (Eds.) Biology of Plant-Microbe Interactions, Vol. 4, International Society for Molecular Plant-Microbe Interactions, St Paul, Minnesota USA, 2004, pp. 394–397.
[53] E. Sergeeva, A. Liaimer, B. Bergman, Evidence for biosynthesis and release of the phytohormone indole-3-acetic acid by cyanobacteria, Planta 215 (2002) 229–238.
[54] A. Liaimer, A. Matveyev, B. Bergman, Isolation of host plant induced cDNAs from *Nostoc* sp. PCC 9229 forming symbiosis with the angiosperm *Gunnera* spp, Symbiosis 31 (2001) 293–307.
[55] C.-M. Wang, M. Ekman, B. Bergman, Expression of *ntcA, glnB, hetR* and *nifH* and corresponding protein profiles along a developmental sequence in the *Nostoc-Gunnera* symbiosis, Mol. Plant-Microbe Interact. 17 (2004) 436–443.
[56] J.S. Pate, P. Lindblad, C.A. Atkins, Pathways of assimilation and transfer of the fixed nitrogen in coralloid roots of cycad-*Nostoc* symbioses, Planta 176 (1988) 461–471.
[57] E. Söderbäck, B. Bergman, The *Nostoc-Gunnera* symbiosis: carbon fixation and translocation, Physiol. Plant. 89 (1993) 125–132.
[58] K.G. Black, R. Parsons, B.A. Osborne, Uptake and metabolism of glucose in the *Nostoc-Gunnera* symbiosis, New Phytol. 153 (2002) 297–305.
[59] J. Wouters, S. Jansson, B. Bergman, The effect of exogenous carbohydrates on nitrogen fixation and *hetR* expression in *Nostoc* PCC 9229 forming symbiosis with *Gunnera*, Symbiosis 28 (2000) 63–76.

Chapter 12

Associative Nitrogen Fixation

Anne Van Dommelen and Jos Vanderleyden

12.1 Introduction

Biological N_2 fixation is the only biotic process that makes atmospheric N_2 available to living organisms. Besides terrestrial N_2 fixation, which is estimated to contribute for $90 \times 10^{12}-140 \times 10^{12}$ g of N per year [1], marine free-living and symbiotic N_2 fixation contribute in great part to the natural input of nitrogen in the N cycle. On the basis of the extrapolated contribution of the best-known marine diazotrophs, the cyanobacterial *Trichodesmium* spp., the marine contribution to the global annual N_2 fixation is estimated at ~80×10^{12} g of N per year. Symbiotic cyanobacteria found in tropical diatoms and planktonic unicellular cyanobacteria fix N_2 at rates comparable to or even in excess of those of *Trichodesmium* spp. Culture-independent assessments of marine diazotroph diversity targeting *nifH* revealed that heterotrophic N_2-fixers actively expressing their nitrogenase are also present in the tropical and subtropical oceans [2].

The process of N_2 fixation is unique to Bacteria and Archaea. The most important crops for the nutrition of the world's population – rice, maize and wheat – do not naturally form specialized symbiotic structures, the

Biology of the Nitrogen Cycle
Edited by H. Bothe, S.J. Ferguson and W.E. Newton

so-called nodules. This does not mean that atmospheric N_2 cannot contribute to the N demand of non-legume crops. Some gramineous crops such as sugarcane, wetland rice and Kallar grass (*Leptochloa fusca* (L.) Kunth) can derive a substantial part of their N demand from biological N_2 fixation [3]. This has opened the search for plant associated N_2-fixing bacteria.

Most diazotrophs associated with plant roots belong to the proteobacteria. They are found in the α-subgroup (species of e.g. *Gluconacetobacter*, *Azospirillum*), in the β-subgroup (species of e.g. *Derxia*, *Azoarcus*, *Burkholderia*, *Herbasprillum*), as well as in the γ-subgroup (species of e.g. *Azotobacter*, *Serratia*, *Klebsiella*, *Pseudomonas*, *Pantoea*). Exceptions are, for example, the N_2-fixing *Paenibacillus* species that have been found in the rhizosphere of many plants including crops and they belong to the Firmicutes.

In contrast to rhizobia in nodules, there is no evidence that plant-associated diazotrophs colonize living plant cytoplasm or occur inside host cells in an organelle called the symbiosome. Some have been detected inside root cortex cells but the plant cell cytoplasm was disintegrated [4]. Plant-associated diazotrophs able to colonize intercellularly the inner tissues of plants without causing any apparent damage or ecological threat to the host are referred to as endophytic (e.g. *Herbaspirillum* species, *Gluconacetobacter* species, *Azoarcus* species, *Burkholderia* species). *Azospirillum* species have a more 'associative' or 'facultative endophytic' lifestyle. They sometimes enter the root tissues, but are also found in large numbers on the root surface. Epiphytic diazotrophic bacteria like *Beijerinckia fluminensis* and *Azorhizophilus paspali* (*Azotobacter paspali*) are mainly isolated from rhizoplane [4], [5].

Nitrogen is not only a limiting nutrient for plants, many insects also feed on a N-poor diet. The finding of effective N_2-fixing bacteria in the gut of termites and fruit flies [6] opens the question of the extent and ecological impact of associative N_2 fixation in animals.

In this section we will give a short overview of the best-studied plant associated N_2-fixing bacteria.

12.2 Non-symbiotic colonization of plants

Studies in well-defined systems have elucidated pathways of colonization of grass roots by diazotrophic endophytes, like *Azoarcus* sp., *Herbaspirillum seropedicae* and *Gluconacetobacter diazotrophicus* [4]. The establishment of these endophytic associations differs remarkably from highly developed root nodule symbioses, in which rhizobia enter the plants through root hairs via infection threads. The colonization of grasses by endophytes is more comparable to the crack entry in more primitive nodule symbioses. One site of primary colonization is the points of emergence of lateral roots, where bacterial microcolonies can easily be detected. Bacterial cells have been found between the cell layers of the lateral root and the cortex of the main root. Entry near lateral roots has also been observed for *Azospirillum* spp. although these bacteria are mainly rhizoplane colonizers. Another entry path of grass endophytes is the root tip at the zone of elongation and differentiation. Since the tissue is not yet differentiated at the root tip, they may reach the site of the developing stele before the formation of the endodermis, which will later constitute a thick-walled boundary. This explains how these bacteria are able to invade xylem vessels [7].

Colonization by *Azospirillum* has also been studied in detail. These associative bacteria not only colonize the sites of lateral root emergence, they also colonize the root hair zone. An important factor in this colonization is the bacterial chemotactic motility. The attachment of *Azospirillum* to plant roots has been shown in short-term *in vitro* binding assays as a biphasic process. In a first step, the bacteria adsorb to the roots as single cells in a rapid, weak and reversible manner. The polar flagellum and outer-membrane proteins are involved in this step. The adsorption phase is followed by an anchoring phase in which bacterial aggregates are formed and the bacteria are firmly and irreversibly anchored to the roots. Extracellular polysaccharides play a role in this step [8], [9].

In contrast to symbiotic systems, associative interactions are less specific. Large and diverse populations of heterotrophic diazotrophs can be isolated from the surface and rhizospheres of different gramineous plants [3]. This does not exclude some 'pickiness' in associative interactions. As will be discussed in Section 12.3.5, the N_2-fixation phenotype is dependent on the plant species and can even differ according to the cultivar. Another example of some kind of specificity is the limited host range found for *Azoarcus* species.

12.3 N_2-fixing bacteria associated with plants

12.3.1 Azospirillum

Azospirilla are one of the earliest discovered and best characterized associative N_2-fixing bacteria. The genus *Azospirillum* owes its name to its N_2-fixing capability (*Azo-*) and the spiral movements of the cell (*-spirillum*) [10]. These short, rod-shaped, slightly curved Gram-negative bacteria were first isolated from soil in the Netherlands in 1925 [11].

Not many reports followed this first isolation until the group of Dr. Johanna Döbereiner 'rediscovered' *Azospirillum* in the mid-1970s. This resulted from a search for associative N_2-fixing bacteria in the rhizosphere of *Digitaria* and *Zea mays* in Brazil. Azospirilla are mainly found in large numbers in tropical soils, but in various temperate zones, and even in tundra and semi-desert sites of the Canadian High Arctic [12] the isolation of *Azospirillum* spp. has been reported. Azospirilla appear to colonize crop plants as well as weeds, annual as well as perennial plants and they have even been isolated from leaves of marine mangroves and from lake or pond water [8].

Presently eight *Azospirillum* species have been described: *A. brasilense*, *A. lipoferum*, *A. halopraeferens*, *A. irakense*, *A. largimobile*, *A. doebereinerae*, *A. oryzae* and *A. amazonense* [8], [13]. *Azospirillum* has been shown to positively influence plant growth, crop yields and N content of the plant. This plant stimulatory effect exerted by *Azospirillum* has been attributed to several mechanisms, including biological N_2 fixation. However, there is still no clear evidence that significant amounts of fixed N_2 contribute to the N content of cereals when inoculated with *Azospirillum*. Available C-source, mineral N concentration and prevailing O_2-tension at the root surface can be limiting factors for significant associative N_2 fixation. Moreover, the NH_4 derived from N_2 fixation is mostly only available for the plant after the bacterial cells die because *A. brasilense* excretes only small amounts of NH_4^+ under diazotrophic growth. The presence of a high-affinity NH_4^+-uptake system and efficient assimilation of NH_4^+ by the glutamine synthetase enable the bacteria to use most of their fixed N for their own growth. Nevertheless, the expression of an NH_4^+-transporter gene of tomato, which was shown to be upregulated by low levels of NH_4^+, responded to root-associated N_2-fixing *Azospirillum* spp. [14]. When NH_4^+ assimilation is reduced by a mutation in the glutamine synthetase, significant levels of NH_4^+ are excreted [15]. These mutants are also superior in enhancing growth of greenhouse winter wheat (A. Van Dommelen *et al.*, unpublished results). Nevertheless, the fact that these mutants can significantly contribute to the N budget of plants, in view of the relatively low number of bacteria (10^5 fresh weight of roots) usually observed, awaits further proofs.

Presently the main plant growth-promoting trait of *Azospirillum* is considered to be its ability to produce plant-growth-regulating substances, in particular auxins, which increase the number of lateral roots and enlarge the root hairs. Given that the concentration of auxin and other phytohormones is suboptimal in roots, this could result in a higher nutrient uptake and an improved water status of the plant under suboptimal conditions for normal plant root development [16]. The potential sustainable yield-enhancing capacity of *A. lipoferum* and *A. brasilense* strains have been exploited by using them in commercial inocula.

12.3.2 Herbaspirillum

Although most of the bacteria in the genus *Herbaspirillum* are N_2-fixers colonizing diverse plants endophytically, also clinical and environmental isolates belong to this genus. This resembles the situation in other species of the β-proteobacteria, where plant-associated or even symbiotic diazotrophs, opportunistic pathogens and potent degraders of pollutants belong to the same genera, like *Burkholderia* (see Section 12.3.7), *Ralstonia* and *Azoarcus* (see Section 12.3.6) [17].

The finding of species with and without N_2-fixation genes within a same genus suggests that these genes have been acquired by horizontal gene transfer. Phylogenetic studies support this hypothesis [18].

Diazotrophic *H. seropedicae* were first isolated from different species of the *Gramineae* family, like maize (*Zea mays*), sorghum (*Sorghum bicolor*), sugarcane (*Saccharum officinarum*) and wild rice (*Oryza sativa*). Later they were also found in association with dicotyledonous plants and could be isolated from root nodules of legumes and in roots of monocots outside the grass family as well as stems of different cultivars of banana (*Musa* spp.) and pineapple (*Ananas comosus* (L.) Merril) [17].

Herbaspirillum spp. are aggressive colonizers of the root interior, establishing themselves not only in the cortex and vascular tissues of roots but also systemically in the whole plant. Using axenic systems of different plants, a significant stimulation of root development due to inoculation by *H. seropedicae* and *H. frisingense* was demonstrated. Histochemical analysis of seedlings of maize, sorghum, wheat and rice grown in vermiculite showed that *H. seropedicae* colonizes root surfaces and inner tissues. Root exudation sites, such as axils of secondary roots and intercellular spaces of the root cortex are firstly colonized, subsequently the vascular tissue is occupied and expression of *nif* genes is observed in roots, stems and leaves. The expression of *nif* genes was also observed in bacterial colonies located in the external mucilaginous root material. Colonization of plant tissue by *H. seropedicae* does not depend on the N_2-fixing ability, since similar numbers of cells were isolated from roots or shoots of the plants inoculated with Nif^+ or Nif^- strains [19].

When colonizing rice seedlings, nitrogenase activity of *H. seropedicae* was only present when a C source was added. In some cultivars, *H. seropedicae* nitrogenase activity could significantly (approximately 30%) enhance the N content as compared to the non-inoculated controls [20].

Presently, the *H. seropedicae* genome is being sequenced by the Brazilian GENOPAR program (http://www.genopar.org/).

12.3.3 Pseudomonas spp. and Azotobacter vinelandii

Azotobacter vinelandii is a Gram-negative, strictly aerobic and widely distributed free-living soil bacterium that has many interesting features, including the ability to grow on a wide variety of carbohydrates, alcohols and organic acids, alginate production and N_2 fixation. Unlike most diazotrophs, *A. vinelandii* is able to fix N_2 in the presence of atmospheric O_2 concentrations. Protection of the nitrogenase from O_2 inactivation is a complex process in which high respiration rates, formation of a complex with a so-called Shethna protein, autoprotection by reduction of O_2 to H_2O_2, morphological protection and the formation of alginate have been proposed to play a role [21], [22]. Phylogenetic data and gene organization data strongly suggest that *A. vinelandii* belongs to the genus *Pseudomonas sensu stricto*. Originally, N_2-fixing ability was considered a major physiological characteristic differentiating *A. vinelandii* from *Pseudomonas* species. The occurrence of N_2 fixation in *Pseudomonas* species has been long debated, but in recent years several genuine *Pseudomonas* strains that can fix N_2 have been identified [18], [23].

P. stutzeri strain A15 (formerly *Alcaligenes faecalis*) is a N_2-fixing *Pseudomonas* repeatedly isolated as a particular predominant diazotrophic strain in the rhizosphere of Chinese paddy field rice. It was shown to express *nifH* in the rice roots. Other N_2-fixing *P. stutzeri* strains have been isolated from sorghum (*Sorghum bicolor*) and caper (*Capparis spinosa*) [24]. Also other plants have been reported to have *Pseudomonas* strains in their rhizosphere.

12.3.4 Gluconacetobacter (formerly Acetobacter) diazotrophicus

This small endophytic bacterium is a Gram-negative rod that exhibits micro-aerobic N_2 fixation that is relatively tolerant towards O_2. It has several unusual physiological properties for a diazotroph, such as tolerance to low pH and to high sugar concentrations. For most diazotrophs, cell growth and biological N_2 fixation are optimal near neutral pH. However, *G. diazotrophicus* does not grow at pH values higher than 7. Optimal growth of *G. diazotrophicus* occurs at pH 5.5 and growth and N_2 fixation have been reported at pH as low as 2.5. *G. diazotrophicus* also grows and fixes N_2 in medium with up to 30% sucrose indicating that it has adapted for growth inside the sugarcane plant. Its sensitivity (both growth and N_2-fixation activity) to even low salt concentrations is partially relieved by higher sucrose levels. Furthermore, this organism has no assimilatory nitrate reductase, so N_2 fixation continues in the presence of NO_3^-, and its nitrogenase is only partially inhibited by added NH_4^+ via an unknown process that does not involve covalent modification (ADP-ribosylation) of its Fe–protein component. These last two properties might have considerable agricultural significance such that biological N_2 fixation might be supplemented by N fertilizer [25].

Although this bacterium is ideally suited to survive and grow within plant tissues, it does not easily colonize intact plants and hardly survives in the soil [26]. *G. diazotrophicus* was considered in early studies as an endophyte associated only with sugarcane and other sucrose-accumulating plants such as sweet potato and Cameroon grass. Later *G. diazotrophicus* was also found in endophytic associations with multiple host plants such as *Coffea arabica*, *Eleusine coracana* and *Ananas comosus*, and in very distant geographical regions [27].

G. diazotrophicus has been found in the roots, stems, and aerial parts of several Australian and Brazilian sugarcane cultivars. Also, the xylem sap and the intercellular spaces of sugarcane stems have been suggested as a possible suitable location for the bacterium because of the low pO_2 within them that should allow nitrogenase expression [28].

In N-deficient conditions, N_2 fixation of *G. diazotrophicus* is responsible for enhanced growth and N content of sugarcane. Colonization is not dependent on the N_2-fixation capacity and when N is not limiting, growth enhancement is observed in plants inoculated with either wild-type or Nif^- mutants, suggesting an additional effect of a plant-growth-promoting factor provided by *G. diazotrophicus* [29].

The genome of *G. diazotrophicus* is presently being sequenced (http://www.riogene.lncc.br/).

12.3.5 Klebsiella

Klebsiella can frequently be isolated from the root surfaces of various plants. *K. pneumoniae*, *K. oxytoca* and *K. planticola* are all capable of fixing N_2 and are classified as associative N_2 fixers. Isolations of N_2-fixing *Klebsiella* were reported from rice leaves, rhizosphere, grassland soil, sweet potato [30], [31]. The *Klebsiella*-type strain, *K. pneumoniae*, is a model system in N_2 fixation, but other strains of this same species are feared pathogens. Plant-growth-promoting strains of *K. pneumoniae* are able to colonize the rhizosphere and interior of seedlings of legumes like alfalfa *(Medicago sativa)* and *Medicago truncatula*, but colonize *Arabidopsis thaliana*, wheat (*Triticum aestivum*) and rice (*Oryza sativa*) in much higher numbers [32].

K. pneumoniae was shown to fix N_2, relieve N-deficiency symptoms and increase total N and N concentration in wheat (*Triticum aestivum* L.). N derived from N_2 fixation was found in plant tissues and in chlorophyll, as shown by the ^{15}N isotope dilution technique. Nitrogenase reductase is produced in the intercellular space of the root cortex. The N_2-fixation phenotype is dependent on the wheat cultivar [33]. For maize, *K. pneumoniae* was not capable of relieving the N-deficiency symptoms of non-fertilized maize in either the field or the greenhouse, although the yield of N-fertilized maize was significantly enhanced [34].

Although *K. pneumoniae* was found to reside in the intercortical layers of the stem and in the root of maize, nitrogenase reductase was only found in the roots when the bacteria were supplied with an exogenous C-source, as shown by immunolocalization with an antibody against purified nitrogenase reductase [35].

12.3.6 Azoarcus

Associative N_2-fixing species of the genus *Azoarcus* were first found in association with Kallar grass (*Leptochloa fusca* (L.) Kunth), an undomesticated C4 plant tolerant to soil salinity, alkalinity and water-logged conditions cultivated in Pakistan. Later they were also isolated from field-grown rice cultivated in Nepal and from resting stages of plant-associated fungi found in rice field soil from Pakistan. *A. indigens*, *A. communis* and *A.* sp. BH72 occur inside roots or on the root surface of Gramineae and have never been isolated from root-free soil so far, except for members of *A. communis* originating from a petroleum refinery oily sludge in France or from a compost biofilter in Canada. This is in contrast to the soil-borne strains of the genus, such as *A. tolulyticus*, *A. toluvorans*, *A. toluclasticus*, *A. evansii*, *A. buckelii* or *A. anaerobius*, which do not originate from living plants but mostly from soil and sediments. Although presently plant-associated *Azoarcus* spp. have been isolated only from a limited number of samples, they may be more widely distributed than assumed. In molecular-ecological studies on root material or fungal spores, *Azoarcus* 16S rDNA genes or *nifH* genes have been retrieved which did not correspond to genes of cultivated strains or species [36].

In situ hybridization studies demonstrated expression of *Azoarcus* nitrogenase in the root cortex of Kallar grass and inside rice roots [4].

N_2 fixation was studied in more detail in *Azoarcus* sp. BH72. Compared to *Azospirillum* spp., *Azoarcus* is more tolerant to O_2. Similarly to *Azospirillum*, the expression of nitrogenase genes is transcriptionally regulated in response to O_2 (fully repressed at 4% O_2 in the headspace) or combined N (repressed by 0.5 mM NH_4^+ or nitrate). A particular trait of N_2 fixation in this strain is the formation of so-called 'diazosomes'. This novel intracytoplasmic membrane stack is formed when cells are shifted down to extremely low O_2 concentrations (30 nM) and is associated with augmented rates and efficiency of N_2 fixation, called 'hyperinduction' [36].

12.3.7 Burkholderia

The genus *Burkholderia* is an extremely heterogeneous group that includes soil bacteria, plant-growth-promoting rhizobacteria, and human and plant pathogens. Strains of *Burkholderia cepacia* can survive within vacuoles of free-living amoebae [37]. *B. vietnamiensis* was the first species of this genus found to fix N_2. *B. vietnamiensis* was discovered in association with roots of rice plants grown in Vietnam and was later found inside maize roots, as well as in the rhizosphere and rhizoplane of maize and coffee plants. *B. vietnamiensis* has attracted interest because of its abilities to promote rice plant growth and to enhance grain yield [27]. *B. brasilensis* and *B. tropicalis* were isolated as N_2-fixing bacteria from banana (*Musa* spp.) and pineapple (*Ananas comosus* (L.) Merril) [38]. Recently, other N_2-fixing *Burkholderia* species were reported: *B. kururiensis*, *B. tropica*, *B. unamae*, *B. silvatlantica* [39] and *B. silvatlantica* [personal communication Estrada-de los Santos].

Although it was long thought that only members of the α-subclass of proteobacteria where able to nodulate leguminous plants, *B. tuberum*, *B. phymatum* and *B. caribensis* strains were found to be N_2-fixing legume symbionts [40]. They possess nodulation genes that have been acquired by horizontal transfer [41]. *Ralstonia taiwanensis* is a related β-proteobacteria found to nodulate *Mimosa* species [40].

12.3.8 Other associative N_2-fixing bacteria

Serratia marcescens strains have been isolated from surface-sterilized roots and stems of different rice varieties. More detailed studies using light and transmission electron microscopy combined with immunogold labeling confirmed an endophytical establishment within roots, stems, and leaves. When inoculated *in vitro* on rice, N_2 fixation was measured only when external carbon (e.g. malate, succinate or sucrose) was added to the rooting medium [42].

Pantoea species are primarily plant pathogens, but *Pantoea* species have also been isolated endophytically and epiphytically from a wide variety of crops: *P. agglomerans* (*Enterobacter agglomerans*, *Erwinia herbicola*) has for example been isolated from corn, cotton and wheat [31]. A *Pantoea* species able to grow over a wide range of temperature, pH and salt concentration was isolated from sugarcane in Cuba [43].

γ-Proteobacteria of the orders *Enterobacteriales* or *Pseudomonadales* were found as exclusive occupants of the nodules in some Mediterranean wild *Hedysarum* legume species. These γ-proteobacteria included *Pantoea agglomerans*, *Enterobacter kobei*, *Enterobacter cloacae*, *Leclercia adecarboxylata*, *Escherichia vulneris* and *Pseudomonas* sp. [44], and expand the group of nodulating bacteria to the γ-division of the proteobacteria.

Besides finding nodulating strains outside of the *Rhizobiales*, rhizobia may also occur as endophytes in the roots of cereals, such as rice, wheat and maize, without nodule formation. They are able to promote the growth of these non-legumes, but this growth promotion is related to mechanisms independent of biological N_2 fixation, such as phosphate-solubilizing ability, and neither root nodules nor N_2 fixation are observed during these interactions [14].

Nitrogen-fixing *Paenibacillus* have been found in the rhizosphere of many plants and important crop species. They include: *P. polymyxa*, *P. macerans*, *P. durus* (synonyms: *Paenibacillus azotofixans*, *Bacillus azotofixans*, *Clostridium durum*), *P. peoriae*, *P. borealis*, *P. brasilensis*, *P. graminis* and *P. odorifer*. Strains belonging to these species are considered to be important for agriculture. They can influence plant growth and health directly by the production of phytohormones, by providing nutrients, by fixing N_2 and/or by the suppression of deleterious microorganisms through antagonistic functions. Besides their contribution as N_2 fixers, many of these strains are also of industrial importance for the production of chitinases, amylases, proteases and antibiotics [45].

Anaerobic clostridia have been found to fix N_2 in a consortium with diverse non-diazotrophic bacteria in various gramineous plants. A major feature of these consortia is that N_2 fixation by the anaerobic clostridia is supported by the elimination of O_2 by the accompanying bacteria in the culture. These consortia are widespread in wild rice species and pioneer plants, which are able to grow in unfavorable locations [46].

With the use of cultivation-independent methods, based on the detection of the *nifH* gene, a remarkable diversity of up to now uncultured diazotrophs was detected in association with roots of grasses [47].

12.4 Conclusion

The number of N_2-fixing plant-associated bacteria identified is still growing, but we are far from having a complete view of the ecological impact of these associations. Although a distinct clear contribution of N_2 fixation has been reported in some endophytic associations, it is not always clear which species is the most important fixer and how interactions of the microbial communities found in association with plants influence the N_2-fixation activity. Another question that deserves future investigation is how fixed N_2 becomes available to the plant. Is it only released upon cell death and disintegration or are their other routes of fixed-N delivery, perhaps stimulated by the environment? For example, accumulation of glutamate is a response to osmotic stress. Glutamate can be accumulated by reducing its further transformation into glutamine by glutamine synthetase. Reducing the level of NH_4^+ assimilation by the glutamine synthetase correlates with NH_4^+ excretion [15]. When rhizobia are in symbiosis with their host plants, their NH_4^+-assimilation is reduced by reducing glutamine synthetase activity. In endophytic associations where substantial amounts of fixed N are transferred to the plant (e.g. sugarcane), it would be interesting to find out which factors influence the release of fixed N_2.

Understanding and optimizing N_2-fixing plant-bacteria associations have promising prospective for sustainable agriculture. Not only would it reduce the energy and pollution cost of the industrial reduction of N_2 from fertilizer industry, but since N_2-fixing bacteria generally also have other plant-growth-promoting traits (better development of root due to plant-growth-factor production, enhanced resistance to pathogens, enhanced mineral uptake through, e.g. P-solubilization) the general plant health, and indirectly human health, will benefit from it.

The challenge remains to compile and integrate all the factors (genetic, biotic, abiotic) that influence associative N_2 fixation and to find ways to control them at low cost.

Finally, the ecological impact and extent of N_2 fixation activity in insects and perhaps higher animals is an interesting route for future investigation [6].

References

[1] P.M. Vitousek, J.D. Aber, R.W. Howarth, G.E. Likens, P.A. Matson, D.W. Schindler, W.H. Schlesinger, D.G. Tilman, Human alteration of the global nitrogen cycle: sources and consequences, Ecol. Appl. 7 (1997) 737–750.
[2] C. Bird, J. Martinez Martinez, A.G. O'Donnell, M. Wyman, Spatial distribution and transcriptional activity of an uncultured clade of planktonic diazotrophic gamma-proteobacteria in the Arabian sea, Appl. Environ. Microbiol. 71 (2005) 2079–2085.
[3] E.K. James, Nitrogen fixation in endophytic and associative symbiosis, Field Crops Res. 65 (2000) 197–209.
[4] B. Reinhold-Hurek, T. Hurek, Life in grasses: diazotrophic endophytes, Trends Microbiol. 6 (1998) 139–144.
[5] J.I. Baldani, V.L.D. Baldani, History on the biological nitrogen fixation research in graminaceous plants: special emphasis on the Brazilian experience, An. Acad. Bras. Cienc. 77 (2005) 549–579.
[6] A. Behar, B. Yuval, E. Jurkevitch, Enterobacteria-mediated nitrogen fixation in natural populations of the fruit fly *Ceratitis capitata*, Mol. Ecol. 14 (2005) 2637–2643.
[7] T. Hurek, B. Reinhold-Hurek, *Azoarcus* sp. strain BH72 as a model for nitrogen-fixing grass endophytes, J. Biotechnol. 106 (2003) 169–178.
[8] A. Hartmann, J.I. Baldani, The genus *Azospirillum*, in: M. Dworkin et al. (Eds.) The Prokaryotes: An Evolving Electronic Resource for the Microbiological Community, Springer-Verlag, New York, 2003, http://link.springer-ny.com/link/service/books/10125/.
[9] S. Burdman, G. Dulguerova, Y. Okon, E. Jurkevitch, Purification of the major outer membrane protein of *Azospirillum brasilense*, its affinity to plant roots, and its involvement in cell aggregation, Mol. Plant Microbe Interact. 14 (2001) 555–561.
[10] J.J. Tarrand, N.R. Krieg, J. Döbereiner, A taxonomic study of the *Spirillum lipoferum* group, with the description of a new genus, *Azospirillum* gen.nov. and two species, *Azospirillum lipoferum* (Beijerinck) comb. nov. and *Azospirillum brasilense* sp. nov., Can. J. Microbiol. 24 (1978) 967–980.
[11] M.W. Beijerinck, über ein *Spirillum*, welches freien Stickstoff binden kann? Centralbl. Bakt. II 63 (1925) 353–357.
[12] P. Nosto, L.C. Bliss, F.D. Cook, The association of free-living nitrogen-fixing bacteria with the roots of High Arctic graminoids, Arct. Alp. Res. 26 (1994) 180–186.

[13] C.H. Xie, A. Yokota, *Azospirillum oryzae* sp. nov., a nitrogen-fixing bacterium isolated from the roots of the rice plant *Oryza sativa*, Int. J. Syst. Evol. Microbiol. 55 (2005) 1435–1438.

[14] E. Somers, J. Vanderleyden, M. Srinivasan, Rhizosphere bacterial signalling: a love parade beneath our feet, Crit. Rev. Microbiol. 30 (2004) 205–240.

[15] A. Van Dommelen, V. Keijers, A. Wollebrants, J. Vanderleyden, Phenotypic changes resulting from distinct point mutations in the *Azospirillum brasilense glnA* gene, encoding glutamine synthetase, Appl. Environ. Microbiol. 69 (2003) 5699–5701.

[16] O. Steenhoudt, J. Vanderleyden, *Azospirillum*, a free-living nitrogen-fixing bacterium closely associated with grasses: genetic, biochemical and ecological aspects, FEMS Microbiol. Rev. 24 (2000) 487–506.

[17] M. Schmid, J.I. Baldani, A. Hartmann, The Genus *Herbaspirillum*, in: M. Dworkin et al. (Eds.) The Prokaryotes: An Evolving Electronic Resource for the Microbiological Community, Springer-Verlag, New York, 2005, http://link.springer-ny.com/link/service/books/10125/.

[18] H. Rediers, J. Vanderleyden, R. De Mot, *Azotobacter vinelandii*: a *Pseudomonas* in disguise? Microbiology 150 (2004) 1117–1119.

[19] L.D.B. Roncato-Maccari, H.J.O. Ramos, F.O. Pedrosa, Y. Alquini, L.S. Chubatsu, M.G. Yates, L.U. Rigo, M.B.R. Steffens, E.M. Souza, Endophytic *Herbaspirillum seropedicae* expresses *nif* genes in gramineous plants, FEMS Microbiol. Ecol. 45 (2003) 39–47.

[20] E.K. James, P. Gyaneshwar, N. Mathan, W.L. Barraquio, P.M. Reddy, P.P. Iannetta, F.L. Olivares, J.K. Ladha, Infection and colonization of rice seedlings by the plant growth-promoting bacterium *Herbaspirillum seropedicae* Z67, Mol. Plant Microbe Interact. 15 (2002) 894–906.

[21] J.H. Becking, The Family *Azotobacteraceae*, in: M. Dworkin et al. (Eds.) The Prokaryotes: An Evolving Electronic Resource for the Microbiological Community, Springer-Verlag, New York, 1999, http://link.springer-ny.com/link/service/books/10125/.

[22] W. Sabra, A.P. Zeng, H. Lnsdorf, W.D. Deckwer, Effect of oxygen on formation and structure of *Azotobacter vinelandii* alginate and its role in protecting nitrogenase, Appl. Environ. Microb. 66 (2000) 4037–4044.

[23] H. Vermeiren, J. Vanderleyden, W. Hai, Colonization and *nifH* expression on rice roots by *Alcaligenes faecalis* A15, in: Malik et al. (Eds.) Nitrogen Fixation with Non-Legumes, Kluwer Academic Publishers, Great Britain, 1998, pp. 167–177.

[24] H. Vermeiren, A. Willems, G. Schoofs, R. de Mot, V. Keijers, W. Hai, J. Vanderleyden, The rice inoculant strain *Alcaligenes faecalis* A15 is a nitrogen-fixing *Pseudomonas stutzeri*, Syst. Appl. Microbiol. 22 (1999) 215–224.

[25] K. Fisher, W.E. Newton, Nitrogenase proteins from *Gluconacetobacter diazotrophicus*, a sugarcane-colonizing bacterium, Biochim. Biophys. Acta. 1750 (2005) 154–165.

[26] R.M. Boddey, S. Urquiaga, B.J.R. Alves, V. Reis, Endophytic nitrogen fixation in sugarcane: present knowledge and future applications, Plant Soil 252 (2003) 139–149.

[27] P. Estrada-De Los Santos, R. Bustillos-Cristales, J. Caballero-Mellado, *Burkholderia*, a genus rich in plant-associated nitrogen fixers with wide environmental and geographic distribution, Appl. Environ. Microbiol. 67 (2001) 2790–2798.

[28] N.A. Tejera, E. Ortega, R. Rodes, C. Lluch, Influence of carbon and nitrogen sources on growth, nitrogenase activity, and carbon metabolism of *Gluconacetobacter diazotrophicus*, Can. J. Microbiol. 50 (2004) 745–750.

[29] M. Sevilla, R.H. Burris, N. Gunapala, C. Kennedy, Comparison of benefit to sugarcane plant growth and 15N2 incorporation following inoculation of sterile plants with *Acetobacter diazotrophicus* wild-type and Nif- mutants strains, Mol. Plant Microbe Interact. 14 (2001) 358–366.

[30] F. Grimont, P.A.D. Grimont, C. Richard, The Genus *Klebsiella*, in: M. Dworkin et al. (Eds.) The Prokaryotes: An Evolving Electronic Resource for the Microbiological Community, Springer-Verlag, New York, 1999, http://link.springer-ny.com/link/service/books/10125/.

[31] C.A. Jr. Asis, K. Adachi, Isolation of endophytic diazotroph *Pantoea agglomerans* and nondiazotroph *Enterobacter asburiae* from sweet-potato stem in Japan, Lett. Appl. Microbiol. 38 (2004) 19–23.

[32] Y. Dong, A.L. Iniguez, E.W. Triplett, Quantitative assessments of the host range and strain specificity of endophytic colonization by *Klebsiella pneumoniae* 342, Plant Soil 257 (2003) 49–59.

[33] A.L. Iniguez, Y. Dong, E.W. Triplett, Nitrogen fixation in wheat provided by *Klebsiella pneumoniae* 342, Mol. Plant Microbe Interact. 17 (2004) 1078–1085.

[34] P.J. Riggs, M.K. Chelius, A.L. Iniguez, S.M. Kaeppler, E.W. Triplett, Enhanced maize productivity by inoculation with diazotrophic bacteria, Aust. J. Plant Physiol. 28 (2001) 829–836.

[35] M.K. Chelius, E.W. Triplett, Immunolocalization of dinitrogenase reductase produced by *Klebsiella pneumoniae* in association with *Zea mays* L., Appl. Environ. Microbiol. 66 (2000) 783–787.

[36] B. Reinhold-Hurek, T. Hurek, The genera *Azoarcus*, *Azovibrio*, *Azospira* and *Azonexus*, in: M. Dworkin et al. (Eds.) The Prokaryotes: An Evolving Electronic Resource for the Microbiological Community, Springer-Verlag, New York, 2004, http://link.springer-ny.com/link/service/books/10125/.

[37] C.L.B. Marolda, B. Hauroder, M.A. John, R. Michel, M.A. Valvano, Intracellular survival and saprophytic growth of isolates from the

Burkholderia cepacia complex in free-living amoebae, Microbiology 145 (1999) 1509–1517.
[38] L. Magalhaes Cruz, E.M. de Souza, O.B. Weber, J.I. Baldani, J. Dobereiner, F.O. Pedrosa, 16S ribosomal DNA characterization of nitrogen-fixing bacteria isolated from banana (*Musa* spp.) and pineapple (*Ananas comosus* (L.) Merril), Appl. Environ. Microbiol. 67 (2001) 2375–2379.
[39] J. Caballero-Mellado, L. Martinez-Aguilar, G. Paredes-Valdez, P.E. Santos, *Burkholderia unamae* sp. nov., an N_2-fixing rhizospheric and endophytic species, Int. J. Syst. Evol. Microbiol. 54 (2004) 1165–1172.
[40] W.M. Chen, L. Moulin, C. Bontemps, P. Vandamme, G. Bena, C. Boivin-Masson, Legume symbiotic nitrogen fixation by beta-proteobacteria is widespread in nature, J. Bacteriol. 185 (2003) 7266–7272.
[41] L. Moulin, A. Munive, B. Dreyfus, C. Boivin-Masson, Nodulation of legumes by members of the β-subclass of Proteobacteria, Nature 411 (2001) 948–950.
[42] P. Gyaneshwar, E.K. James, N. Mathan, P.M. Reddy, B. Reinhold-Hurek, J.K. Ladha, Endophytic colonization of rice by a diazotrophic strain of *Serratia marcescens*, J. Bacteriol. 183 (2001) 2634–2645.
[43] F.G. Loiret, E. Ortega, D. Kleiner, P. Ortega-Rodes, R. Rodes, Z. Dong, A putative new endophytic nitrogen-fixing bacterium *Pantoea* sp. from sugarcane, J. Appl. Microbiol. 97 (2004) 504–511.
[44] Y. Benhizia, H. Benhizia, A. Benguedouar, R. Muresu, A. Giacomini, A. Squartini, Gamma proteobacteria can nodulate legumes of the genus *Hedysarum*, Syst. Appl. Microbiol. 27 (2004) 462–468.
[45] M.R. Coelho, I. von der Weid, V. Zahner, L. Seldin, Characterization of nitrogen-fixing *Paenibacillus* species by polymerase chain reaction-restriction fragment length polymorphism analysis of part of genes encoding 16S rRNA and 23S rRNA and by multilocus enzyme electrophoresis, FEMS Microbiol. Lett. 222 (2003) 243–250.
[46] K. Minamisawa, K. Nishioka, T. Miyaki, B. Ye, T. Miyamoto, M. You, A. Saito, M. Saito, W. Barraquio, N. Teaumroong, T. Sein, T. Tadashi, Anaerobic nitrogen-fixing consortia consisting of clostridia isolated from gramineous plants, Appl. Environ. Microbiol. 70 (2004) 3096–3102.
[47] Z. Tan, T. Hurek, B. Reinhold-Hurek, Effect of N-fertilization, plant genotype and environmental conditions on *nifH* gene pools in roots of rice, Environ. Microbiol. 5 (2003) 1009–1015.

Chapter 13

Measuring N_2 Fixation in the Field

Jonathan P. Zehr and Joseph P. Montoya

13.1 Introduction

Biological N_2 fixation is an important source of N in aquatic and terrestrial environments [1, 2], since N_2 is unavailable to most organisms. Denitrification results in the loss of N to N_2 and must be balanced by N_2 fixation globally in order to support life and habitability of the planet. The N-cycle has been substantially perturbed by anthropogenic activities, and currently there is concern over how the N-cycle may be changing and the implications for the future [2].

Natural biological N_2 fixation is important in terrestrial systems, on the order of 128 Tg N yr^{-1} (Tg = 10^{12} g or 10^6 metric tons) [2], which is distributed throughout agricultural, forested, and grassland habitats [3]. In marine environments, N_2 fixation occurs at rates that are roughly equivalent to those in the terrestrial environment [2]. Marine N_2 fixation occurs in localized habitats such as microbial mats and salt marshes, but was believed to be of minor importance in the water column of the open ocean until only recently. Recent biological and geochemical studies demonstrated that N_2 fixation may be the primary source of new N supporting

Biology of the Nitrogen Cycle
Edited by H. Bothe, S.J. Ferguson and W.E. Newton

primary production in nutrient-poor open ocean ecosystems [4–6]. In concert with the increasing appreciation of the role of N_2 fixation in supporting oceanic production, the number of known marine N_2-fixers has expanded greatly in recent years through application of molecular approaches that target the *nif* genes encoding the enzymes of N_2 fixation (e.g. [7, 8]).

Field measurements are important for determining when and where N_2 fixation occurs, and to determine the relative contribution of different sources of N in supporting biological production. Measuring or estimating N_2 fixation can be done using several methods, including a mass balance approach, stable isotopic methods using ^{15}N, and the C_2H_2-reduction technique [3]. Among N_2-fixers, different species are capable of fixing N_2 under a broad range of environmental conditions and in a variety of habitats [9, 10]. Relatively recently, molecular approaches have been developed to identify and quantify N_2-fixing microorganisms in the environment, without the need for cultivation [11]. Here we discuss the common geochemical (^{15}N natural abundance measurements) and experimental (C_2H_2-reduction reduction and $^{15}N_2$ fixation) approaches to quantifying the rate of N_2 fixation in natural marine ecosystems. We also present the molecular techniques for identifying and assaying the activity of individual nitrogenase phylotypes. In combination, these diverse approaches promise to show the links between N_2 fixation rates and biological diversity. This chapter focuses primarily on the application of these techniques in marine systems, but also provides an overview of the application of these methods to terrestrial environments.

13.2 Rate measurements

13.2.1 Acetylene reduction assay

The C_2H_2-reduction assay (ARA) is commonly used to quantify nitrogenase activity in both marine and freshwater systems [12–14]. When present at relatively high concentration (e.g., $pC_2H_2 > 0.1$ atm), C_2H_2 is preferentially reduced by nitrogenase to ethylene (C_2H_4), which is easily quantified by flame ionization gas chromatography (Figure 3.1). Small gas chromatographs are easily transported and may be used in the field with limited logistical support. In comparison to the $^{15}N_2$ fixation assay, the C_2H_2-reduction procedure is relatively inexpensive and can provide information on rates in near real-time.

ARA is commonly carried out in glass or polycarbonate bottles equipped with a septum seal. C_2H_2, either from a commercial supplier or generated as needed [15], is added to a final partial pressure of at least 0.1 atm. The size of the incubation vessel and the quantity of seawater incubated depend on the abundance of diazotrophs and their activity. For marine diazotrophs, a liquid:gas phase ratio of about 1:1 is commonly used [16], but a higher liquid/gas ratio can be used to improve the overall sensitivity of the assay. The limit of detection by C_2H_2-reduction is determined by both the analytical sensitivity of the gas chromatograph used and the background concentration of C_2H_4 present in the C_2H_2. For a volume-optimized incubation, the C_2H_2-reduction assay can detect a rate on the order of 0.5 nmol N_2 L^{-1} h^{-1} [17]. Although the standard C_2H_2-reduction assay is less sensitive than the $^{15}N_2$ fixation method in oligotrophic systems, recent work on continuous flow systems and laser photoacoustic techniques for quantifying C_2H_4-concentration promise increased sensitivity and responsiveness [18].

After addition of C_2H_2, the experimental bottles are typically incubated under in situ or simulated in situ conditions. A sample of the headspace is withdrawn periodically for the measurement of ethylene concentration by flame ionization gas [15]. Because the headspace of the incubation bottle is sampled and analyzed, the gas:liquid ratio within the bottle and the solubility of C_2H_4 must both be known in order to quantify the total production of C_2H_4 by C_2H_2-reduction. The Bunsen (solubility) coefficient for C_2H_4 varies with temperature and salinity, and use of a single, average value can lead to significant bias in rate measurements [19]. Both tabular and functional summaries of the temperature and salinity dependence of the Bunsen coefficient of C_2H_4 are available [20]. At any time point, the quantity of C_2H_4 produced can be calculated from the response (peak height) of the gas chromatograph [15]:

$$C_2H_4 \text{ formed} = \frac{\text{peak height}_{\text{sample}}}{\text{peak height}_{\text{standard}}} \times \text{conc.std (nmol/ml)} \times \text{GPV} \times \text{SC} \quad (1)$$

where:

GPV = gas-phase volume of assay vessel in mL

SC = solubility correction for C_2H_4 in the aqueous phase = $1 + (\alpha A/B)$

α = Bunsen coefficient for C_2H_4 at the appropriate temperature and salinity

A = volume of the aqueous phase

B = volume of the gas phase (or GPV)

Once the total quantity of C_2H_4 produced is calculated, its regression as a function of time provides a direct estimate of the rate of C_2H_2-reduction (i.e., nitrogenase activity) within the experimental system.

The C_2H_2-reduction assay involves a substrate analog, so converting a rate of C_2H_2-reduction to a rate of N_2 fixation requires an estimate of the reduction ratio (C_2H_2 reduction:N_2 reduction). The theoretical reduction ratio is 3:1, reflecting the number of reducing equivalents required by nitrogenase to reduce C_2H_2 to C_2H_4 [2] relative to the number required to reduce N_2 to NH_4^+ [6]. In fact, a ratio of 4:1 may be more appropriate in view of the overall stoichiometry of the nitrogenase reaction: $8H^+ + 8e^- + N_2 \rightarrow 2NH_3 + H_2$ (Figure 3.1). H_2-formation by nitrogenase is much lower in the presence of C_2H_2, resulting in a relatively greater yield of C_2H_4 [20]. Given the potential variation in reduction ratio, direct calibration through $^{15}N_2$ fixation assays provide the most reliable estimates of N_2 fixation based on C_2H_2-reduction. In a recent compilation of field experiments on *Trichodesmium* spp. in the North Atlantic, a mean reduction ratio was 3.58 ± 0.21 (mean $\pm$ SE, $N = 178$) in parallel C_2H_2-reduction and $^{15}N_2$ fixation assays [4]. These experiments spanned a range of conditions, including environmentally relevant ranges of light intensity and thus provide a robust estimate of the reduction ratio in this important marine diazotroph. The availability of $^{15}N_2$ and facilities for measuring ^{15}N abundance in organic matter make it increasingly feasible to calibrate C_2H_2-reduction assays using parallel incubations with ^{15}N tracer.

13.2.2 $^{15}N_2$ fixation

Direct tracer methods provide an unambiguous measure of the movement of N from one biologically active pool to another. The radioisotope of N (^{13}N) has a very short half-life (ca. 10 min) and the requirement for a cyclotron prevents its use in most field applications [21]. In contrast, the stable isotope of N, ^{15}N, is readily available in a variety of forms, including $^{15}N_2$ and can be used for direct measurements of N_2 fixation activity in the field. The $^{15}N_2$ fixation assay provides greater sensitivity than C_2H_2-reduction at low biomass and activity levels [17], though rates cannot generally be measured in real time in the field given the need for access to an isotope ratio mass spectrometer to quantify the abundance of ^{15}N. The sensitivity of the $^{15}N_2$ fixation assay varies with biomass concentration and is greater than that of the ARA in oligotrophic systems [17].

Stable isotope tracer experiments differ from radioisotope-based approaches in one important respect: they are ratio measurements in which the abundance of the rare isotope (^{15}N) is normalized to that of the common isotope (^{14}N). For N, this means that ^{15}N tracer experiments are normalized to the standing stock of N in a particular pool and hence produce biomass-specific rate estimates, normally symbolized as V [T^{-1}]. In contrast, radioisotope experiments effectively involve the tracing and counting of atoms, producing a direct measure of mass flux, ρ [MT^{-1}]. Specific rates (V) and mass fluxes (ρ) are simply related by the standing stock of N (B): $\rho = BV$, but this conversion introduces another source of error associated with the biomass measurement itself.

Tracer assays of N_2 fixation provide estimates of the net rate of incorporation of $^{15}N_2$ into biomass. In contrast, ARA provides an estimate of the overall (gross) activity of nitrogenase (Figure 3.1). The high reduction ratios (C_2H_2-reduction:$^{15}N_2$ fixation) sometimes measured in *Trichodesmium* may reflect release of fixed N in dissolved form. The importance of this extracellular release is not clear, and continuous culture studies often show a good balance between C_2H_2-reduction and growth rates [22]. These systems provide the best opportunity for resolving the importance of excretion of recently fixed N since biomass production can be measured independently of both C_2H_2-reduction and $^{15}N_2$ fixation.

In a typical experiment, N_2- and CO_2 fixation rate measurements are carried out in polycarbonate bottles equipped with septum caps. A gastight syringe is used to add $^{15}N_2$ (99 atom%, Cambridge Isotope) to a final ^{15}N enrichment of 2–5 atom%. After incubation for 2–36 h, incubations are terminated by filtration through a pre-combusted GF/F filter, and the filter is then dried at 60°C and stored over desiccant until analyzed by continuous-flow isotope ratio mass spectrometry (CF-IRMS). Biomass specific N_2 fixation rates, normalized to organic N, can be calculated by isotope mass balance as previously described [17]:

$$V(t^{-1}) = \frac{\mathrm{AP}_{\mathrm{PN}_{\mathrm{final}}} - \mathrm{AP}_{\mathrm{PN}_{\mathrm{initial}}}}{\mathrm{AP}_{\mathrm{N}_2} - \mathrm{AP}_{\mathrm{PN}_{\mathrm{initial}}}} \times \frac{1}{\Delta t} \tag{2}$$

where

AP = atom% ^{15}N of the sample (e.g., PN_{final}) or substrate (i.e., N_2) pool

Δt = length of the incubation

$AP_{PN_{initial}}$ can often be approximated using the global average abundance of ^{15}N (0.366 atom%), but a direct measurement is necessary to quantify low rates with high precision.

As noted earlier, calculation of a volumetric rate of N_2 fixation requires multiplication of the N-specific rate by the concentration of PN. Since N_2 fixation rates are customarily expressed in terms of fixation of molecular N_2, a factor of 2 is also introduced:

$$\mathrm{N_2\text{-}\,fixation\ rate\,(mol\,N_2}, t^{-1}) = \frac{\mathrm{AP}_{\mathrm{PN}_{\mathrm{final}}} - \mathrm{AP}_{\mathrm{PN}_{\mathrm{initial}}}}{\mathrm{AP}_{\mathrm{N}_2} - \mathrm{AP}_{\mathrm{PN}_{\mathrm{initial}}}} \times \frac{1}{\Delta t} \times \frac{\mathrm{PN}_{\mathrm{final}}}{2} \tag{3}$$

One advantage of the $^{15}N_2$ fixation tracer assay is the potential for measuring rates of N_2– and CO_2 fixation in a single incubation simply by adding $H^{13}CO_3^-$ at the same time as the $^{15}N_2$. The isotope mass balance approach can then be used to calculate CO_2 fixation rates from ^{13}C incorporation into organic matter. In a pure culture of diazotrophs, this approach allows direct determination of the stoichiometry of organic matter production.

13.2.3 Mass balance

In terrestrial systems, in theory it is possible to estimate N_2 fixation by net change in total N with time. This method is based on the fact that total N will accumulate through N_2 fixation, but the difficulty is that all inputs and outputs must be known [3]. A major problem is the inherent spatial temporal variability of soil systems that makes it difficult to determine N_2 fixation rates with high degrees of certainty by mass balance [2]. Correlative relationships [3] at large spatial scales are often used to estimate habitat and ecosystem level N_2 fixation rates [2].

The physical and biological dynamics of marine ecosystems make it very difficult to estimate N_2 fixation through a straightforward mass balance approach. On the other hand, deviations from the robust N:P stoichiometry of deep water nutrients provide a useful, integrative index of the impact of N_2 fixation and denitrification on the water column N-budget. A parameter, N*, was defined that quantifies the deviations from the average oceanic NO_3^-:PO_4^{3-} stoichiometry [23]. High N* values in the upper thermocline of the Sargasso Sea and other oligotrophic waters reflect the impact of local N_2 fixation in adding fixed N to the water column [6, 23]. Such nutrient budgets provide one of the most robust estimates of basin and global N_2 fixation.

13.2.4 Stable isotope budgets

The natural abundance of the stable N isotope, ^{15}N, can provide an in situ measure of the impact of N_2 fixation on the local N budget. Marine diazotrophs produce organic matter with a $\delta^{15}N$ (relative to atmospheric N_2) of about -1 to -2‰ [24, 25], which is strongly depleted relative to the isotopic composition of deepwater nitrate, which averages about 4.5‰ globally [26, 27]. This isotopic contrast provides the opportunity for evaluating the relative importance of upwelled NO_3^- and N_2 fixation in providing the N required by primary producers through simple isotopic mixing models using the $\delta^{15}N$ of subsurface NO_3^- (typically ca. 4.5‰) and diazotrophic biomass (ca. -2‰).

Particles and zooplankton from oligotrophic regions of the ocean typically have a low $\delta^{15}N$ [25, 28–31]. One potential complication is that this isotopic signature could, in principle, arise through trophic processes, which tend to preferentially retain ^{14}N in the upper water column while exporting ^{15}N in sinking organic matter [32], but the spatial distribution and magnitude of the low $\delta^{15}N$ signature both provide strong evidence that N_2 fixation is its primary source [25, 31]. Recent advances in compound-specific N isotope analysis provided additional evidence that N_2 fixation is the primary source of the low $\delta^{15}N$ characteristic of organic matter in the tropical and subtropical oceans [33]. In such regions, the contribution of N_2 fixation to biomass, estimated as PN, can be calculated with a simple isotope mixing model [27]:

$$\%\ \text{Diazotroph N} = 100 \times \left(\frac{\cdot\, ^{15}N_{\text{particles}} - \cdot\, ^{15}NO_3^-}{\cdot\, ^{15}N_{\text{diazotroph}} - \cdot\, ^{15}NO_3^-} \right) \quad (4)$$

where
$\delta^{15}N_{particles} = \delta^{15}N$ of bulk PN
$\delta^{15}N_{diazotroph} = -2‰$
$\delta^{15}NO_3^- = 4.5‰$

13.3 Detecting N_2 fixation potential and N_2-fixing microorganisms

13.3.1 General

Although cultivation has been used to isolate N_2-fixing microorganisms over decades, it has been realized that cultivation yields only a small percentage of organisms that are present in the environment. Cultivation-independent approaches using molecular biological and immunological techniques can be used to assay the potential for N_2 fixation and identify the organisms involved.

Immunological techniques have been used primarily in aquatic systems to identify microorganisms that contain nitrogenase [34]. Antibodies raised to the Fe protein or the MoFe protein can be used in cellular immunoassays to identify N_2-fixing cells. This has been done with aquatic cyanobacteria [34]. The advantage of this approach is that it detects cells that have expressed the nitrogenase protein. There are limitations to this approach, however. It is difficult to permeabilize some cells, and in the case of the Fe protein the active or inactive forms of the protein cannot be distinguished.

13.3.2 Gene diversity

Nitrogenase is composed of two multisubunit proteins, encoded by *nifHDK* (conventional nitrogenase containing Mo) or *vnfHDGK/anfHDGK* (V- or Fe-containing proteins, respectively). The *nifH/vnfH/anfH* gene has been particularly well conserved through evolution making it an easy target for DNA–DNA hybridization or polymerase chain reaction (PCR) amplification methods. A number of PCR primer sets have been developed for *nifH* and *nifD* to amplify nitrogenase from cultivated microorganisms and the environment [35–38]. The phylogeny of *nifH* is roughly concordant with 16S rRNA phylogeny, making it useful for identifying related taxa from their *nifH* sequences (Figure 13-2). Nitrogenase genes amplified from the environment can be tentatively identified by phylogenetic analysis, although they cannot be linked with rRNA ribotypes. The amplified *nif* gene sequences are sometimes termed phylotypes. One cluster of the *nifH* phylogeny deviates from rRNA phylogeny, in that spirochaetes, green sulfur bacteria, sulfate-reducing bacteria, and some Archaea form a deeply branching group usually called Cluster III (Figure 13-2).

nif PCR amplification products from the environment contain the diverse array of nitrogenase gene sequences represented in the diazotrophic

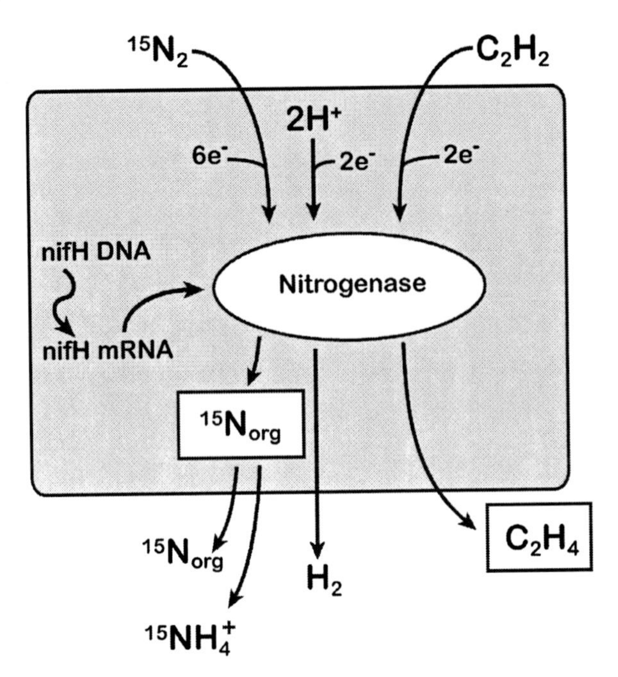

Figure 13-1 Schematic of pools and fluxes involved in the $^{15}N_2$ fixation and acetylene reduction assays. The pools measured in these assays are enclosed in boxes. Note that the $^{15}N_2$ fixation assay measures the net rate of incorporation into biomass, while the acetylene reduction assay measures the total (gross) activity of nitrogenase. Two electrons are required to reduce a molecule of acetylene to ethylene and six electrons are needed to reduce a molecule of N_2 to organic nitrogen, resulting in a 3:1 reduction ratio. The hydrogenase reaction of nitrogenase consumes additional reducing power under N_2 fixing conditions, resulting in an overall reduction ratio of about 4:1. Note the important role of the *nif* gene (DNA) and *nif* mRNA in producing nitrogenase and therefore regulating N_2 fixation activity.

community. However, it must be remembered that most, if not all, PCR primer sets bias for certain types of sequences, and the presence of sequences in the amplification products may or may not reflect the numerical abundance of microorganisms in the environment.

The amplification products can be characterized by a number of methods to determine the individual sequence types and to assess diazotroph diversity. By amplifying, cloning, and sequencing *nifH* genes, or by characterizing the amplification products by DGGE, TRFLP, or some other method, the diversity and identity of N_2-fixing microorganisms can be determined [11]. Recently, macro- and microarrays have been

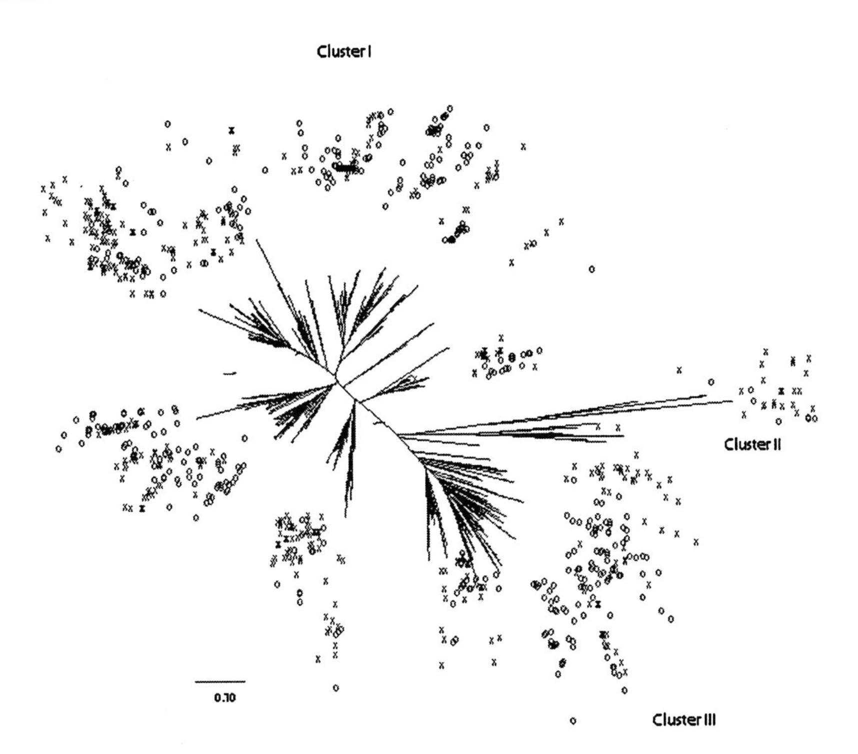

Figure 13-2 Phylogenetic distribution of *nifH* genes recovered from terrestrial (soil) and aquatic (oceans, lakes, microbial mats, and salt marsh) environments, showing broad phylogenetic diversity of prokaryotic N_2-fixing microorganisms in the environment. X = terrestrial, O = aquatic.

developed that can be used to characterize *nifH* phylotypes from amplification products [39].

Diazotroph communities in many environments are diverse (Figure 13-2). The characterization of nitrogenase gene diversity is useful for determining the types of microorganisms that could be involved in N_2 fixation, but not all organisms that have the *nif* genes express them [11]. In fact, many of the nitrogenase genes do not appear to be expressed in many environments [11].

13.3.3 Gene expression

The significance of *nif* gene diversity can be investigated by determining the extent to which the genes are expressed in messenger RNA (mRNA). This technique relies on converting RNA extracted from the environment into cDNA, which is then amplified by PCR. The detected gene sequences, which can be detected by cloning and sequencing, DGGE, or microarray techniques, then represent microorganisms who are expressing the nitrogenase genes [40–42]. Although the expression of the *nif* genes are highly regulated [43], and mRNA synthesis should be indicative of nitrogenase protein synthesis and activity, in photoheterotrophic growth of *Rhodobacter* and *Rhodospirillum*, nitrogenase genes are expressed in the presence of fixed N and the enzyme is used to balance redox [44]. However, it has been shown that expression can be correlated with nitrogenase activity in *Azotobacter* in soil [38].

The RT-PCR approach has been used in aquatic systems and terrestrial environments. In the open ocean, new diazotrophs were discovered that were shown to contribute substantially to the N-budget [8]. Nitrogenase gene expression has also been detected in salt marshes [45] and microbial mats [46]. The use of gene expression assays help to identify the major players in N_2 fixation, even though expression itself is insufficient evidence for N_2 fixation.

References

[1] P.M. Vitousek, R.W. Howarth, Nitrogen limitation on land and in the sea: how can it occur? Biogeochemistry 13 (1991) 87–115.

[2] J. Galloway, F. J. Dentener, D.G. Capone, E.W. Boyer, R.W. Howarth, S.P. Seitzinger, G.P. Asner, C.C. Cleveland, P.A. Green, E.A. Holland, D.M. Karl, A.F. Michaels, J.H. Porter, A.R. Townsend, C.J. Vorosmarty, Nitrogen cycles: past, present and future, Biogeochemistry 70 (2004), 153–226.

[3] C.C. Cleveland, A.R. Townsend, D.S. Schimel, H. Fisher, R.W. Howarth, L.O. Hedin, S.S. Perakis, E.F. Latty, J.C. von Fischer, A. Elseroad, M.F. Wasson, Global patterns of terrestrial biological nitrogen (N_2) fixation in natural ecosystems, Global Biogeochem. Cy. 13 (1999), 623–645.

[4] D.G. Capone, J.A. Burns, J.P. Montoya, A. Subramaniam, C. Mahaffey, T. Gunderson, A.F. Michaels, E.J. Carpenter, Nitrogen fixation by *Trichodesmium* spp.: an important source of new nitrogen to the tropical and subtropical North Atlantic Ocean, Global Biogeochem. Cy. 19 (2005) doi:10.1029/2004GB002331.

[5] D. Karl, A. Michaels, B. Bergman, D. Capone, E. Carpenter, R. Letelier, F. Lipschultz, H. Paerl, D. Sigman, L. Stal, Dinitrogen fixation in the world's oceans, Biogeochemistry 57/58 (2002) 47–98.

[6] C. Deutsch, N. Gruber, R.M. Key, J.L. Sarmiento, A. Ganachaud, Denitrification and N_2 fixation in the Pacific Ocean, Global Biogeochem. Cy. 15 (2001) 483–506.

[7] J.P. Zehr, M.T. Mellon, S. Zani, New nitrogen fixing microorganisms detected in oligotrophic oceans by the amplification of nitrogenase (*nifH*) genes, Appl. Environ. Microbiol. 64 (1998), 3444–3450.

[8] J.P. Zehr, J.B. Waterbury, P.J. Turner, J.P. Montoya, E. Omoregie, G.F. Steward, A. Hansen, D.M. Karl, Unicellular cyanobacteria fix N_2 in the subtropical North Pacific Ocean, Nature 412 (2001) 635–638.
[9] G. Stacey, H.J. Evans, R.H. Burris, Biological Nitrogen Fixation, Chapman and Hall, New York, 1992.
[10] E.W. Triplett, Prokaryotic Nitrogen Fixation: A Model System for the Analysis of a Biological Process, Horizon Scientific Press, Wymondham, Norfolk, England, 2000.
[11] J.P. Zehr, B.D. Jenkins, S.M. Short, G.F. Steward, Nitrogenase gene diversity and microbial community structure: a cross-system comparison, Environ. Microbiol. 5 (2003) 539–554.
[12] D. Capone, Benthic nitrogen fixation, in: R.T. Blackburn, J. Sørensen (Eds.) Nitrogen Cycling in Coastal Marine Environments, John Wiley & Sons Ltd, New York, 1988, pp. 85–125.
[13] W.D.P. Stewart, G.P. Fitzgerald, R.H. Burris, In situ studies on nitrogen fixation using the acetylene reduction technique, PNAS USA 58 (1967) 2071–2078.
[14] R.W. Hardy, R.D. Holsten, E.K. Jackson, R.C. Burns, The acetylene-ethylene assay for nitrogen fixation: laboratory and field evaluation, Plant Physiol. 43 (1968) 1185–1207.
[15] D.G. Capone, J.P. Montoya, Nitrogen fixation and denitrification, in: J. Paul (Ed.) Marine Microbiology, Academic Press, London, 2001, pp. 501–515.
[16] D.G. Capone, Marine nitrogen fixation: what's the fuss? Curr. Opin. Microbiol. 4 (2001) 341–348.
[17] J.P. Montoya, M. Voss, P. Kahler, D.G. Capone, A simple, high-precision, high-sensitivity tracer assay for N_2 fixation, Appl. Environ. Microbiol. 62 (1996) 986–993.
[18] M. Staal, S.T. Lintel-Hekkert, F. Harren, L. Stal, Nitrogenase activity in cyanobacteria measured by the acetylene reduction assay: a comparison between batch incubation and on-line monitoring, Environ. Microbiol. 3 (2001) 343–351.
[19] E. Breitbarth, M.M. Mills, G. Friedrichs, J. LaRoche, The Bunsen gas solubility coefficient of ethylene as a function of temperature and salinity and its importance for nitrogen fixation assays, Limnol. Oceanogr.-Meth. 2 (2004) 282–288.
[20] J. Postgate, Nitrogen Fixation, 3rd ed., The Press Syndicate of the University of Cambridge, Cambridge, UK, 1968.
[21] D.G. Capone, Microbial nitrogen cycling, in: R.R. Christian, S. Newell (Eds.) Manual of Environmental Microbiology, Section IV: Aquatic Environments, ASM Press, Washington, DC, 1996, pp. 334–342.
[22] C.M. Holl, J.P. Montoya, Interactions between nitrate uptake and nitrogen fixation in the marine diazotroph, *Trichodesmium* J. Phycol. 41 (2005) 1178–1183.

[23] N. Gruber, J.L. Sarmiento, Global patterns of marine nitrogen fixation and denitrification, Global Biogeochem. Cy. 11 (1997) 235–266.
[24] E.J. Carpenter, H.R. Harvey, B. Fry, D.G. Capone, Biogeochemical tracers of the marine cyanobacterium *Trichodesmium*, Deep-Sea Res. Pt I, Oceanog. Res. 44 (1997) 27–38.
[25] J.P. Montoya, E.J. Carpenter, D.G. Capone, Nitrogen fixation and nitrogen isotope abundances in zooplankton of the oligotrophic, North Atlantic Limnol. Oceanogr. 47 (2002) 1617–1628.
[26] K.-K. Liu, I.R. Kaplan, The eastern tropical Pacific as a source of ^{15}N-enriched nitrate in seawater off southern California, Limnol. Oceanogr. 34 (1989) 820–830.
[27] D.M. Sigman, M.A. Altabet, R. Michener, D.C. McCorkle, B. Fry, R.M. Holmes, Natural abundance-level measurement of the nitrogen isotopic composition of oceanic nitrate: an adaptation of the ammonia diffusion method, Mar. Chem. 57 (1997) 227–242.
[28] T. Saino, A. Hattori, ^{15}N natural abundance in oceanic suspended particulate matter, Nature 283 (1980) 752–754.
[29] E. Wada, A. Hattori, Natural abundance of ^{15}N in particulate organic matter in the North Pacific Ocean, Geochim. Cosmochim. Acta 40 (1976) 249–251.
[30] J.P. Montoya, P.H.Wiebe, J.J. McCarthy, J.J. Natural abundance of ^{15}N in particulate nitrogen and zooplankton in the Gulf Stream region and Warm-Core Ring 86A, Deep-Sea Res. 39 (Suppl. 1) (1992) S363–S392.
[31] J.E. Dore, J.R. Brum, L.M. Tupas, D.M. Karl, Seasonal and interannual variability in sources of nitrogen supporting export in the oligotrophic subtropical North Pacific Ocean, Limnol. Oceanogr. 47 (2002) 1595–1607.
[32] D.M. Checkley Jr., C.A. Miller, Nitrogen isotope fractionation by oceanic zooplankton, Deep-Sea Res. 36 (1989) 1449–1456.
[33] J.M. McClelland, C.M. Holl, J.P. Montoya, Nitrogen sources to zooplankton in the Tropical North Atlantic: Stable isotope ratios of amino acids identify strong coupling to N_2-fixation, Deep-Sea Res. I 50 (2003) 849–861.
[34] C.A. Currin, H.W. Paerl, G.K. Suba, R.S. Alberte, Immunofluorescence detection and characterization of N_2-fixing microorganisms form aquatic environments, Limnol. Oceanogr. 35 (1990) 59–71.
[35] J.P. Zehr, L.A. McReynolds, Use of degenerate oligonucleotides for amplification of the *nifH* gene from the marine cyanobacterium *Trichodesmium* spp., Appl. Environ. Microbiol. 55 (1989) 2522–2526.
[36] M.J. Church, B.D. Jenkins, D.M. Karl, J.P. Zehr, Vertical distributions of nitrogen-fixing phylotypes at Station ALOHA in the oligotrophic North Pacific Ocean, Aquat. Microb. Ecol. 38 (2005) 3–14.

[37] T. Ueda, Y. Suga, N. Yahiro, T. Matsuguchi, Genetic diversity of N_2-fixing bacteria associated with rice roots by molecular evolutionary analysis of a *nifD* library, Can. J. Microbiol. 41 (1995) 235–240.
[38] H. Burgmann, F. Widmer, W.W. Sigler, J. Zeyer, mRNA extraction and reverse transcription-PCR protocol for detection of *nifH* gene expression by *Azotobacter vinelandii* in soil, Appl. Environ. Microbiol. 69 (2003) 1928–1935.
[39] B.D. Jenkins, G.F. Steward, S.M. Short, B.B Ward, J.P. Zehr, Fingerprinting diazotroph communities in the Chesapeake Bay by using a DNA macroarray, Appl. Environ. Microbiol. 70 (2004) 1767–1776.
[40] T. Egener, T. Hurek, B. Reinhold-Hurek, (1999) Endophytic expression of *nif* genes of *Azoarcus* sp strain BH72 in rice roots Mol, Plant Microbe Interact. 12 (1999) 813–819.
[41] S. Zani, M.T. Mellon, J.L. Collier, J.P. Zehr, Expression of *nifH* genes in natural microbial assemblages in Lake George, NY detected with RT-PCR, Appl. Environ. Microbiol. 66 (2000) 3119–3124.
[42] M.J. Church, B.D. Jenkins, C. Short, D.M. Karl, J.P. Zehr, Temporal patterns of nitrogenase (*nifH*) gene expression in the oligotrophic North Pacific Ocean, Appl. Environ. Microbiol. 71 (2005) 5362–5370.
[43] M.J. Merrick, R.A. Edwards, Nitrogen control in bacteria, Microbiol. Rev. 59 (1995) 604–622.
[44] M.A. Tichi, F.R. Tabita, Interactive control of *Rhodobacter capsulatus* redox-balancing systems during phototrophic metabolism, J. Bacteriol. 183 (2001) 6344–6354.
[45] M.M. Brown, M.J. Friez, C.R. Lovell, Expression of *nifH* genes by diazotrophic bacteria in the rhizosphere of short form *Spartina alterniflora*, FEMS Microbiol. Ecol. 43 (2003) 411–417.
[46] E.O. Omoregie, L.L. Crumbliss, B.M. Bebout, J.P. Zehr, Determination of nitrogen-fixing phylotypes in *Lyngbya* sp and *Microcoleus chthonoplastes* cyanobacterial mats from Guerrero Negro, Baja California, Mexico, Appl. Environ. Microbiol. 70 (2004) 2119–2128.

Part III

Other Reactions of the Nitrogen Cycle

Chapter 14

Biochemistry and Molecular Biology of Nitrification

Stuart J. Ferguson, David J. Richardson and Rob J.M. van Spanning

14.1 Introduction

There can be no argument that the least understood steps of the N-cycle are those of nitrification, the aerobic oxidations of NH_4^+ to NO_2^- and of NO_2^- to NO_3^-. Classically these reactions are regarded as being catalysed by chemolithoautotrophs exemplified by *Nitrosomonas* and *Nitrobacter* species, respectively. There is more recent evidence that these nitrification reactions can, to some extent, also be catalysed by heterotrophs.

The lithotrophs are undoubtedly important in natural environments and it is with these organisms that we start. Before doing so, we note that the reason why the biochemistry and molecular biology of nitrification is poorly understood is almost certainly related to the difficult problem of growing large enough quantities of cells from which to prepare vesicular membranes and purified proteins. These organisms grow in culture to very low cell densities.

Biology of the Nitrogen Cycle
Edited by H. Bothe, S.J. Ferguson and W.E. Newton

14.2 The general energetic problem faced by lithotrophic nitrifiers

Nitrosomonas and *Nitrobacter* use the energy released from the aerobic oxidation of NO_2^- or ammonia for cell growth. NAD(P)H is required for various reductive biosynthesis steps, starting with the capture of CO_2 by the Calvin cycle. Neither aerobic oxidation of NH_4^+ nor that of NO_2^- provides sufficient reducing power to reduce $NAD(P)^+$ to NAD(P)H. This apparent barrier to reductive biosynthesis is overcome by an often-misunderstood process known as reversed electron transport. Furthermore, although we can be fairly certain of its occurrence in the cells of *Nitrosomonas* and *Nitrobacter*, it is very difficult to demonstrate this process experimentally.

Nitrosomonas and *Nitrobacter* depend on a chemiosmotic mechanism of energy transduction. This means that, as electrons flow to O_2 in these organisms, an H^+-electrochemical gradient (also known as H^+ motive force, PMF) is set up over the cytoplasmic membrane. This gradient not only drives the synthesis of ATP, but also simultaneous reversed electron transfer in the case of *Nitrosomonas* and *Nitrobacter*. Electrons enter the Nitrobacter respiratory chain from the NO_2^-/ NO_3^- couple, with an E_0' of 420 mV, towards the 'oxygen end'. The majority of electrons flow to O_2 with concomitant generation of the H^+-electrochemical gradient. But a minority flow 'in reverse', so as to reduce $NAD(P)^+$ to NAD(P)H for which E'_0 is −320 mV. This reversed electron transport requires an energy source under all feasible physiological concentrations of NO_2^-, NO_3^-, $NAD(P)^+$ and NAD(P)H. The driving force is widely accepted to be the H^+-electrochemical gradient. This can be understood if it is recalled that 'normal' electron transport from NADH to O_2, in a heterotrophic bacterium or mitochondrion, involves the translocation of H^+ outward across the membrane as the electrons pass through the H^+-translocating NADH dehydrogenase and/or cytochrome bc_1 complex. All that is needed, therefore, is that the electron flow is reversed in one or both of these complexes so that electrons reach NAD(P). It is sometimes asked why the NAD(P)H so produced does not immediately reoxidise with the electrons flowing back to O_2. The answer lies in energetics; the ratio of $NAD(P)H/NAD(P)^+$ will be low because of the steady-state consumption of NAD(P)H for reductive biosynthesis and electron flow from NO_2^- to O_2 will sustain the proton motive force. As we shall see, many of the special biochemical features of *Nitrosomonas* and *Nitrobacter* need to be understood in the context of the ability of the electron transport system to catalyse reversed electron transfer.

14.2.1 Nitrobacter

The energy yielding reaction used by *Nitrobacter* for growth is deceptively simple:

$$2\,NO_2^- + O_2 \rightarrow 2\,NO_3^-$$

What needs to be understood is the molecular nature of the NO_2^--oxidising enzyme, how it feeds electrons into the respiratory chain system and how in turn O_2 is reduced to water. This process must generate an H^+-electrochemical gradient across the cytoplasmic membrane.

There is undoubtedly a cytochrome oxidase enzyme of the aa_3-type. This can be asserted on the basis of both isolation of the protein and knowledge of its gene sequence obtained from both a specific cloning project [1], as well as from the more recent genome sequence. The latter also shows that *Nitrobacter* posesses the necessary enzymes for the production of the *a*-type heme. Cytochrome aa_3 has been shown to act as a H^+ pump in many different types of cells. Key residues needed for this pumping are present in the *Nitrobacter* protein. Experimentally, there is conflicting evidence as to whether the aa_3-type cytochrome *c* oxidase is a H^+ pump or not. When the purified protein was reconstituted into phospholipid vesicles, no H^+ pumping could be detected, although a parallel work with the enzyme from another source did show H^+ pumping [2]. However, it has been notoriously difficult to obtain well-sealed cytochrome oxidase vesicles that exhibit H^+ pumping. Thus the demonstration of H^+ pumping by intact cells fed with electrons from the non-physiological donor ascorbate can be taken as support for the H^+ pumping activity [3]. Of course, this conclusion relies upon aa_3-type cytochrome *c* oxidase being the only oxidase. The recent genome sequence shows that there are genes for *bo*- and *bd*-type oxidases. The conditions under which these are expressed are not known, and as both *bo*- and *bd*-type oxidases are, by analogy with other organisms, quinol oxidases, their role in oxidising NO_2^-, if any, is not clear. Reduction of ubiquinone by NO_2^- would be endergonic. For the purposes of the analysis, in this chapter we continue to assume that an aa_3-type oxidase catalyses most, if not all, electron transport to O_2.

aa_3-type cytochrome c oxidase has a Cu_A domain within the globular region of a subunit that in bacteria extends into the periplasm. Cu_A is the 'electron receiver' and in most systems its electron donor is a *c*-type monoheme cytochrome with redox potential in the region of +250 mV. Such a *c*-type cytochrome, with α-band maximum at 550 nm, has been purified from *Nitrobacter*. Its sequence shows that it is related to mitochondrial cytochrome *c* and thus is very likely an electron donor to aa_3-type cytochrome *c* oxidase. However, further studies showed that a membrane-bound cytochrome c_{550} was also present with mid-point potential +270 mV, and was able to mediate electron flow between preparations of NO_2^--oxidase and cytochrome aa_3 when each was incorporated into phospholipid vesicles [4].

It is when we turn to the NO_2^--oxidation step that very critical questions about *Nitrobacter* have to be addressed. The first is to recall that E_0'; for the NO_2^-/NO_3^- couple is +420 mV. Thus even with a NO_2^-/NO_3^- ratio of 100, the reaction will have a potential of approximately 300 mV. At this substrate-to-product ratio electron transfer from NO_2^- to one of the cytochromes c_{550} would be possible. However, as *Nitrobacter* operates aerobically, and the next step of the N-cycle, NO_3^--reduction, functions anaerobically, it is probable that *Nitrobacter* has evolved to function in the presence of significant accumulated NO_3^- concentrations. On the other hand, since NO_2^- is toxic, its concentration can be expected to be minimised. Thus assumption of equimolar NO_3^- and NO_2^-, a condition that corresponds (in energetic terms) to the E_0' value, probably overestimates the reducing power available to *Nitrobacter*. The properties of the NO_2^--oxidizing enzyme are clearly of critical importance. Two research groups have reported preparations of this enzyme. In 1983 Tanaka et al. [5] described a three-subunit *Nitrobacter winogradskyi* enzyme cytochrome a_1c_1. There were three subunits with estimated molecular weights 55K, 29K and 19K, with the 29K subunit containing the covalently bound heme *c* centre. It was this preparation that was used to assay the ability of the enzyme to mediate electron transfer from NO_2^- to O_2. The name a_1 is now outdated. The authors concluded that the protein contained heme *a* on the basis of pyridine hemochrome spectra and the designation a_1 served to distinguish it from the aa_3 oxidase. Later this enzyme preparation was described as also containing Mo and FeS centres [6]. A preparation of the enzyme from *Nitrobacter hamburgensis* lacked the a_1 heme [7] while a subsequent preparation [8] also lacked the *c*-heme. This latter preparation contained molybdenum and iron with two subunits of molecular weights 115K and 65K. We can now be essentially certain that these two subunits are related to the NarGH subunits of respiratory NO_3^--reductases. Although attempts to make the latter type of enzyme oxidise NO_2^- have failed, one can imagine that only relatively minor changes are needed in order to generate appropriate redox centres in an NO_2^--oxidase. The genome sequence clearly shows two reading frames, designated NorA and NorB on the basis of earlier partial sequence information [9], which must be this enzyme. Redox potentiometry on membrane fractions from *Nitrobacter* can be deduced to be rich in this NO_2^--oxidase, suggesting operating potentials of approximately +350 mV [10, 11].

The NarG and NarH subunits of respiratory NO_3^--reductase are known from X-ray crystallography to be globular polypeptides that are attached to the membrane via the transmembrane *b*-type cytochrome (with two hemes) subunit. These two globular subunits are also known to be located at the cytoplasmic side of the cell membranes. For the *Nitrobacter* NO_2^--oxidase two absolutely critical questions are (a) the location of the NorA or NorB subunits and (b) the nature of putative transmembrane subunit to which NorA and NorB are attached. The location issue has been controversial. Bioenergetic arguments, to which we shall return below, have suggested a location at the cytoplasmic surface, but immunolabelling studies have indicated the opposite. Usually, the location can be discerned on the basis of the presence or absence of an N-terminal export signal sequence. Given that NorA is a molybdoprotein, export would be indicated by a Tat signal sequence. Inspection of the N-terminal region of NorA suggests such a sequence is absent, although one can discern what appears to be an evolutionarily degraded Tat sequence. The analysis does, therefore, suggest that the NorA and NorB subunits are located at the cytoplasmic side. As *c*-type cytochromes and the Cu_A centre of cytochrome oxidase are at the periplasmic side of the membrane, the reader will immediately see that we need to identify a possible transmembrane route for electrons. The initial preparation of NO_2^--oxidase was argued to contain both *a*- and *c*-type heme. *c*-type heme has never been observed in the transmembrane region of a protein and although a predicted diheme *c*-type cytochrome is located close to the Nor genes, this seems unlikely to be an agent for electron transfer across the membrane. A *c*-type cytochrome is often indicated by a visible absorption spectrum or heme-staining band on a gel. Other kinds of covalently attached heme, as for instance seen in the cytochrome *bf* complex of thylakoids, could give similar signals.

It is possible that one or two molecules of *a*-type heme are sandwiched between transmembrane helices but this would need experimental support. The open reading frames adjacent to NorA or NorB give no clue as to cofactor content, but there is an open reading frame that appears to code for several transmembrane helices. These helices are predicted to contain conserved histidine residues that could act as heme ligands analogously to the NarI subunit of NO_3^--reductase (see Chapter 2). Whether this is true or not, the electrons have to pass from the transmembrane segment to globular domains in the periplasm. This is unlikely to be direct if we use NarI as a model. There could be an associated *c*-type cytochrome subunit that has been called c_1 in the original preparations. One might speculate that this is the diheme *c*-type cytochrome, the gene for which is found adjacent to the *nor*A and *nor*B genes.

A cytoplasmic site of NO_2^--oxidation has several implications. First, it implies that there must be (a) transport system(s) for the import and export of NO_2^- and NO_3^- into and out of the cell. This is the opposite situation to that which applies during NO_3^- respiration by the NarGH enzyme (see Chapter 2). It has been proposed that NarK proteins fulfil this role. Two NarK proteins are predicted from the *Nitrobacter* genome; a speculation is that one could be for NO_2^- import and one for NO_3^- export.

The second consequence of a cytoplasmic location for NO_2^--oxidation is that the electrons will be 'pulled' across the membrane by the membrane potential that is positive at the external surface. Thus, if the NO_2^--oxidase contains redox centres with E_0' values in the range 450 $\pm$ 50 mV, then the expected potential, +150 mV outside, will contribute to electron transfer to cytochrome *c*. Such behaviour can readily explain why oxidation of NO_2^- is inhibited by protonophores in both cells and membrane vesicles; loss of the membrane potential would result in a lower steady-state concentration reduced cytochrome *c* and hence a slower rate of electron transfer to O_2.

There are profound bioenergetic consequences of both the location of the active site of NO_2^--oxidase and the possible H^+ pumping activity of cytochrome oxidase [12]. Figure 14-1A shows how the H^+-electrochemical gradient could be generated with a cytoplasmic location for the NO_2^--oxidase and a H^+ pumping activity in cytochrome oxidase. The effective charge and H^+ translocation stoichiometry per pair of electrons flowing is two. The initial outward movement of the electrons through the NO_2^--oxidase is effectively compensated by two of the four positive charges per two electrons that are translocated outward by cytochrome oxidase. It follows that if the cytochrome oxidase were unable to pump H^+, in other words if the inward movement of electrons to meet H^+ at the O_2 reduction site was the sole function of cytochrome oxidase, then no gradient would be generated. Such a scheme can thus be eliminated from consideration. On the other hand, as Figure 14-1B also shows, a non-H^+-pumping cytochrome oxidase would be compatible with a periplasmic site of NO_2^--oxidation. The positive charge translocation stoichiometry would be two per $2e$ with a non-H^+-pumping oxidase and four per $2e$ with a H^+-pumping oxidase. Clearly, although a cytoplasmic location of the NO_2^--oxidase is currently indicated, this needs to be confirmed.

Either of the schemes in Figure 14-1 explains H^+-electrochemical gradient generation limited to electron flow from NO_2^- to O_2, but not how electrons are delivered to $NAD(P)^+$. The genome sequence of *Nitrobacter* shows that a cytochrome bc_1 complex and a H^+-translocating NADH ubiquinone oxidoreductase are both present. Thus it is very probably that a small fraction of the reduced cytochrome c_{550}, soluble or membrane-bound, is oxidised by the cytochrome c_1 of the bc_1 complex instead of by cytochrome oxidase. The electrons would pass through the complex to generate ubiquinol from ubiquinone. This is opposite of the usual direction in which the bc_1 complex functions. It will go backwards if the cyt *c* (reduced)/cyt *c* (oxidised) and ubiquinone/ubiquinol ratios are sufficiently high so that the H^+-electrochemical gradient is able to drive H^+ into the bacterial cytoplasm from the periplasm via the activity of the bc_1 complex. Similarly, provided the ubiquinol/ubiquinone and NAD^+/NADH ratios are appropriate, the reversal of the normal direction of H^+ movement through the NADH dehydrogenase will also lead to reversed electron flow to produce NADH. In other words, compared with a 'normal' heterotrophic bacterial respiratory chain, the free-energy difference between the NADH/NAD^+ and cyt *c* (reduced)/cyt *c* (oxidised) couples must be much lower than usual, so that a H^+-electrochemical gradient of the usual magnitude ~200 mV would not allow the normal outward movement of H^+ if electrons tried to move from NADH to cytochrome *c*. The generation of

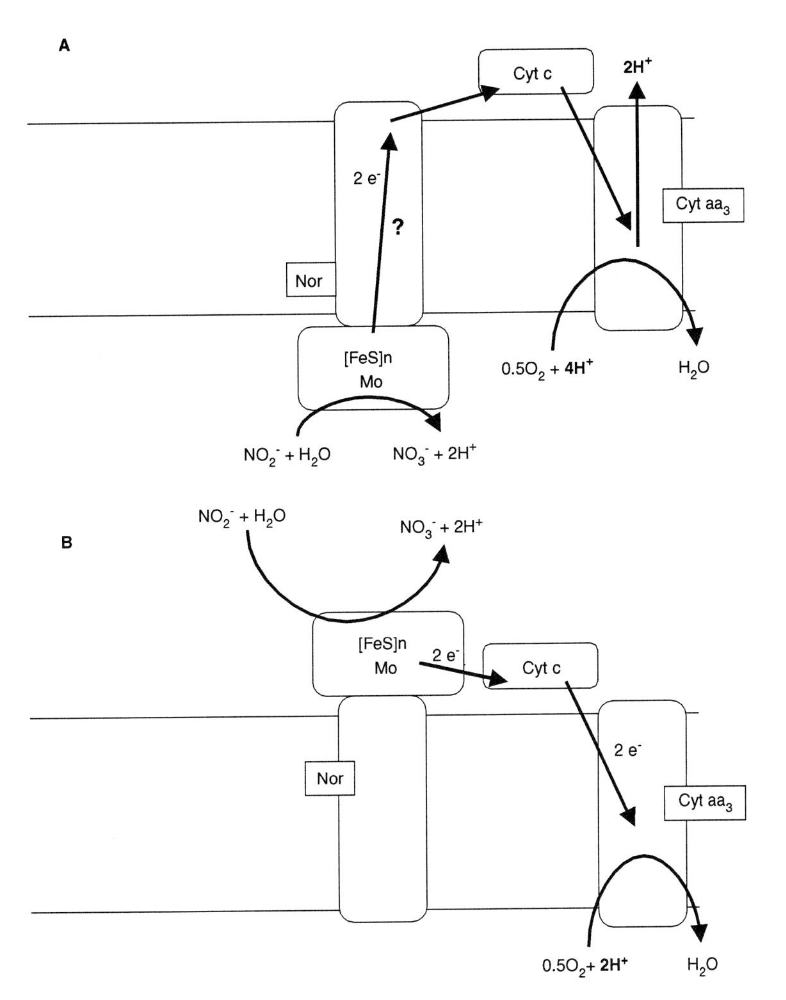

Figure 14-1 Possible topological schemes for nitrite oxidation in *Nitrobacter*. (A) Nitrite is oxidised at the cytoplasmic side of the membrane at an active site containing Mo and electrons pass via several FeS centres ([FeS]$_{n}$). The route for transfer of electrons across the membrane via a subunit of the NO_2^--oxidase (Nor) is unknown (?). A periplasmic *c*-type cytochrome is believed (see text) to transfer electrons to the aa_3 oxidase. The latter acts as a proton pump and hence half the protons taken up at the cytoplasmic side are used for formation of water and the other half are translocated. This scheme requires transport of NO_2^- into, and of NO_3^- out of, the cell (not shown). (B) Nitrite is oxidised at the periplasmic side of the membrane. This scheme does not require a proton-pumping cytochrome oxidase although such a function could be added. Although seemingly simpler than the scheme in (A), the available evidence favours scheme (A) (see text).

NADPH from NADH probably depends upon a transhydrogenase enzyme for which there is evidence from the genome.

14.2.2 Nitrosomonas

The oxidation of ammonia to NO_2^- by *Nitrosomonas* is not a straightforward process (see [13] for recent review). The first step is conversion of NH_3 (or $NH^+{}_4$ cations) into NH_2OH (Figure 14-2). This reaction requires molecular O_2 plus two electrons,believed to be supplied by ubiquinol. The ammonia mono-oxygenase (AMO) responsible for this reaction has never been purified, but it has been identified by a suicide-inhibitor labelling method. From the sequences of the two subunits of the enzyme it is recognised to be very similar to a Cu-dependent methane mono-oxygenase for which a crystal structure has recently been obtained [14]. The structure of

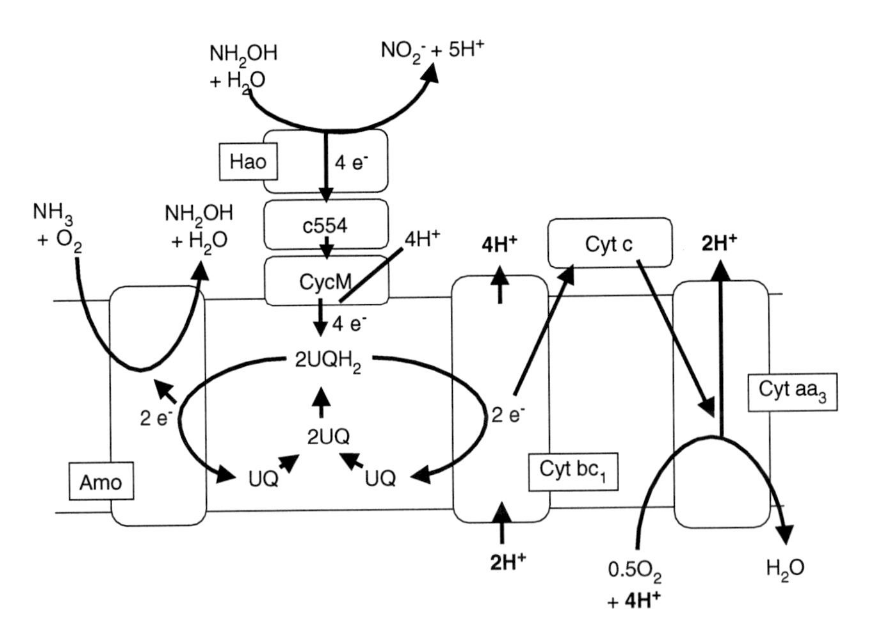

Figure 14-2 A scheme for electron transport pathways in *N. europaea*. It is assumed that electrons derived from NH_2OH are delivered from NH_2OH dehydrogenase (Hao) to ubiquinone via cytochrome c_{554} and cytochrome c_m which is a member of the NapC family (see text). It is assumed that cytochrome c_m catalyses ubiquinol formation with concomitant uptake of H^+ from the periplasmic side. The positioning of the active site of Amo on the periplasmic side is by analogy with methane mono-oxygenase. Protons released upon the putative oxidation of UQH_2 by Amo are not shown but probably are released to the periplasm.

CH_4-mono-oxygenase shows many transmembrane helices along with Cu atoms bound in more globular regions at the periplasmic side of the membrane. Unfortunately, the exact arrangement of the Cu ions at an active site is not clear but at least this structure of methane mono-oxygenase provides a framework for future understanding of the AMO. The idea that ubiquinol provides electrons for the AMO is supported by the fact that partially purified preparations of the enzyme can use duroquinol as electron donor.

Assuming that ubiquinol is the source of electrons for ammonia mono-oxygenase, the mechanism for ubiquinone reduction must next be identified. This is thought to be linked to the oxidation of NH_2OH to NO_2^-, a process that occurs in the periplasm. This is a four-electron reaction that is catalysed by a multi-heme protein of the *c*-type. The electrons are believed to pass on to a tetraheme cytochrome known a cytochrome c_{554}. From there the path of the electrons is not firmly established. However, there is another tetraheme *c*-type cytochrome that is anchored to the membrane via a transmembrane helix and clearly belongs to the NapC/NirT group of proteins that are widely implicated in transferring electrons from ubiquinol or menaquinol to periplasmic reductases [15]; in one case, a respiratory NO_2^--reductase from *Wolinella succinogenes*, there is experimental evidence for this. Given this precedent, and the similarity between the NH_2OH oxidoreductase and the NO_2^--reductase, it seems very likely that this NapC-type protein, sometimes called cytochrome c_{m552} in this context [13], catalyses electron transfer from cytochrome c_{554} to ubiquinone.

It seems that ubiquinol must be principally oxidised via two routes, at almost equal rates. Thus half of the ubiquinol molecules must be oxidised by the ammonia mono-oxygenase that, as explained above, requires two electrons for each molecule of ammonia converted into NH_2OH. The other pathway of ubiquinol oxidation is through electron transfer to O_2 with concomitant generation of the H^+-electrochemical gradient. The exact pathway for this electron transfer has been uncertain but it is now thought likely to be via a cytochrome bc_1 complex to an aa_3-type oxidase. This view is based on information from the genome sequence, which shows the presence of the cytochrome bc_1 complex as well as cytochrome aa_3 as the sole oxidase [13]. *Nitrosomonas europaea* has two distinct loci for the expression of cytochrome *c* oxidases, which terminate the aerobic respiratory network. Both clusters contain the genes encoding the major catalytic subunit of the oxidase and a smaller subunit where electrons from cytochrome *c* enter the complex. Both copies of the small subunit have a stretch of amino acids diagnostic for binding of a dinuclear copper site (Cu_A). In that respect, both types of oxidase resemble the aa_3-type cytochrome *c* oxidases found in mitochondria of eukaryotes and in many bacterial species. *N. europaea* does not have the potential to express a so-called cbb_3-type oxidase, which is a uniquely bacterial type of oxidase with a relatively high affinity for O_2 (K_m of about 5 nM). Electron transport to O_2 from NH_2OH is strongly inhibited by classical inhibitors of the bc_1 complex [16]. Until recently, the cytochrome bc_1 was not recognised to be present in *Nitrosomonas* and so many schemes for electron flow have in the past been shown with direct electron transfer to the oxidase; this would give a lower H^+ translocation stoichiometry and now seems unlikely.

A third route for oxidation of ubiquinol is via reversed electron transfer through a NADH-ubiquinone oxidoreductase complex, a process that generates the NAD(P)H. NADPH would be formed from NADH through the activity of a H^+-motive-force-dependent transhydrogenase enzyme required for reductive biosynthesis. As explained in the introductory part of this chapter, this process is driven by the H^+ motive force that is generated by electron transfer from ubiquinol to O_2; both the cytochrome bc_1 complex and cytochrome aa_3 contribute to this generation. It has been estimated that about 5% of ubiquinol is oxidised by this reversed electron transfer pathway.

14.3 Denitrification reactions catalysed by nitrifiers

Given that ammonia oxidation by nitrifiers is an obligatory aerobic process because of the need for molecular O_2 in conversion of NH_3 to NH_2OH, it is not immediately apparent why *Nitrosomonas* species, e.g. *N. europaea*, should contain NO_2^-- and NO-reductases. However, under some circumstances at the oxic/anoxic boundary these proteins are expressed and cause nitrifiers to generate N_2O. *Nitrosomonas* contains the Fnr system for responding to anaerobiosis [17]. Remarkably, neither the genes for *norCBQD* gene cluster for N_2O-reductase nor the *nirK* gene for NO_2^--reductase are preceded by a typical recognition site for the Fnr transcription factor. The *nirK* gene is the last in a cluster of four genes. The first gene encodes a blue Cu protein that contains the residues involved in the ligation of type-I and type-II Cu-centres. Accordingly, its EPR (Electron Paramagnetic Resonance) spectrum showed signatures diagnostic for type-I and type-II copper centres. The protein has been isolated and showed NO_2^--reductase activity with reduced cytochrome c_{552} as electron donor, indicating that it might also be involved in the production of NO. It has homology to the C-terminal part of a periplasmic protein (CopA) involved in the sequestering of copper in *Pseudomonas syringae*. The second and third genes encode di- and mono-heme *c*-type cytochromes, respectively, which have been isolated and characterized by Whittaker et al. [13]. They co-purified with NirK during all purification steps. The cytochromes have a so far unresolved role in tolerance towards nitric oxide. The gene cluster is preceded by an *nsrR* gene, which is transcribed divergently and encodes a NO_2^--sensitive regulator involved in transcriptional repression of the *nirK* gene cluster [18]. The exact circumstances when denitrification occurs are not altogether clear. On the one hand, inactivation of the *nirK* gene of *N. europaea* resulted in a decreased NO_2^- tolerance of cells growing in aerobic batch cultures, indicating that the physiological role of NO_2^--reductase in *N. europaea* under these growth conditions is for the detoxification of NO_2^-. On the other hand, NO_2^--respiration and subsequent denitrification increases under conditions of low O_2 as more electron donors become available. Many reports describe the increase of N_2O, and sometimes N_2, with decreasing O_2 concentrations. It has been suggested that the distribution of electrons between the terminal oxidase, the AMO and the denitrifying enzymes on the one hand, and the resulting N loss on the other hand, is determined by the competition between AMO and the terminal oxidase for O_2 and for electrons from the respiratory network. Electron flow from ubiquinol to NO_2^- or NO will involve H^+-motive-force generation only at the cytochrome bc_1 complex step. Two charges per 2*e* will be translocated compared with six charges per 2*e* if O_2 is the electron acceptor. However, this will be sufficient to drive some reversed electron transport alongside ATP synthesis. Taking all these considerations together, it appears that denitrification in *N. europaea* serves both a role in NO_2^- and NO management and in respiration-coupled energy transduction when O_2 is limited. N_2O-reductase is absent from *N. europaea* which explains why some nitrifiers can release N_2O.

14.4 Heterotrophic nitrification

Aerobic oxidation of NH_3 to NO_2^- and of NO_2^- to NO_3^- is not restricted to chemolithotrophs such as *Nitrosomonas* and *Nitrobacter* sp. Little is known of hetrotrophic nitrification but the denitrifier *Paracoccus denitrificans* provides an example of a heterotrophic nitrifier [19]. There is evidence that a similar ammonia mono-oxygenase may be present as in chemolithotrophs but that a distinct NH_2OH-oxidase must be used. Heterotrophic nitrifiers, unlike autotrophic nitrifiers, are incapable of using nitrification to support growth. It has therefore been proposed that heterotrophic nitrification is not linked to the generation of a H^+ motive force and that the electron acceptors for a 'NH_2OH oxidase' in heterotrophic nitrifiers have more positive potentials than the cytochrome bc_1 complex, for example cytochrome c_{550} and pseudoazurin. *P.* denitrificans has an AMO-like enzyme activity in membranes that can extract two electrons from the quinol pool and use these to convert NH_3 to NH_2OH using molecular O_2 in a manner that is envisaged to be similar to that of autotrophic nitrifiers. The NH_2OH will be released into the periplasm. An NH_2OH-oxidase from *P. denitrificans* has been isolated from the periplasm and found to have different properties from the NH_2OH

oxidoreductase from *N. europaea*. The enzyme from *P. denitrificans* has a molecular mass of 20 kDa and contains non-heme, rather than heme, iron. A non-heme NH_2OH-oxidase has also been partly purified from the Gram-positive heterotrophic nitrifier *Arthrobacter globiformis*, indicating that the non-heme-iron NH_2OH-oxidase might be widespread among heterotrophic nitrifiers. Under anaerobic conditions, N_2O is the product of the *P. denitrificans* NH_2OH-oxidase. This is probably the product of a $2e^-$ oxidation of NH_2OH to yield nitroxyl radicals that can then dimerize to form N_2O. By contrast, under aerobic conditions NO_2^- can be formed, presumably as a result of an oxidation of nitroxyl to NO_2^- by O_2. In *N. europaea*, NO_2^- is the product of NH_2OH oxidoreductase even in the absence of O_2. As discussed earlier, four electrons are passed to cytochrome c_{554} as NH_2OH is oxidised to NO_2^-. Two of these electrons are passed into the quinol pool and then used by ammonia mono-oxygenase. The other two electrons are used in the electron transport chain, either to generate a H^+ motive force, via the cytochrome bc_1 complex to an oxidase, or to generate NADH via reverse electron transfer to NADH. Therefore more than two electrons from the NH_2OH oxidoreductase must reduce cytochrome c_{554}, so that growth can be coupled with autotrophic nitrification. If the product of NH_2OH-oxidase in heterophic nitrification is nitroxyl and the reaction to NO_2^- involves reaction with O_2, then only two electrons are produced per NH_2OH. This would not allow for autotrophic growth. Instead it is proposed in *P. denitrificans* that the electrons are transferred to the cytochrome c_{550}/pseudoazurin pool. These two electrons can then be passed from cyt c_{500}/peudoazurin to the non-H^+-motive NO_2^-, N_2O or NO reductases. This then allows for further reduction of the NO_2^- generated from NH_2OH via aerobic denitrification reactions that are not coupled to energy conservation. Thus the overall result is to withdraw two electrons from the quinol pool and dissipate these without energy conservation. This is an example of redox balancing and reflects the proposed role for heterotrophic nitrification in energy spillage, rather than energy conservation. Heterotrophic nitrifiers are believed to be widely distributed in soils. However, most of these organisms cannot as yet be cultured when performing heterotrophic nitrification. Although the enzymology, molecular biology and ecology of heterotrophic nitrification remains poorly understood, it is hoped that the emerging genome sequences of some heterotrophic nitrifiers will begin to help to redress this gap in our knowledge on nitrification.

References

[1] G. Berben, *Nitrobacter winogradskyi* cytochrome *c* oxidase genes are organised in a repeated gene cluster, Antonie van Leeuwenhoek 69 (1996) 305–315.

[2] N. Sone, Y. Yanagita, K. Hon-nami, Y. Fukumori, T. Yamanaka, Proton-pump activity of *Nitrobacter agilis* and *Thermus thermophilus* cytochrome *c* oxidase, FEBS Lett. 155 (1983) 150–154.
[3] H.-G. Wetzstein, S.J. Ferguson, Respiration-dependent proton translocation and the mechanism of proton motive force generation in *Nitrobacter winogradskyi*, FEMS Microbiol. Lett. 30 (1985) 87–92.
[4] T. Nomoto, Y. Fukumori, T. Yamanaka, Membrane-bound cytochrome c is an alternative electron donor for cytochrome aa_3 in *Nitrobacter winogradskyi*, J. Bacteriol. 175 (1993) 4400–4404.
[5] Y. Tanaka, Y. Fukumori, T. Yamanaka, Purification of cytochrome a_1c_1 from *Nitrobacter agilis* and characterisation of nitrite-oxidation system of the bacterium, Arch. Microbiol. 135 (1983) 265–271.
[6] M. Fukuoka, Y. Fukumori, T. Yamanaka, The *Nitrobacter winogradskyi* cytochrome a1c1 is an iron-sulfur molybdoenzyme having hemes-a and hemes-c, J. Biochem. 102 (1987) 525–530.
[7] H. Sundermeyer-Klinger, W. Meyer, B. Warninghoff, E. Bock, Membrane-bound nitrite oxidoreductase of *Nitrobacter*: evidence for a nitrate reductase system, Arch. Microbiol. 140 (1984) 153–158.
[8] M. Meincke, E. Bock, D. Kastram, P.M.H. Kroneck, Nitrite oxidoreductase from *Nitrobacter hamburgensis*: redox centers and their catalytic role, Arch. Microbiol. 158 (1992) 127–131.
[9] K. Kirstein, E. Bock, Close genetic relationship between *Nitrobacter hamburgensis* nitrite oxidase and *Escherichia coli* nitrate reductases, Arch. Microbiol. 160 (1993) 447–453.
[10] W.J. Ingledew, P.J. Halling, Paramagnetic centers of the nitrite oxidising bacterium *Nitrobacter*, FEBS Lett. 67 (1976) 90–93.
[11] T. Yamanaka, Y. Fukumari, The nitrite oxidising system of *Nitrobacter winogradskyi*, FEMS Microbiol. Rev. 54 (1993) 259–270.
[12] S.J. Ferguson, Is a proton pumping cytochrome oxidase essential for energy-conservation in *Nitrobacter*? FEBS Lett. 146 (1982) 239–243.
[13] A.B. Hooper, D. Arcerio, D. Bergmann, M.P. Hendrich, The oxidation of ammonia as an energy source, in: D. Zannoni (Ed.) Respiration in Archaea and Bacteria, vol. 2: Diversity of Prokaryotic Respiratory Systems, Springer, Berlin, 2004, pp. 121–147.
[14] R.L. Lieberman, A.C Rosenzweig, Crystal structure of a membrane-bound metalloenzyme that catalyses the biological oxidation of methane, Nature 434 (2005) 177–182.
[15] M.D. Roldan, H.J. Sears, M.R. Cheesman, S.J. Ferguson, A.J. Thomson, B.C. Berks, D.J. Richardson, Spectroscopic characterisation of a novel multiheme c-type cytochrome widely implicated in bacterial electron transport, J. Biol. Chem. 273 (1998) 28785–28790.
[16] M. Whittaker, D.J. Bergmann, D. Arciero, A.B. Hooper, Electron transfer during the oxidation of ammonia by the chemolithotropic

bacterium *Nitrosomonas europaea*, Biochim. Biophys. Acta 1459 (2000) 346–355.
[17] I. Schmidt. R.J.M. van Spanning, M.S.M. Jetten, Denitrification and ammonia oxidation by Nitrosomonas europaea wild type and nirK- and NorB- deficient mutants, Microbiology 150 (2004) 4107–4114.
[18] H.J.E. Beaumont, S.I. Lens, W.N.M. Reijnders, R.J.M. van Spanning, Expression of nitrite reductase in *Nitrosomonas europaea* involves NsrR, a novel NO_2^--sensitive transcription repressor, Mol. Microb. 54 (2004) 148–158.
[19] D.J. Richardson, J.M. Wehrfritz, A. Keech, L.C. Crossman, M.D. Roldan, H.J. Sears, C.S. Butler, A. Reilly, J.W.B. Moir, B.C. Berks, S.J. Ferguson, A.J. Thomson, S. Spiro, The diversity of redox proteins involved in bacterial heterotrophic nitrification and aerobic denitrification, Biochem. Soc. Trans. 26 (1998) 401–408.

Chapter 15

The Ecology of Nitrifying Bacteria

Jim I. Prosser

15.1 Introduction

Nitrification is the oxidation of reduced forms of inorganic and organic N to NO_3^-. It therefore plays a central role in the global N-cycle, providing an important link between decomposition of organic matter, releasing NH_4^+, and denitrification, for which it provides the essential electron acceptor. Unlike ammonification and denitrification, nitrification is carried out by a relatively restricted number of organisms. The most studied process involves the sequential oxidation of inorganic NH_4^+ to NO_3^-, via two distinct functional groups of proteobacteria, autotrophic NH_3 oxidisers and autotrophic NO_2^- oxidisers. However, this relatively simple view of nitrification has been challenged by increasing understanding of the diversity of autotrophic nitrifiers, discovery of the Anammox process (see Chapter 17) and the more recent discovery of NH_3-oxidising archaea. As a consequence, we can distinguish four microbial groups carrying out NH_3 oxidation: NH_3-oxidising betaproteobacteria, gammaproteobacteria, planctomycetes and archaea. In addition, a number of heterotrophic bacteria and fungi oxidise reduced forms of organic N to NO_3^-. NO_2^--oxidising

Biology of the Nitrogen Cycle
Edited by H. Bothe, S.J. Ferguson and W.E. Newton

bacteria have been studied less intensively and appear to be restricted to the proteobacteria.

Increased knowledge of the diversity of nitrifiers has been accompanied by greater awareness of their metabolic versatility, and its consequences for their ecosystem function. Recent technical advances now enable investigation of populations of nitrifying bacteria and their activities in natural environments, providing the potential for greater understanding of the mechanisms driving nitrifier diversity and establishing links with physiological diversity and ecosystem function. This chapter therefore begins with a description of the microbial groups carrying out nitrification and their distribution in major habitats. This is followed by a discussion of the influence of nitrifier physiology, physiological diversity and environmental factors on their community structure, environmental distribution, activities and ecosystem function.

15.2 Nitrifying microorganisms

15.2.1 Autotrophic ammonia-oxidising bacteria

NH_3 oxidation in most environments is considered to be dominated by Gram-negative autotrophic NH_3 oxidisers. These organisms were originally placed within a single taxonomic group based on their ability to grow autotrophically, gaining energy from oxidation of NH_3. Five genera of autotrophic NH_3 oxidisers were described on the basis of phenotypic characters: cell morphology, intracellular membrane structure, motility, tolerance to high concentrations of NH_3 or salt and urease activity. Biochemical, physiological and limited genetic studies have focused on the type strain for the *Nitrosomonas* genus, *Nitrosomonas europaea,* and full genome sequences are now available for *N. europaea* and *Nitrosospira multiformis.*

Classification was revised following phylogenetic analysis of 16S rRNA genes of cultivated NH_3 oxidisers and the key functional gene, *amo*A, which encodes the active subunit of ammonia monooxygenase (see Chapter 15) [1, 2]. This demonstrated two major divisions, gammaproteobacterial (AOG) and betaproteobacterial (AOB) NH_3 oxidisers. The former are represented by a small number of *Nitrosococcus* strains, e.g. *Nitrosococcus oceani*. Most cultivated strains, however, belong to the betaproteobacteria and 16S rRNA gene analysis places them in seven major lineages within a monophyletic group (Figure 15-1) consisting of: *Nitrosospira*, *Nitrosomonas europaea/Nitrosococcus mobilis*, *Nitrosomonas communis, Nitrosomonas marina, Nitrosomonas oligotropha*, *Nitrosomonas cryotolerans* and *Nitrosomonas* sp. Nm143.

Phylogenetic analysis of AOB has been informed by molecular characterisation of natural communities of NH_3 oxidisers, which enabled, for the

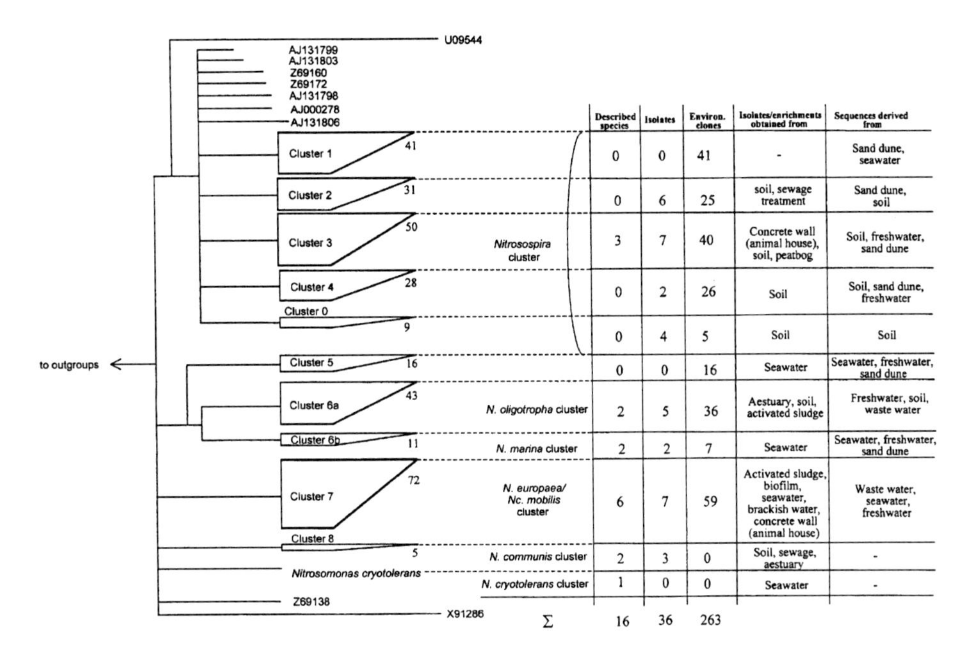

Figure 15-1 Phylogenetic classification of betaproteobacterial NH_3 oxidisers, based on 16S rRNA gene sequences. Heights of triangles represent the numbers of sequences in each cluster, from Ref. [7].

first time, investigation of their population and community ecology. Traditional approaches to community analysis were prevented by slow growth of NH_3 oxidisers in laboratory culture, consequent difficulties in isolating and identifying pure cultures and biases associated with cultivation. Sequence analysis of 16S rRNA and *amo*A genes amplified from extracted nucleic acids predicted or confirmed the major lineages and indicates the existence of several subgroups, or clusters, within the *Nitrosospira* lineage. Support for these subgroups is less strong than that for *Nitrosomonas* lineages but differences in distribution and physiological characteristics have been observed, suggesting that they are useful in detecting different ecotypes. In addition, environmental clone sequences suggest the existence of two novel clusters, *Nitrosospira* cluster 1 and *Nitrosomonas* cluster 5, which are not currently represented in laboratory cultures. Confirmation of their role in NH_3 oxidation requires physiological characterisation of pure, or at least enrichment cultures, but their abundance in some environments suggests their ecological importance. Molecular data also suggest an additional *Nitrosospira* cluster, *Nitrosospira* cluster 0.

15.2.2 NH_3-oxidising archaea

NH_3-oxidising archaea (AOA) were recently discovered through a soil metagenomic study, involving analysis of large sections of the chromosome of sufficient length to contain sets of genes encoding pathways [3]. Archaea are traditionally associated with extreme environments but analysis of 16S rRNA environmental clones has demonstrated their abundance in a wide range of 'non-extreme', mesophilic environments. Their role in NH_3 oxidation was indicated by the presence of a 16S rRNA gene sequence representative of the crenarchaeal Group 1.1b sub-lineage on the same genomic fragment as gene sequences similar to, but distinct from, bacterial *amo*A and *amo*B genes. Primers specific for AOA *amo*A genes have since shown their expression to be common in terrestrial and marine environments. In addition, an NH_3-oxidising, mesophilic crenarchaeote, *Nitrosopumilus maritimus*, has been isolated from a marine aquarium [4]. This strain belongs to the crenarchaeal Group 1.1a lineage, grows chemolithoautotrophically, with NH_3 as a sole energy source, at similar rates to other autotrophic NH_3 oxidisers and possesses *amo*A, B and C sequences similar to those obtained from metagenomic studies. Phylogenetic analysis of translated protein sequences from *amo*A gene sequences amplified from soil and marine environments indicates at least five distinct crenarchaeal *amo* lineages, associated with the different environments from which they were obtained [5]. It is too early to assess the significance of AOA for nitrification in natural environments. However, quantitative PCR techniques indicate that they may be 1–2 orders of magnitude more abundant than bacterial NH_3 oxidisers. Potentially, therefore, they may be major contributors to global nitrification.

15.2.3 NO_2^--oxidising bacteria

In most environments, nitrification is limited by oxidation of NH_3 and NO_2^- rarely accumulates. As a consequence, NO_2^- oxidisers have been studied less frequently and biochemical and physiological studies have focused on *Nitrobacter*. Phenotypic classification, based mainly on cell morphology and membrane ultrastructure, led to description of four genera of NO_2^- oxidisers. *Nitrobacter* (rod- or pear-shaped cells) and *Nitrococcus* (coccoid cells) possess intracytoplasmic membranes and are related to photosynthetic bacteria. *Nitrospina* (rods or spherical cells) lacks intracytoplasmic membranes, while *Nitrospira* grows as helical cells or vibrios. Analysis of 16S rRNA genes placed these four genera in four phyla within the proteobacteria. Most information is available on *Nitrobacter* (alphaproteobacteria) and *Nitrospira*, which forms a distinct phylum within the betaproteobacteria. The remaining genera are based on only a few cultivated organisms: two cultivated strains of *Nitrospina gracilis* (deltaproteobacteria) and one strain of *Nitrococcus mobilis* (gammaproteobacteria).

Knowledge of NO_2^- oxidiser communities and their diversity is being extended by analysis of environmental clones. Relatively few 16S rRNA gene clone sequences are available for *Nitrobacter*, and its phylogeny is based mainly on cultivated strains. The *Nitrobacter* genus appears to contain a narrow range of organisms while numerous *Nitrospira* 16S rRNA genes have been obtained from environmental clones, and few from cultures. These indicate that *Nitrospira* represents a large, highly diverse group, with lineages with no cultivated representatives. More cultures are required to provide confidence that all function as autotrophic NO_2^- oxidisers.

15.2.4 Heterotrophic nitrifiers

Although heterotrophic nitrification has been studied less than autotrophic nitrification, many heterotrophic bacteria and fungi can oxidise organic N compounds. Detailed physiological studies on a small number of strains indicate that bacterial heterotrophic nitrifiers employ similar mechanisms to bacterial NH_3 oxidisers, while heterotrophic nitrification by fungi involves reaction of organic N compounds with hydroxyl radicals produced in the presence of hydrogen peroxide and superoxide. Heterotrophic nitrification does not appear to generate energy for growth. However, the greater physiological diversity of heterotrophic nitrifiers increases the range of environments and environmental conditions in which nitrification is possible.

15.3 Community structure of nitrifiers in natural environments

15.3.1 Molecular characterisation of NH_3 oxidiser communities

A range of primers have been developed for the amplification of 16S rRNA genes and *amo*A genes from (particularly betaproteobacterial) autotrophic NH_3 oxidisers and have been used to target DNA, rRNA and mRNA, as appropriate [6, 7]. Amplification products are analysed in several ways.

1. Amplicons can be cloned, sequenced and subjected to phylogenetic analysis, providing detailed information on the phylogenetic relationships between community members and with other NH_3 oxidisers. This approach can generate estimates of the relative abundance of different phylogenetic groups but requires sequencing of large numbers of clones and comparison of communities requires sequencing of replicate clone libraries.
2. Sequences can also be compared with those in databases, to provide rapid identification.
3. Comparison of relative abundances of different groups within communities is usually attempted using fingerprinting techniques, including

denaturing or temperature gradient gel electrophoresis (DGGE, TGGE) and terminal restriction fragment length polymorphism (T-RFLP).
4. 16S rRNA gene probes have been used for direct detection of NH_3 oxidisers in environmental samples using fluorescence in situ hybridisation (FISH). This has been particularly valuable for studies of nitrification in wastewater treatment systems, where cell concentrations are high. It is less useful in freshwater and marine systems, where cell concentrations are lower, and in soil, where visualisation is made difficult by particulate material. However, a major advantage of this approach is the lack of requirement for PCR amplification and the ability to investigate spatial distribution and organisation of cells.
5. Microarray-based techniques are being developed with the potential for more rapid and comprehensive characterisation of NH_3 oxidiser communities, through hybridisation with extracted nucleic acids.

These basic molecular methods are also being used in combination with others. For example, FISH can be used in combination with microautoradiography to demonstrate in situ incorporation of ^{14}C–CO_2 by target NH_3 oxidisers. Finer scale information can be obtained using stable isotope probing, in which ^{12}C- and ^{13}C-labelled nucleic acids are separated and subjected to molecular analysis after incubation of environmental samples with CO_2, or other potential substrates, labelled with ^{13}C.

Use of these techniques inevitably introduces bias, e.g. associated with cell lysis, nucleic acid extraction efficiency and purification, PCR amplification and choice of primer. These must be considered when designing experiments and interpreting data. However, molecular techniques represent a major advance on cultivation-based methods and it is now possible to study NH_3 oxidiser diversity and community composition.

15.3.2 General features

Molecular surveys of a wide range of environments have generated several common findings. All betaproteobacterial NH_3 oxidisers in pure or enrichment cultures fall within a monophyletic group that contains no cultivated non-NH_3 oxidiser. There is considerable diversity within environmental sequences falling within this monophyletic group and identical environmental clones are rarely found, unless large numbers are sequenced or short stretches analysed. Sequences are rarely identical to those of pure cultures but identity has occasionally been found with AOB strains in enrichment cultures from the same source and replicate enrichments may contain AOB with identical sequences. Phylogenetic analysis of environmental sequences suggests the existence of novel lineages of NH_3 oxidisers, although confirmation of their role in NH_3 oxidation requires analysis of cultivated representatives or in situ demonstration of NH_3 oxidation.

15.3.3 Distribution of NH_3 oxidisers in major environments

All bacterial soil NH_3 oxidisers appear to belong to the betaproteobacteria, although few attempts have been made to detect gammaproteobacteria sequences. Nitrosomonads appear easier to isolate from soil, but soil clone libraries are dominated by *Nitrosospira*-like sequences, particularly in acid soils. Soils of higher pH, approaching neutrality, also contain members of the *N. communis* and *N. oligotropha* lineages.

Representatives of the *Nitrosomonas marina* lineage have been isolated from marine environments and molecular techniques indicate widespread distribution of *Nitrosospira* and *Nitrosomonas*. These generally form phylogenetic groups which are distinct from those found in terrestrial and freshwater NH_3 oxidisers and the *Nitrosomonas* 143 lineage is restricted to marine environments or estuaries. Other lineages contain sequences from all environments. In addition, marine environments contain sequences that fall within *Nitrosomonas* cluster 5 and *Nitrosospira* cluster 1 lineages, which contain no cultivated representatives.

The gammaproteobacterial NH_3 oxidiser, *Nitrosococcus oceani*, has been isolated from marine environments, but initial attempts to detect gammaproteobacteria sequences failed because of inadequate PCR primers. Improved primers indicate widespread distribution but it is not clear whether they are more abundant than betaproteobacterial NH_3 oxidisers. There is, however, some evidence of biogeography, with different geographical distributions of *amo*A and 16S rRNA genes.

NH_3 oxidiser communities in freshwater environments are dominated by members of the *Nitrosomonas oligotropha* lineage and *Nitrosospira* clusters 0, 2, 3 and 4. Isolates have been obtained from eutrophic waters, including *N. europaea, N. eutropha* and *N. nitrosa* and sequences of other *Nitrosomonas* lineages, including cluster 5, and of the *N. marina* lineage have been detected in freshwater environments.

Nitrosomonads, from non-marine lineages, are generally more abundant than nitrosospiras in activated sludge and fixed-film wastewater treatment plants and *N. eutropha, N. europaea* and *Nitrosococcus mobilis* have been isolated. The relatively high ammonium concentrations and fast throughput of material in wastewater treatment plants might be expected to increase selective pressure and some systems are dominated by single NH_3 oxidiser phylotypes. Others contain high abundance of several populations and these differences may influence stability and efficiency of these systems.

15.4 Factors influencing the ecology of nitrifying bacteria

15.4.1 General

Physiological studies of NH_3- and NO_2^--oxidising bacteria have been limited to relatively few strains, which have limited relevance to those in natural communities. For example, most work on NH_3 oxidisers has been carried out on *N. europaea*, which was originally isolated from soil, but molecular surveys indicate that soils are dominated by nitrosospiras, rather than nitrosomonads. *N. europaea* has advantages for laboratory study, in that it grows relatively fast, but many of its characteristics do not reflect those of environmental strains, particularly those in soil and marine environments.

Despite difficulties in isolating pure cultures of NH_3 oxidisers, the lack of availability of isolates and the difficulties in physiological characterisation, there is sufficient information to indicate links between phylogeny and physiological properties of ecological importance [8]. Most noticeably, there is a clear distinction between marine and non-marine strains, but other examples will be provided below.

15.4.2 Enrichment and isolation of pure cultures of nitrifying bacteria

Enrichment of autotrophic nitrifiers from natural environments is relatively straightforward. Samples are inoculated into neutral-pH, mineral salts medium, containing NH_3 or NO_2^- and growth is indicated by acidification of the medium (for NH_3 oxidisers) and disappearance of NH_3 or NO_2^-, or production of NO_2^- and NO_3^-, respectively. Growth usually occurs after aerobic incubation for several weeks and numbers of heterotrophic contaminants can be reduced by successive subculturing. However, contaminants grow faster than nitrifiers and isolation of pure cultures requires end-point dilution methods or isolation of colonies on solid media. Isolation is therefore difficult, takes several months and is frequently unsuccessful. In addition, the process reduces the probability of isolating organisms that are representative of natural strains. Nevertheless, in the absence of in situ, cultivation-independent methods for assessing physiology of specific phylotypes, characterisation of cultivated representatives of these groups is necessary to determine physiological and functional diversity.

Assessment of growth as changes in total cell concentration or biomass follows traditional methods, but is made difficult by the small size of cells and low biomass yields, respectively. Viable cell enumeration by dilution plate methods is not practicable due to slow growth on solid media. Even after incubation for several weeks, colonies can only be visualised using low-power microscopy. Viable cell enumeration of pure cultures and environmental populations therefore requires use of the most probable number (MPN) method.

Most probable number enumeration of nitrifiers involves multiple inoculation of tubes containing liquid inorganic medium, plus ammonium or NO_2^-, with dilutions of cell suspensions of environmental samples. Viable counts are based on statistical analysis of the number of tubes showing growth after incubation. The probabilistic nature of this method generates intrinsic variability, and estimated numbers will depend on the length of the incubation period. However, an additional major factor is selectivity of the growth media used. This is exemplified by a study of NH_3 oxidisers in agricultural soils [9]. MPN incubations contained sequences related to *Nitrosomonas*, which was not detected by molecular methods. In addition, incubations in media containing high ammonium concentration were dominated by *Nitrosomonas*-like sequences and *Nitrosospira* cluster 3 phylotypes, cultivated representatives of which are tolerant to high NH_3 concentration. In contrast, MPN incubations on low NH_3 medium were dominated by *Nitrosospira* cluster 3 and cluster 2 phylotypes that are sensitive to high NH_3. In addition, MPN counts were significantly lower (1–2 orders of magnitude) than those determined using cultivation-independent, competitive PCR. The results therefore demonstrate the selectivity of laboratory growth media and the NH_3 preferences of different phylogenetic groups. They also demonstrate that communities are functionally diverse and that phylogenetic groups with important physiological characteristics, such as NH_3 tolerance, can be present at relatively low abundance, undetectable by standard molecular techniques.

15.4.3 Measurement of nitrification activity

Substrates and products of nitrification, NH_3, NO_2^- and NO_3^-, are relatively easy to measure in laboratory systems and environmental samples. This facilitates measurement of nitrification rate, particularly in pure culture studies where analysis is not complicated by other N-cycling processes. Zero-order kinetics are observed at non-limiting substrate concentrations and rates are proportional to biomass concentration. This measure of 'potential nitrification' is used to estimate nitrifier biomass in environmental samples. At lower, limiting substrate concentrations nitrification follows saturation kinetics, generating values for activity saturation constants. During longer incubations (days to weeks), growth occurs, with exponential increases in product concentration. This allows estimation of maximum specific growth rates, at non-limiting substrate concentration, or saturation constants for growth during substrate-limited growth, typically determined in chemostats.

These kinetics are complicated by several factors in natural environments. For example, in soil specific growth rate and cell activity will be determined by local NH_3 concentration, which will vary significantly. Nitrification will continue until NH_3 is fully utilised, or until it is limited by acidification of the microenvironment in which the organisms are active. As a consequence, NH_3 depletion or NO_2^- and NO_3^- production, averaged over large volumes, will result from many small pockets or bursts of activity within the soil, masking true kinetics of growth. Changes in NO_3^- concentration in bulk soil are frequently described by the logistic equation. However, one study [10] demonstrated that the kinetics of NO_3^- production was determined by the distribution of lag times prior to onset of nitrification in soil aggregates. The amount of NO_3^- produced in each aggregate was then 'self-limited' by acidification. A second important factor is the influence of other processes, i.e. assimilation, immobilisation and denitrification, on changes in NH_3 and NO_3^- concentrations. Accurate measurement of kinetics therefore requires comparison of rates in the presence and absence of inhibitors of these processes, or of nitrification, and/or use of ^{15}N-based techniques.

Cellular oxidation rates have traditionally been determined in laboratory culture, with subsequent comparison to those measured in environmental samples based on rates of product formation and MPN counts. This approach is now being enhanced by improved methods for cell enumeration, particularly FISH and quantitative PCR. In one study, real-time PCR was used to quantify NH_3 oxidiser abundance in microcosms supplied with different concentrations of NH_3 and in field samples. Rates of NH_3 oxidation were also measured and used to determine cell yield and cell activities [11]. Similar approaches, involving enumeration by FISH, have been used to determine cell activities in NH_3 oxidiser aggregates in activated sludge. It is also possible to combine microsensor analysis of NH_3, NO_2^-, NO_3^- and O_2 concentrations in biofilms with FISH, microautoradiography (MAR) and local quantification of ^{14}C bicarbonate uptake (β-microimaging) to determine the contributions and activities of different members of NH_3 oxidiser communities. Development of these cultivation-independent techniques greatly increases the ability to investigate the physiology of nitrifying bacteria in situ, reducing reliance on cultivated organisms when assessing physiological diversity.

15.4.4 Maximum specific growth rate of nitrifiers

In laboratory culture, NH_3 and NO_2^- oxidisers have maximum specific growth rates in the ranges 0.02–0.088 h^{-1} and 0.018–0.058 h^{-1}, respectively, equivalent to cell doubling times in the ranges 35–8 h and 39–12 h [12]. Growth yields are approximately 1 g biomass per mol NH_3 or NO_2^-. These optimal growth characteristics have implications for analysis of natural populations. The throughput of wastewater treatment processes is limited by maximum specific growth rates of NH_3 oxidisers, which are lower than those of other organisms essential for wastewater treatment. Strategies are therefore adopted to retain biomass, on filters or in biofilms, or to re-inoculate, e.g. through sludge return in activated sludge processes. Maximum specific growth rates also indicate the minimum time required for changes in community structure in response to environmental change. Even under ideal conditions, it may take several days for a minority population to increase abundance and influence nitrification rates.

15.4.5 Affinity of nitrifying bacteria for ammonia and nitrite

Saturation constants for growth range from 0.05 to 0.07 mM NH_4^+ and 0.015 to 0.178 mM NO_2^-, for NH_3 and NO_2^- oxidisers, respectively. Saturation constants for activity, measured on cell suspensions, are higher: 0.12–14 mM NH_4^+ and 1.6–3.6 mM NO_2^-, respectively. Thus, NH_3 is unlikely to limit growth or activity of AOB in wastewater treatment systems, but will limit soil and aquatic NH_3 oxidation, except in microenvironments with locally high concentrations. NO_2^--concentrations are close to detection levels in the majority of natural environments, and saturation constants therefore predict substrate-limited growth and activity of NO_2^- oxidisers. However, their greater biomass and potential for heterotrophic growth ensure that NO_2^- is used rapidly and rarely accumulates. It must be emphasised, however, that these growth parameters are difficult to obtain experimentally and quoted figures are based on relatively few studies.

There is a general view that nitrosomonads have higher maximum specific growth rates and lower substrate affinity than nitrospiras, and that the two groups are adapted to feast–famine and fast–famine existences, respectively. This view is too simplistic; for example, some *Nitrosospira* strains can grow relatively rapidly at high NH_3 concentrations, while some nitrosomonads are inhibited by high NH_3, but differences in substrate affinities have been used to explain community differences. For example, cultivated representatives of the *Nitrosomonas oligotropha* lineage have been reported with high affinity for NH_3, explaining their dominance in some riverine environments, where NH_3 concentrations are low.

15.4.6 Inhibition of nitrifiers by high substrate concentration

NH_3 and NO_2^- oxidisers are inhibited by high concentrations of their respective substrates and difference in tolerance to high NH_3 was used for phenotypic classification of NH_3 oxidisers. High concentrations in soil result from application of NH_3-based fertilisers and from local increases in NH_3 due to decomposition of organic matter, faecal material and urine patches. Prolonged fertilisation can lead to changes in NH_3 oxidiser communities, although it is often difficult to distinguish effects of NH_3 concentration and associated management activities, such as liming and ploughing. For example, there is evidence that ploughing reduces spatial heterogeneity in AOB communities and reduces diversity. In addition, the impact of changes in diversity may be less that those arising through changes in total biomass and cell concentration.

There is also evidence that NO_2^- oxidiser communities are influenced by long-term NH_4^+ fertilisation and ploughing [13]. Interestingly, fertilisation in this case did not influence NH_3 oxidiser communities or *Nitrobacter*, which showed low diversity across all soil plots. However, fertilisation led to changes in *Nitrospira* communities, with characterisation of novel *Nitrospira* phylogenetic groups. These may be derived from different ecotypes, but most physiological studies of NOB have been carried out on *Nitrobacter*, and little is known of *Nitrospira* physiology or physiological diversity. The study does, however, suggest a role for *Nitrospira* in soil NO_2^- oxidation and further investigation of physiological diversity is of interest, particularly given the role of NO_2^- oxidisers in denitrification.

In aquatic environments, there is evidence for changes in NH_3 oxidiser communities with depth in lakes, particularly when stratified, and between sediments and overlying water, suggesting that these may represent distinct ecotypes. In some cases, these changes may result from differences in substrate affinity and sensitivity.

Tolerance to high NH_3 concentrations has been shown to provide a link between NH_3 oxidiser community structure and dynamics, physiological diversity and the dynamics of soil nitrification [14]. Sheep grazing on grassland soil leads to locally high concentrations of NH_3, through rapid hydrolysis of urea in urine patches. Soil microcosms, in which unfertilised soil was amended with synthetic sheep urine (SSU), demonstrated significant variability in the length of the lag phase prior to NO_3^- production. In microcosms containing unfertilised soil, this variability was due to the nature of the initial NH_3 oxidiser community. Molecular analysis demonstrated the presence of two subgroups of *Nitrosospira* cluster 3, clusters 3a and 3b, with occasional detection of *Nitrosomonas*. In microcosms dominated initially by cluster 3b strains, lag phases were short. In those dominated initially by cluster 3a, lag phases were long (several weeks) and the onset of NO_3^- production followed increases in relative abundance of, and eventual dominance by *Nitrosospira* cluster 3b strains (Figure 15-2). This therefore demonstrated a link between nitrification dynamics and the diversity and community structure of AOB in this soil. The link with physiological diversity was shown by determining NH_3 tolerance, in laboratory batch cultures, of pure culture and enrichments (obtained from the same grassland site) representative of clusters 3a and 3b. These showed cluster 3b strains, which led to rapid nitrification, to be tolerant to NH_3 concentrations typical of sheep urine patches, while 3a strains were inhibited. In microcosms constructed using fertilised soils no clear relationship was found between initial community structure and lag phase, or in changes in relative abundances of different groups during incubation. This is probably due to the higher AOB biomass in these soils, through prolonged fertilisation, such that all strains, including NH_3 tolerant strains, will be more abundant and capable of producing detectable NO_3^- levels without a significant lag period.

Although NO_2^- rarely accumulates in natural environments, local gradients in NO_2^- concentration may occur and investigation of NO_2^- oxidisers in wastewater treatment processes indicates physiological diversity in *Nitrospira* with respect to NO_2^- concentration. FISH was used to detect and distinguish NO_2^- oxidisers belonging to two *Nitrospira* lineages on

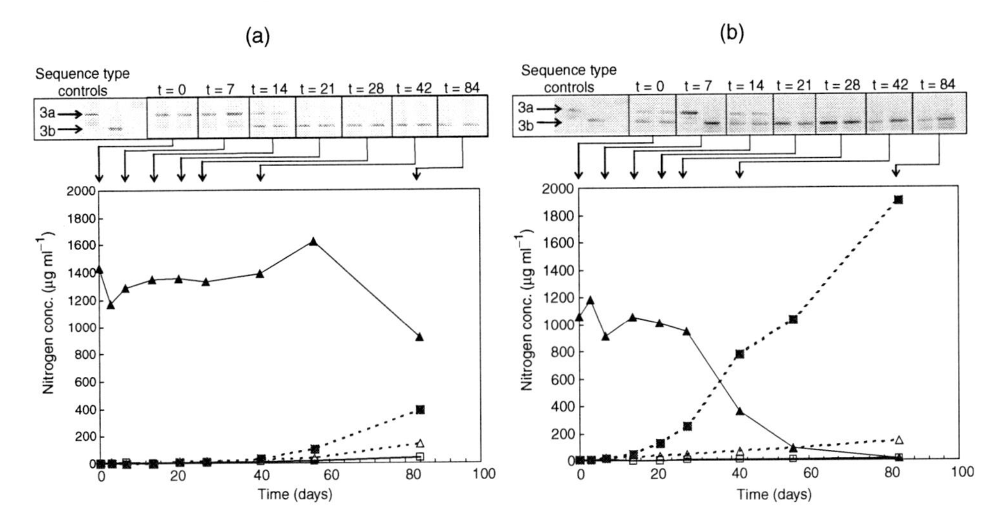

Figure 15-2 The influence of synthetic sheep urine on NH_3 oxidiser communities in soil microcosms. Ammonium (▲, △) and NO_2^- + NO_3^- (■, □) concentrations and NH_3 oxidiser sequence types were measured in control microcosms (□, △) and microcosms amended with synthetic sheep urine (■, ▲). DGGE profiles of NH_3 oxidiser communities from microcosms established with unimproved (a) and improved (b) soil and amended with urea at $t = 0$. Triplicate soil microcosms were sampled destructively at each time point and mean values plotted, from Ref. [14].

aggregates in activated sludge. Detailed quantitative image analysis provided evidence that the two *Nitrospira* lineages were located at different distances from NH_3 oxidiser microcolonies, which were also detected using FISH. This suggested differences in NO_2^- concentration preference, which were confirmed in incubation studies.

15.4.7 Transformation of carbon compounds by nitrifying bacteria

Assimilation of organic carbon has been demonstrated in a number of cultivated NH_3 oxidisers, but does not appear to confer any significant growth advantage. Recently, however, analysis of the *N. europaea* genome correctly predicted the potential for utilisation of fructose as an alternative source of carbon, although ammonia oxidation is still required for energy production [15]. Ammonia oxidisers may also be important degraders of organic pollutants in oligotrophic environments. This results from broad substrate specificity of the ammonia monooxygenase, enabling oxidation of, e.g., methanol, bromoethane, ethylene, propylene, cyclohexane, benzene and phenol.

In contrast to autotrophic NH_3 oxidisers, NO_2^- oxidisers range from strains which grow best autotrophically to those which are more heterotrophic in nature, and many can grow mixotrophically, on both NO_2^- and organic compounds. This is one explanation for their greater abundance, than NH_3 oxidisers, and for the lack of accumulation of NO_2^- in natural environments. Organic compounds also provide the substrate for denitrification by NOB under anaerobic conditions.

AOB are also capable of oxidising CH_4 and CO, and may therefore contribute to reductions in these two greenhouse gases. NH_3 oxidisers and CH_4 oxidisers share many characteristics. Both possess membrane-bound monooxygenase enzymes, and amino acid sequences confirm significant similarities, but also distinct phylogenetic groups corresponding to 16S rRNA gene phylogenies. Both enzymes are capable of oxidising both NH_3 and CH_4 and have common inhibitors. Ammonia oxidisers have higher oxidation rates and lower saturation constants for NH_3 than CH_4, with the converse true for CH_4 oxidisers, and neither appears to grow on the major substrate of the other group. As a consequence, NH_3 oxidisers are not believed to contribute significantly to global methane oxidation. Nevertheless, a stable isotope probing study provided some evidence for assimilation of CH_4 and methanol by NH_3 oxidisers [16]. Grassland soils were incubated with ^{13}C-labelled CH_4 and CH_3OH, and NH_3 oxidizer 16S rRNA and *amo*A sequences were detected in ^{13}C-labelled nucleic acids. This may, however, have resulted from assimilation of $^{13}CO_2$ released by methanotrophs and methylotrophs. In addition, NH_3 and CH_4 suppress activity of methane and NH_3 oxidisers, respectively, and there is evidence that high levels of NH_3 may suppress activity of MOB.

15.4.8 Nitrification at low oxygen concentrations

NH_3-oxidising planctomycetes oxidise NH_3 through the Anammox process, but autotrophic NH_3 oxidation by AOB also occurs at low O_2 concentrations. *Nitrosomonas eutropha* can oxidise NH_3 under anaerobic conditions, using NO_2 or N_2O_4 as an electron acceptor. This generates nitric oxide and hydroxylamine, which is converted to NO_2^-, releasing N_2-gas. In addition, several NH_3 oxidisers can maintain NH_3 oxidation at low O_2 concentrations and can denitrify, generating NO, N_2O and N_2-gas. Under these conditions, 2.5–10% of NH_3 oxidised can be converted to N-oxides. The significance of these processes for their growth and survival, and for ecosystem processes, is unclear but merits further study, given the importance of N_2O as a greenhouse gas. Their importance is likely to be greatest at oxic/anoxic interfaces, such as sediment–water interfaces, biofilms or on soil particles, as they require substrates produced under oxic conditions (NO_2 and NO_2^-).

15.4.9 Starvation mechanisms and biofilm formation

Intermittent supply of NH_3 in terrestrial and aquatic environments necessitates efficient starvation survival mechanisms for ecological success, and the ability to respond rapidly to increased substrate supply. However, starved *N. europaea* cells exhibit significant lag periods before growth after re-supply of NH_3. Lag periods increase with the length of the starvation period, presenting significant ecological disadvantages in natural environments, where competition for NH_3 will be great. However, many natural communities grow on solid surfaces or particles and biofilm formation may accelerate recovery from starvation [17]. When *N. europaea* was grown as biofilms on sand particles, and then starved for several weeks, growth after addition of NH_3 was immediate, with no detectable lag period. This rapid recovery may be due to a quorum-sensing response, as lag periods for starved cell suspensions could be reduced by addition of the homoserine lactone signalling molecule, *N*-(3-oxo-hexanoyl homoserine lactone (OHHL), with NH_3. There is also evidence that degradation of rRNA is slow in starved NH_3 oxidiser cells, increasing the potential for rapid recovery.

Surface attachment and biofilm formation by nitrifiers alters other growth characteristics and resistance to environmental stress. Cell activity of NH_3 and NO_2^- oxidisers is greater when attached to charged surfaces that attract NH_4^+ or NO_2^- ions, respectively. This is of obvious relevance to the soil environment, where NH_4^+ is strongly adsorbed to clay minerals. Biofilm growth is also one explanation for protection of soil nitrifiers from inhibitors of NH_3 oxidation. Commercial inhibitors, such as N-Serve, are added with ammonium-based fertilisers to reduce losses arising from leaching of NO_3^-. They are also of value experimentally where specific inhibition of NH_3 oxidation is required. Biofilm formation increases tenfold the concentration required for inhibition of nitrification. The mechanisms for protection of biofilm communities from stress are not known but nitrifiers, like other bacteria, produce significant quantities of extracellular polymeric material when colonising surfaces.

15.4.10 Formation of cell aggregates by nitrifying bacteria

Decomposing particulate material is an ideal habitat for nitrifiers in aquatic systems, as it provides a region of high, localised ammonium and NO_2^- flux and concentration and the potential for microcolony formation and interactions between NH_3 oxidisers, NO_2^- oxidisers and other organisms. In marine ecosystems, the bacterial communities in particulate material, 'marine snow', differ significantly from those in the surrounding, bulk water. There is also evidence that it selects for different communities of NH_3-oxidising bacteria. In one study, NH_3 oxidiser 16S rRNA gene sequences from particles in Mediterranean Sea samples were dominated by representatives of *Nitrosomonas*, while *Nitrosospira* sequences dominated environmental clone libraries in surrounding sea water. This conforms with the feast:famine/fast:famine distinction between *Nitrosomonas* and *Nitrosospira* (see above).

In activated sludge systems complex and highly active communities develop on loose flocculated material, where FISH analysis shows NH_3 oxidisers as dense, discrete microcolonies containing several thousand cells, often associated with NO_2^- oxidisers. Combined use of FISH and microautoradiography demonstrates that most NH_3 oxidisers in activated sludge systems are active and that they can incorporate pyruvate. Aggregate formation will be promoted by production of extracellular polymeric material and this characteristic, and tolerance to heavy metals, has been used to explain isolation of a strain of *N. oligotropha* from sewage.

15.4.11 Influence of salt concentration on nitrifiers

The influence of salt concentration on growth was an important trait in phenotypic classification of NH_3-oxidising bacteria. Ammonia oxidisers isolated from marine environments have a requirement for, or tolerance of salt at seawater concentrations, while *Nitrosococcus halophilus*, isolated from salt lagoons and salt lakes, prefers even higher salt concentrations. Salinity may also be a selective factor in wastewater treatment plants, due to high salt concentration of industrial waste material. Some cultured representatives of the *N. europaea/N. communis* lineage are tolerant to salt but are not isolated from, or detected in marine samples, and other factors must be involved in selection. Conversely, *Nitrosospira* isolates are not tolerant to salt, but are detected in marine samples, suggesting that they may not be autochthonous or that representative strains have not yet been isolated.

Salinity influences community structure in estuaries. One study showed *N. oligotropha / ureae* and sequences distantly related to Nm143 in freshwater regions but reduction in diversity in marine sites, with dominance by *Nitrosospira* sequences [18]. In another study, DGGE analysis indicated dominance by sequences associated with *Nitrosomonas oligotropha*, *Nitrosomonas* sp. Nm143 and *Nitrosospira* cluster 1 in freshwater, brackish and marine regions, respectively (Figure 15-3) [19]. Stable isotope probing with ^{13}C-bicarbonate indicated that active organisms were related to *Nitrosomonas* sp. Nm143, *Nitrosomonas cryotolerans* and the NO_2^- oxidiser *Nitrospira marina*. *Nitrosospira* cluster 1-like sequences were not detected in ^{13}C-labelled DNA. This suggests that incubation conditions did not favour its growth and activity, or that it is not an autotrophic NH_3 oxidiser. Changes in NH_3 oxidiser communities have also been investigated in the River Seine from wastewater effluent sites to the estuary. Effluent inoculated River water inoculated with NH_3 oxidisers from effluent and communities downstream of effluent outlet were dominated by *Nitrosomonas oligotropha*- and *N. ureae*-related clones. These strains survived and remained dominant during passage down the estuary, until changes occurred in the estuary and replacement by marine NH_3 oxidisers as salinity increased.

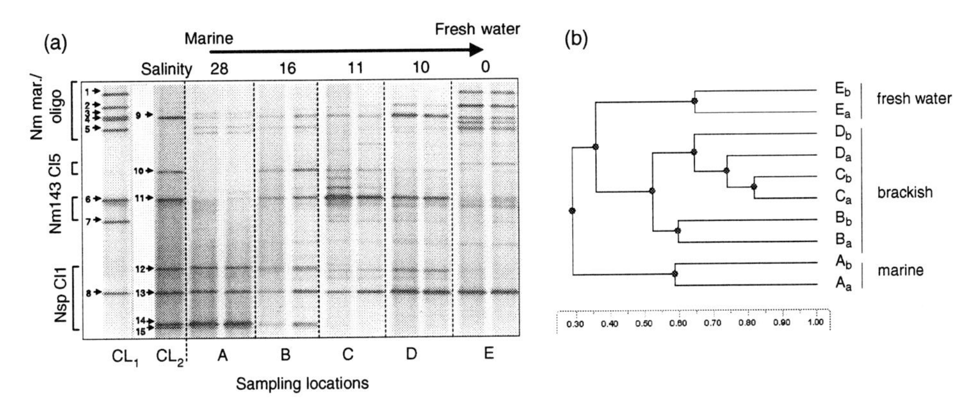

Figure 15-3 Changes in betaproteobacterial NH_3-oxidiser communities in duplicate sediment samples along the Ythan estuary, N.E. Scotland, with salinities (‰) ranging from close to seawater (A) to freshwater (E). betaproteobacterial NH_3 oxidiser-like 16S rRNA gene sequences were amplified from extracted DNA using specific NH_3 oxidiser primers, nested with bacterial primers. The first two lanes (CL1, CL2) contain clone sequences, representative of dominant DGGE bands which were putatively identified by comparison with database sequences. (a) DGGE analysis of amplification products; (b) UPGMA (unweighted pair group method with arithmetic mean) dendrogram assessing the similarity of DGGE profiles illustrated in (a), from Ref. [5].

15.4.12 Influence of pH on nitrifier growth and activity

Laboratory strains of both NH_3 and NO_2^- oxidisers grow optimally in the pH range 7.5–8. Specific growth rate decreases sharply as pH falls below 7 and, with the exception of a few acidophilic NO_2^- oxidiser strains, growth in laboratory batch culture in liquid medium does not occur below pH 6.5. For NH_3 oxidisers, this is due to increasing ionisation of NH_3, which enters the cell by diffusion, to NH_4^+, whose uptake requires active transport. Laboratory growth of NH_3 oxidisers at low pH is possible, but only under particular situations that will pertain in natural environments. For example, the pH minimum for growth can be reduced by 1–1.5 pH units if cells are attached to surfaces or within a biofilm or if growing cells in chemostat culture are subjected to gradual decreases in pH. In addition, some strains can hydrolyse urea, which enters cells by diffusion. Growth on urea appears to be unaffected at pH values down to 3 (Figure 15-4) [20]. However, nitrification of urea may not be complete, as a proportion of NH_3 produced intracellularly diffuses out of the cells, ionises and becomes unavailable for further growth.

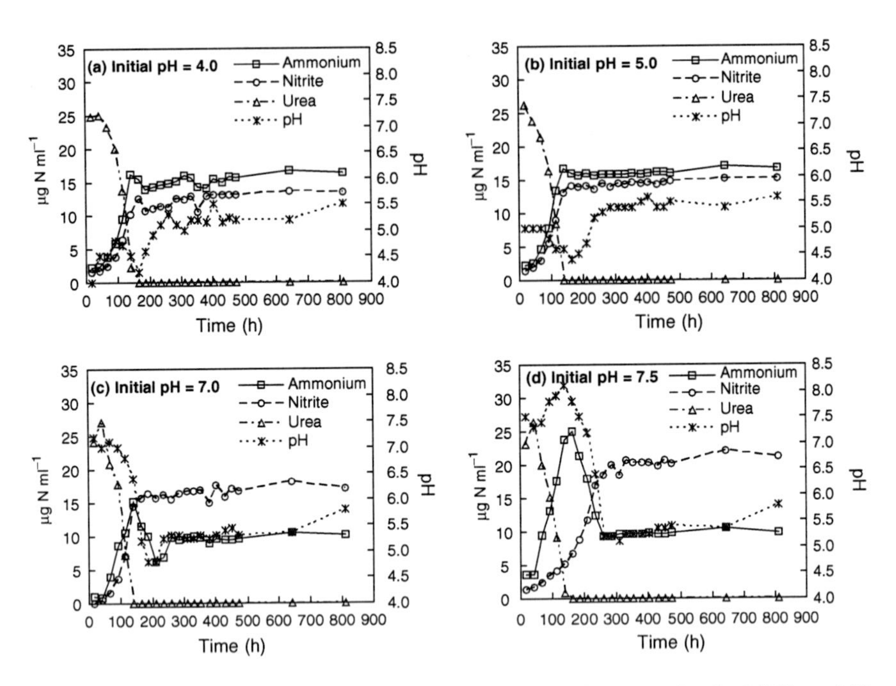

Figure 15-4 pH-independent growth on urea of a ureolytic NH_3 oxidiser, *Nitrosospira* NpAV, which was unable to grow on NH_3 at pH values below 7. Urea, ammonium and NO_2^- concentrations and pH were measured during growth of *Nitrosospira* NpAV in liquid batch culture on poorly buffered medium containing urea, at initial pH values of (a) 4, (b) 5, (c) 7 and (d) 7.5, from [3].

Oxidation of NH_3 is accompanied by a decrease in pH and low pH may limit nitrification in poorly buffered environments or in acid soils. Nevertheless, nitrification has been reported in soils with pH values of 3.5. This may result from activity of heterotrophic nitrifiers, as fungal heterotrophic nitrifiers can grow at low pH. Several attempts have been made to obtain autotrophic NH_3 oxidisers from acid soils using media of different pH. Enrichment, and occasionally isolation of AOB is possible, suggesting a role for autotrophic NH_3 oxidisers in acid soils, but cultivated strains grow only at neutral pH. Nitrification in acid soils may therefore be due to growth of neutrophilic strains, in biofilms or with urease activity.

Evidence for selection of different groups of NH_3 oxidisers in acid soils comes from analysis of AOB 16S rRNA gene sequences amplified from soils maintained at a range of pH values for prolonged periods. In one example, the relative abundance of different AOB phylotypes was determined in soil plots maintained for 36 years at pH values in the range 3.9–6.6. *Nitrosospira* cluster 2 sequences dominated in acid soils but decreased in relative abundance as pH increased, to be replaced by

Nitrosospira cluster 3 sequences. Other groups detected, *Nitrosospira* cluster 4 and *Nitrosomonas* cluster 6a, were less abundant and showed less obvious trends with pH. Other studies of acid soils indicate that *Nitrosospira* cluster 2 strains may be selected at low pH, although it would be dangerous to generalise about physiological characteristics of these groups without more extensive study of cultivated representatives. Unfortunately, these soils, like many others, were dominated by *Nitrosospira* strains, for which laboratory cultivation studies are few, and it is not possible to determine which physiological characteristics led to their greater abundance in acid soils.

15.5 Conclusions

Nitrification plays a central role in the global cycling of N and is of considerable importance economically, in agriculture and wastewater treatment processes, and contributes to important ecosystem processes, such as N_2O production and methane oxidation. The major recent advance in nitrifier ecology has been the ability to investigate nitrifier communities using cultivation-independent techniques. This has greatly increased our understanding of the diversity of bacterial nitrifiers and has led to the discovery of archaeal NH_3 oxidisers. There is also increased awareness of the metabolic versatility of nitrifying bacteria, and the consequences for their contributions to ecosystem processes. A combination of laboratory-based studies on cultivated nitrifiers and, increasingly, cultivation-independent, in situ techniques for analysis of physiology is now providing information on the links between nitrifier diversity and community structure, physiological diversity and ecosystem function. This chapter provides examples of their use to determine the function of uncultivated organisms and the influence of environmental change on microbial community structure. However, the existence of phylogenetic groups with no cultured representatives and the discovery of archaeal NH_3 oxidisers illustrate the exciting developments in this area and the challenges that remain in gaining a mechanistic understanding of the factors controlling the ecology of nitrifying bacteria and predicting their impact on ecosystem functions.

References

[1] U. Purkhold, A. Pommerening-Röser, S. Juretschko, M.C. Schmid, H.P. Koops, M. Wagner, Phylogeny of all recognized species of ammonia oxidizers based on comparative 16S rRNA and *amo*A sequence analysis: Implications for molecular diversity surveys, Appl. Environ. Microbiol. 66 (2000) 5368–5382.

[2] H.P. Koops, U. Purkhold, A. Pommerening-Röser, G. Timmermann, M. Wagner, The lithoautotrophic ammonia-oxidizing bacteria, in: M. Dworkin (Ed.) The Prokaryotes: an Evolving Electronic Resource for the Microbiological Community, Springer-Verlag, New York, 2003.
[3] A.H. Treusch, S. Leininger, C. Schleper, A. Kietzin, H.-. Klenk, S.C. Schuster, Novel genes for nitrite reductase and Amo-related proteins indicate a role of uncultivated mesophilic crenarchaeota in nitrogen cycling, Environ. Microb. 7 (2005) 1985–1995.
[4] M. Könneke, A.E. Bernhard, J.R. De La Torre, C.B. Walker, D.A. Stahl, J.B. Waterbury, Isolation of an autotrophic ammonia-oxidizing marine archaeon, Nature 437 (2005) 543–546.
[5] G.N. Nicol, C. Schleper, The role of Archaea in ammonia oxidation, Trends Microbiol. 14 (2006) 207–212.
[6] J.I. Prosser, T.M. Embley, Cultivation-based and molecular approaches to characterisation of terrestrial and aquatic nitrifiers, Antonie Van Leeuwenhoek 81 (2002) 165–179.
[7] G.A. Kowalchuk, J.R. Stephen, Ammonia-oxidizing bacteria: a model for molecular microbial ecology, Ann. Rev. Microbiol. 55 (2001) 485–529.
[8] H.P. Koops, A. Pommerening-Röser, Distribution and ecophysiology of the nitrifying bacteria emphasizing cultured species, FEMS Microbiol. Ecol. 37 (2001) 1–9.
[9] C.J. Phillips, E.A. Paul, J.I. Prosser, Quantitative analysis of ammonia oxidising bacteria using competitive PCR, FEMS Microbiol. Ecol. 32 (2000) 167–175.
[10] J.A.E. Molina, Components of rates of ammonium oxidation in soil, Soil Soc. Amer. J. 49 (1985) 603–609.
[11] Y. Okano, K.R. Hristova, L.E. Jackson, B. Gebreyesus, D. Lebauer, K.M. Scow, C.M. Leutenegger, R.F. Denison, Application of real-time PCR to study effects of ammonium on population size of ammonia-oxidizing bacteria in soil, Appl. Environ. Microbiol. 70 (2004) 1008–1016.
[12] J.I. Prosser, Autotrophic nitrification in bacteria, Adv. Microb. Physiol. 30 (1989) 125–181.
[13] T.E. Freitag, L. Chang, C.D. Clegg, J.I. Prosser, Influence of inorganic nitrogen-management regime on the diversity of nitrite oxidizing bacteria in agricultural grassland soils, Appl. Environ. Microbiol. 71 (2005) 8323–8334.
[14] G. Webster, T.M. Embley, T.E. Freitag, Z. Smith, J.I. Prosser, Links between ammonia oxidiser species composition, functional diversity and nitrification kinetics in grassland soils, Environ. Microbiol. 7 (2005) 676–684.
[15] N.G. Hommes, L.A. Sayavedra-Soto, D.J. Arp, Chemolithoorganotrophic growth of *Nitrosomonas europaea* on fructose, J. Bact. 185 (2003) 6809–6814.

[16] S. Radajewski, G. Webster, D.S. Reay, S.A. Morris, P. Ineson, D.B. Nedwell, J.I. Prosser, J.C. Murrell, Identification of active methylotroph populations in an acidic forest soil by stable isotope probing, Microbiol. 148 (2002) 2331–2342.

[17] S.E. Batchelor, M. Cooper, S.R. Chhabra, L.A. Glover, G.S.A.B. Stewart, P. Williams, J.I. Prosser, Cell density-regulated recovery of starved biofilm populations of ammonia-oxidizing bacteria, Appl. Environ. Microbiol. 63 (1997) 2281–2286.

[18] A.E. Bernhard, T. Donn, D.A. Stahl, A.E. Giblin, Loss of diversity of ammonia-oxidizing bacteria correlates with increasing salinity in an estuary system, Environ. Microbiol. 7 (2005) 1289–1297.

[19] T.E. Freitag, L. Chang, J.I. Prosser, Changes in the community structure and activity of betaproteobacterial ammonia-oxidising sediment bacteria along a freshwater-marine gradient. Environ. Microbiol. 8 (2006) 684–696.

[20] S.A.Q. Burton, J.I. Prosser, Autotrophic ammonia oxidation at low pH through urea hydrolysis, Appl. Environ. Microbiol. 67 (2001) 2952–2957.

Chapter 16

Anammox

Huub J.M. Op den Camp, Mike S.M. Jetten and Marc Strous

16.1 Anammox: Discovery and introduction

This chapter describes the microbiological investigation of anaerobic NH_4^+ oxidation (the anammox process). During the last 10 years the anammox process evolved from a largely unexplored part of the biological N cycle to general textbook knowledge. The last few years, it has become clear that anammox is a major player in the global N cycle.

The anammox process is the oxidation of NH_4^+ to N_2 with NO_2^- as the electron acceptor (equation (16.1)). The Gibbs free energy change associated with this reaction is even higher than for aerobic NH_4^+ oxidation and could support autotrophic growth, as was first noted by Engelbert Broda in 1975 [1]. The anammox process was actually discovered 10 years later in a denitrifying pilot plant for the treatment of wastewater from the Gist Brocades yeast factory in Delft [2]. The discovery was important for two reasons. Firstly, the anammox process is very attractive to wastewater treatment (see Section 16.9). Secondly, for almost a century, the picture of

---

Biology of the Nitrogen Cycle
Edited by H. Bothe, S.J. Ferguson and W.E. Newton

the N cycle had generally been considered complete: ammonia could not be oxidised under anoxic conditions:

$$NH_4^+ + NO_2^- \rightarrow N_2 + 2H_2O \ (\Delta G_0 = -357 \text{ kJ/mol}) \qquad (16.1)$$

After its discovery, initially NO_3^- was believed to be the electron acceptor. Only after the role of NO_2^- was recognised, enrichment and study of the responsible microorganism became possible [3, 4]. Another important advance made in these studies was the use of electron microscopy and cytochrome spectra to determine the degree of enrichment of the undefined mixed culture. This was possible because it was found that the putative anammox bacterium was morphologically unusual (hence could be recognised on electron micrographs). An unusual peak appeared at 468 nm in reduced cytochrome spectra of enrichment cultures. Later, this peak was shown to belong to one of the key catabolic enzymes, similar but not identical to nitrifier hydroxylamine oxidoreductase (HAO) [5]. However, all attempts to isolate the responsible microorganism failed. This failure could be explained in part by the instability of the anammox activity. In all experiments with anammox biomass described in the above citations, at least 80% of anammox activity was lost after taking the biomass from the reactors.

Isolation of a responsible microorganism is important because microbiology can only progress when experimental results can be interpreted without ambiguity. Therefore, Kochs's postulate is still one of the most important paradigms: 'To prove that a process is a microbiological process one must isolate the responsible microorganism in pure culture and show that it is still capable of the process'. Isolation would enable a better understanding of the process.

In retrospect, anammox research had several discrete experimental phases: (I) the application of the sequencing batch reactor to anammox enrichment culture [6], leading to the production of large quantities of anammox biomass of relatively defined microbial composition; (II) the physiological study of this biomass [7]; (III) the development of the cell purification technique, the miniaturised activity test and the discovery of the importance of cell density, leading to the evidence that a single microorganism was responsible for anammox [8]; (IV) the molecular ecological work to prove the hypothesis that the anammox bacterium might be a planctomycete (based on ultrastructural evidence) [8]; (V) the development of a molecular toolbox to study the significance of anammox on a global scale [9]; (VI) environmental genomics resulting in a complete genome of the anammox bacterium *Kuenenia stuttgartiensis* and increase of knowledge on the metabolic pathways [10] and (VII) application of the anammox process in wastewater treatment [11].

16.2 Enrichment of anammox biomass

Since the application of molecular techniques to ecology, both molecular biologists and microbiologists have argued that the available conventional cultivation techniques have given a distorted view of the microbial reality. What causes this distortion is the large difference between successful cultivation in the laboratory and survival or growth in the natural habitat. The truth is that the great microbiologists, such as Winogradsky, Beyerinck, Perfiliev and Kluyver, have known this all along. They knew very well that most habitats cannot be reproduced using an agar plate or shake flask and once the (a)biotic conditions become too different, the microbes from those habitats can no longer occupy their ecological niche and cannot be recovered through cultivation attempts. Winogradsky already pointed out the existence of the zymogenous and autochtonous flora. However, at that time no technique other than the microscope could prove this point. Only with the advent of molecular microbiological techniques, i.e. molecular probing, the hard proof of their intuition has been presented. Today it is also argued that many bacteria would be 'uncultivable'. We surmise that they are cultivable!

The principle of selective enrichment, as first developed by Beyerinck, demands that to isolate a particular microorganism, the organism first has to be provided with what might be described as the complement of its ecological niche. This means engineering a cultivation-environment with the right dynamic supply of substrates and nutrients and eliminating the products (via mass transport and dilution or other microorganisms), and to control the pH and temperature. Presently available microbiological techniques are not designed to deal with very slowly growing microorganisms. The sequencing batch reactor (SBR) was applied and optimised for the enrichment and quantitative study of anammox bacteria [6]. The specific maximum anammox activity of the biomass in the reactor and the specific absorption at 468 nm of cell-free extract prepared from the culture were selected as independent parameters to determine the degree of enrichment. The SBR was shown to be a powerful experimental setup with the following strong points: (1) efficient biomass retention (only approximately 10% of the growing biomass was washed out), (2) homogeneous distribution of substrates, products and biomass aggregates over the reactor, (3) reliable operation for more than a year and (4) stable conditions under substrate-limiting conditions. Figure 16-1 shows the exponential increase in N load and the resulting accumulation of anammox biomass. The biomass-specific activity in the reactor was constant during day 246 to 330 at 20 ± 6 nmol NH_4^+/mg protein/min. Also NO_3^- is produced in the reactor (Figure 16-1): part of the NO_2^- is oxidised to NO_3^- to generate reducing equivalents for CO_2 fixation (anammox microorganisms are autotrophs) and therefore is a measure for biomass growth. Apart from N_2, only traces of other gaseous N compounds (N_2O, NO, NO_2) were produced.

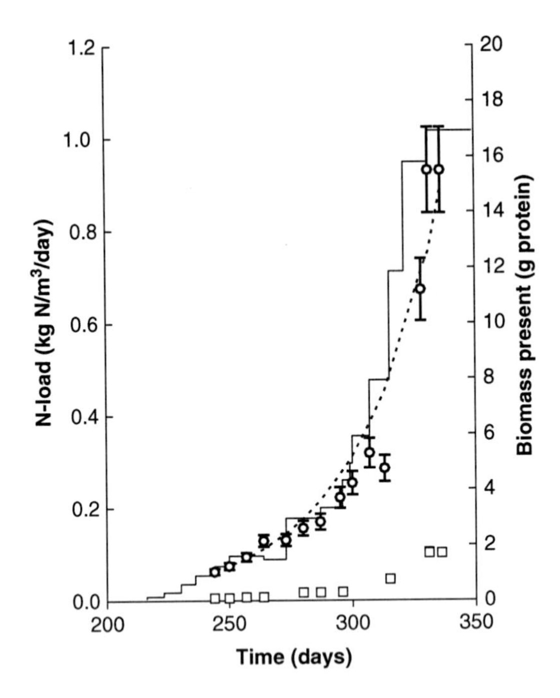

Figure 16-1 Performance of the SBR. The line shows the reactor load. Open circles indicate the measured amount of biomass. The dashed line represents the amount of biomass calculated from the stoichiometry (equation 16.2) and assuming a biomass retention efficiency of 90%. Open squares indicate nitrate concentrations.

More than 10 mass balances over the SBR were compiled and all showed that C and N were completely recovered and the degree of reduction balance was closed within 5%. With these mass balances, the conversion of N and C compounds could be calculated and used to establish the stoichiometry of the anammox process (at NO_2^- limitation, in the presence of 5 mM surplus NH_4^+ and NO_3^-) (see equation (16.2)).

The persisting stable and strongly selective conditions of the SBR led to a high degree of enrichment (74% of the desired microorganism; see also Section 16.3).

$$1\ NH_4^+ + 1.32\ NO_2^- + 0.066\ HCO_3^- + 0.13\ H^+ \rightarrow 1.02\ N_2 + 0.26\ NO_3^- + 0.066\ CH_2O_{0.5}N_{0.15} + 2.03\ H_2O \quad (16.2)$$

16.3 Physiology of anammox aggregates from the SBR

The SBR approach made possible the determination of several important physiological parameters such as the biomass yield (0.066 ± 0.01 C-mol/mol NH_4^+), the maximum specific NH_4^+ consumption rate (45 ± 5 nmol/mg protein/min) and the maximum specific growth rate (0.0027 h^{-1}, doubling time 11 days) [6, 7]. The experimentally found biomass yield correlates well with the theoretically predicted biomass yield. The low activity of anaerobic NH_4^+-oxidising microorganisms not necessarily results from a K-strategy (high substrate affinity as a microbial specialism, usually associated with low maximum growth rates – as opposed to R-strategy, high maximum growth rate and low substrate affinity). Anammox may simply be a kinetically difficult metabolic strategy.

Anammox biomass was found to be active at temperatures between 20 and 43°C (optimum 40 ± 3°C). The pH range was 6.7–8.3 (optimum 8). Anammox is reversibly inhibited by oxygen (complete inhibition already at 1 μM O_2). The affinity for the substrates NH_4^+ and NO_2^- was very high (affinity constants less than 10 μM). The anammox process was inhibited by NO_2^- concentrations higher than 20 mM. When the NO_2^- concentration was more than 5 mM for a longer period (12 h), the anammox activity was completely lost. However, the activity could be restored by addition of trace amounts (±50 μM) of the anammox intermediate, hydrazine. Table 16-1 shows the most important physiological parameters of anammox compared with those of aerobic NH_4^+ and NO_2^- oxidation.

16.4 Identification of the key player

16.4.1 General

The dominant morphotypical cells (74%) from the anammox SBR enrichment culture were coccoid shaped and conspicuously refractile. Attempts to isolate the anammox microorganism from the SBR using classical microbiological techniques (dilution series, obtaining colonies from single cells on a range of media) were unsuccessful. The main obstacle seems to be the extremely slow growth of the organisms (see Section 16.10). However, it appeared to be possible to purify the morphotypical cells by three subsequent steps: (1) biofilm disruption by mild sonication, (2) separation of single cells from remaining biofilm fragments by centrifugation and (3) purification of the single cells by Percoll density gradient centrifugation [8, 12]. Finally, cells were recovered as a broadband in the bottom part of the gradient. The cell numbers of the morphotypical cells increased to 99.7% of the total cell number. The purified cells converted NH_4^+ and NO_2^- and produced NO_3^- in the usual anammox stoichiometry (Figure 16-2, see Section 16.2) and the activity was almost as high as the 74% enriched aggregates. Cell density was a crucial factor; a cell density of at least 4 g protein/l (10^{10}–10^{11} cells/ml) was necessary. Further, cells were only active when 50–100 μM hydrazine or hydroxylamine was added at the start of the experiment. Purified cells incorporated $^{14}CO_2$, strictly dependent on anammox activity. No label was incorporated in the absence of either NH_4^+ or NO_2^-.

Table 16-1 Physiological parameters of the chemolithoautotrophs of the N cycle.

Parameter	Units	Nitrite oxidation in nitrification	Nitrate oxidation in nitrification	Anammox
Gibbs free energy change	kJ/mol substrate	–275	–74	–357
"Activation energy"	kJ/mol substrate	68	44	70
Biomass yield	C-mol/mol substrate*	0.08	0.02	0.07
Aerobic maximum specific conversion rate	μmol/mg protein/min	0.2–0.6	0.07–2.6	0
Anaerobic maximum specific ammonium conversion rate	μmol/mg protein/min	0.002	–	0.05
Maximum growth rate	h^{-1}	0.04	0.04	0.003
Minimum doubling time	Days	0.73	0.71	10.6
Affinity constant for ammonium	μM	5–2600	–	5
Affinity constant for nitrite	μM	–	14–500	<5
Affinity constant for oxygen	μM	10–50	1	–
Temperature range	°C	Up to 42	Up to 47	Up to 43
pH range		4 – 8.5	4–10	6.7–8.3

*For anammox, one C-mol ($CH_2O_{0.5}N_{0.15}$) was equal to 15 g protein.

The dominant 16S rRNA gene sequence in clone libraries obtained from DNA and RNA extracted from the purified cells was used to design specific gene probes for fluorescence in situ hybridization (FISH). The probes were shown to hybridise specifically with the purified morphotypical cells and with the same cells in the SBR enrichment culture. The 16S rRNA gene sequence grouped deep inside the order Planctomycetales (Figure 16-3). The combination of all these steps ultimately led to the final identification of *Candidatus* Brocadia anammoxidans – the chemolithoautotrophic bacterium responsible for anammox, without the need for conventional isolation. The planctomycete identity of the anammox bacterium was surprising because

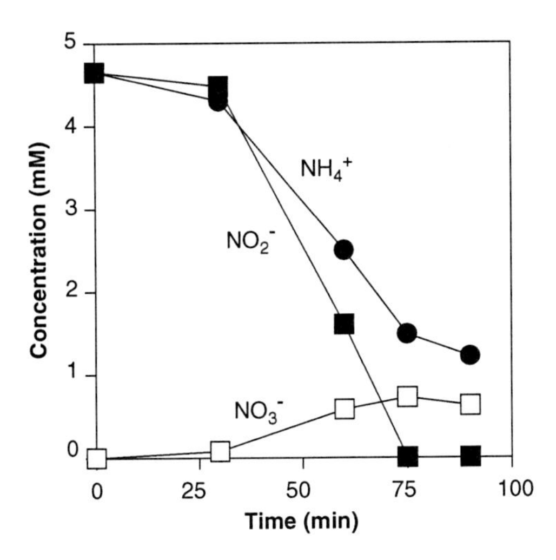

Figure 16-2 Anammox activity of 99.5% pure cell suspensions.

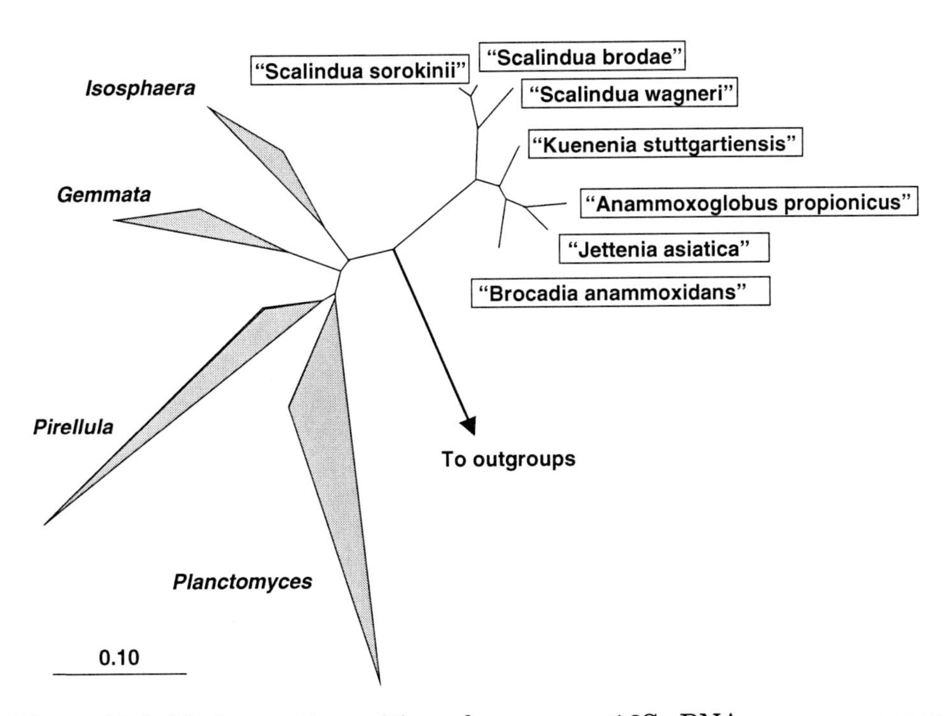

Figure 16-3 Phylogenetic position of anammox 16S rRNA gene sequences (boxes). Bar represents 10% estimated sequence divergence.

the planctomycetes were known as aerobic heterotrophs. In fact, the discovery does not disagree with current views on bacterial evolution: some microbiologists believe that respiration (in the broad sense – including denitrification) evolved from phototrophy (the 'conversion hypothesis') and that the proteobacteria have evolved from a common phototrophic ancestor. Since the common ancestors of the bacterial phyla cannot be discriminated phylogenetically, it follows that chemolithoautotrophy can be expected in every bacterial phylotaxonomic position. Other microbiologists would not believe in a too rigid relationship between phylogeny and metabolism in the first place.

16.4.2 The planctomycetes

The order Planctomycetales is one of the many currently defined (more than 70) distinct divisions that form the domain Bacteria. Before 16S rRNA gene-based phylogeny overturned microbiological taxonomy in the 1980s, the relationships between the various budding, prostecate and stalked bacteria were relatively obscure. Planctomycetes were described for the first time in the 1920s. Owing to their conspicuous morphology and sometimes-high cell numbers, they were observed microscopically in freshwater samples. However, the first planctomycete was isolated in pure culture as late as 1973. Because the number of isolated species remained low, the ecological and evolutionary significance of the planctomycetes went unnoticed until very recently [13].

16.4.2.1 Physiology

Because so few planctomycetes have been isolated in pure culture, and these bacteria have only been studied for their biochemical composition and morphology, we hardly know anything about their physiological significance. The strategy for selective enrichment and isolation of planctomycetes encompasses the use of dilute (heterotrophic) media, (micro-)oxic conditions, antibiotics and *N*-acetylglucosamine as C and N source. However, on the basis of the large differences found among the planctomycetes regarding morphology, biochemical composition and the large phylogenetic distances between them, the planctomycetes presumably occupy a very wide range of possible ecological niches (physiological strategies), such as for example anammox. The planctomycetes presently isolated are considered (micro-)aerobic oligotrophic heterotrophic microorganisms, although the genus *Pirellula* also includes facultative NO_3^- reducers and some planctomycetes are facultative fermenters.

Sequencing of the genome of *Rhodopirellula baltica* was finished in 2003 [14]. With 7.145 Mb this is the largest circular bacterial genome sequenced so far. Surprisingly, all genes required for heterolactic acid fermentation, the key genes for the interconversion of C1 compounds and 110 sulfatases were present in this aerobic heterotrophic bacterium. Sequencing of the genome of *G. obscuriglobus* is still in progress.

16.4.2.2 Phylogeny

The planctomycetes are known for the large phylogenetic distances that separate the individual 16S rRNA gene sequences. On the basis of 16S rRNA phylogeny, biochemical composition and morphology, the genera *Planctomyces* (floating fungus), *Pirellula* (pear-shaped), *Gemmata* (bud-bearer) and *Isosphaera* (spheres of equal size) have been placed in a distinct Order, Planctomycetales, and Family, Planctomycetaceae. Recently, a detailed study of members of the genus *Pirellula* resulted in the definition of two new genera, *Rhodopirellula* and *Blastopirellula*.

16.4.2.3 Morphology

Planctomycetes divide by budding and often have appendages such as stalks or holdfasts. The cell walls of planctomycetes do not contain peptidoglycan, like in Gram-positive or Gram-negative bacteria. Instead, these bacteria have proteinaceous cell walls, like in archaea. These cell walls contain 'crateriform structures', pit-like surface structures.

Planctomycetes have a very distinctive ultrastructure – unique among the prokaryotes. All species have internal membrane structures that separate the cytoplasm into different compartments. On electron micrographs, the structure and electron transparency of the cytoplasm differ for each compartment. Furthermore, the chromosome is present in only one compartment and (ribosomal) RNA is distributed unequally. Therefore, ultrastructural organisation must be an active process for these bacteria. Apparently, they invest energy to transport proteins, RNA and metabolites through their internal membranes and must somehow benefit from their distinct cellular organisation. The nucleoid itself is usually dense and fibrillar, morphologically comparable to the nucleoids of primitive eukaryotes and very uncharacteristic for bacteria.

16.4.2.4 Ecology

Molecular ecology has indicated that the planctomycetes are one of the bacterial divisions represented in almost every ecosystem investigated. Planctomycetes were isolated from sea and freshwater ecosystems, from oligotrophic and polluted habitats, from alkaline (pH 11.6) and acid (pH 4.2) environments, and were found to be meso- and thermophilic (up to 55°C). Molecular studies, e.g. direct sequencing of environmental DNA, indicated that the ecological importance of planctomycetes is still being underestimated.

16.5 Ultrastructure of *Brocadia anammoxidans*

Brocadia anammoxidans has many features in common with previously investigated planctomycetes. Crateriform structures in the cell wall, extracellular appendages, reproduction by budding, the presence of internal compartments and a dense, fibrillar nucleoid were all consistent with observations made previously for *Gemmata obscuriglobus*, *Pirellula marina* and *Pirellula staleyi* [15].

Cells of *B. anammoxidans* were found to have an internal, membrane-bound compartment that we have termed 'anammoxosome'. This anammoxosome did not contain any ribosome or chromosome. Instead, the anammoxosome cytoplasm contained electron-dense inclusions (diameter 20 nm) and was filled with fibre-like structures of unknown function or composition. Apart from the anammoxosome, the cells were organised similar to *Pirellula*, with an intracytoplasmic membrane separating the pirellulosome compartment (containing most or all of the RNA, DNA and the anammoxosome) from a fringing, peripheral compartment. The cell wall appeared to be similar to the cell walls of the other planctomycetes, a typical trilamellar structure.

16.6 Lipids of anammox bacteria

The membranes of anammox bacteria were shown to be composed mainly of so-called ladderane lipids [16]. These lipids contain three to five linearly concatenated cyclobutane moieties with *cis* ring junctions, which occur as fatty acids, fatty alcohols, alkyl glycerol monoethers, dialkyl glycerol diethers and mixed glycerol ether/esters. Comprehensive MS and relative retention time data for all the encountered ladderane membrane lipids are reported, allowing the identification of ladderanes in other bacterial cultures and in the environment. The occurrence of ladderane lipids seems to be a unique feature of the anammox bacteria. This is consistent with their proposed biochemical function, namely as predominant membrane lipids of the so-called anammoxosome (see Section 16.5).

16.7 Significance of anammox on a global scale

16.7.1 General

From a different perspective, some studies have reported N losses in natural ecosystems under anaerobic conditions. However, it is very difficult to discriminate between anammox and other microbial processes (such as denitrification or assimilation) from the black-box data supplied by these studies. The N cycle is easily masked by the carbon cycle, because autotrophic nitrifiers are no true primary producers. Usually, ammonia is a by-product of the biological food chain (ammonification). To study the ecological significance of anammox in man-made and natural ecosystems, a molecular toolbox with a combination of rRNA-based and non-rRNA-based methods is necessary [17–19].

16.7.2 The toolbox

In environmental samples, PCR amplification with general 16S rRNA gene-targeted primers and subsequent phylogenetic analysis of the product is the method of choice to detect previously undescribed organisms. However, planctomycetes such as anammox organisms are still underrepresented in general 16S rRNA gene clone libraries. A more directed PCR approach with the primer Pla46F (a planctomycete-specific forward primer) together with a general eubacterial reverse primer increased relative amounts of planctomycete 16S rRNA gene sequences.

FISH is an excellent tool to collect both qualitative and quantitative data of anammox bacteria in environmental samples [9]. It can also be used to validate the findings of clone libraries. It should be kept in mind that probes specific for planctomycetes or more general eubacterial probes do not always hybridise with the 16S rRNA of anammox organisms or any other environmental clone with a phylogenetic position between the described planctomycete genera and anammox. Probe design will improve as more validated anammox sequences become available. In the future, anammox bacteria might also be detected in environmental samples via immunofluorescence with antibodies raised against anammox-specific proteins. The genomics project on *K. stuttgartiensis* (see Section 16.8) will reveal more anammox-specific genes and proteins, which could be exploited as specific targets for antibodies in future experiments.

Tracer experiments with ^{15}N-labelled NH_4^+ and NO_2^- are the methods of choice for the detection of anammox activity. Under anoxic conditions, labelled $^{15}NH_4^+$ reacts uniquely, in a 1:1 ratio with unlabelled $^{14}NO_2^-$, to form $^{29}N_2$; ($^{14}N^{15}N$) via the anammox reaction. A lot of effort was put into development and use of this method by the research groups of Dalsgaard and Thamdrup. In addition, very sensitive biosensors for online NO_2^- monitoring have become available for the sensitive detection of anammox activity in reactor systems or sediments.

Anammox bacteria have unique lipids (see Section 16.6) that can be used as biomarkers for the presence of anammox cells in the environment. The ladderane lipids are especially suitable, but anammox bacteria also produce characteristic, branched fatty acids and hopanoids. Anammox bacteria are the first strict anaerobes that have been shown to biosynthesise the latter bacterial membrane rigidifiers. Lipids from anammox bacteria are characterized by substantially lower ^{13}C content than their carbon source. The ^{13}C content of ladderane and other lipids is approximately 45% depleted compared to their carbon source, whereas lipids from other autotrophic organisms generally are 20–30% depleted. The isotopic composition of anammox lipids in environmental samples can thus be an additional confirmation of their origin [20].

The conversion of hydroxylamine to hydrazine is a unique reaction catalysed by anammox bacteria and can also be used specifically to detect anammox activity in environmental samples. Because this assay requires rather high anammox cell numbers, it can only be used in samples where anammox bacteria form a substantial part (10–20%) of the microbial population. An alternative anammox activity can be directly assessed by measuring the simultaneous consumption of NH_4^+ and NO_2^- under anoxic conditions.

16.7.3 Anammox in environmental samples

Anammox bacteria are not restricted to wastewater treatment systems. The application of the molecular toolbox described above resulted in the detection of anammox bacteria in more than 30 natural freshwater and marine ecosystems all over the world. Until now, the anammox bacteria belong to five *Candidatus* genera: Brocadia, Kuenenia, Scalindua, Anammoxoglobus and Jettenia (Figure 16-3). These five genera are monophyletic and branch off deep inside the planctomycete lineage of descent. All five genera share the same metabolism, and have a similar ultrastructure, indicating that the capability for anaerobic NH_4^+ oxidation seems to have evolved only once. The ecosystems where anammox was found were anoxic sediments and anoxic water columns characterized by (dynamic) nutrient gradients. In these ecosystems, e.g. Black Sea, Golfo Dulce and the Benguela upwelling system, anammox accounted for 20–100% of total N_2 production [17–19, 21]. The strong N deficit in some ocean ecosystems was until now attributed to denitrification, but the recent findings have showed unequivocally that the anammox bacteria are major players in the global N cycle.

16.8 Biochemistry and the role of environmental genomics

The biochemical mechanism of anaerobic NH_4^+ oxidation is not well understood. In the first paper on this subject, hydrazine and hydroxylamine were proposed as catabolic intermediates. The anammox enrichment culture was shown to produce hydrazine in the presence of excess hydroxylamine. Thus far two dominant proteins, e.g. a hydroxylamine oxidoreductase and a small *c*-monheme cytochrome *c*-552, were purified from crude extracts of *B. anammoxidans* and *K. stuttgartiensis* [5, 22]. The first enzyme was shown to be present exclusively inside the anammoxosome by immuno-labelling. The current hypothesis for the biochemistry of the anammox reaction is depicted in Figure 16-4. Since many chemolithoautotrophs (i.e. nitrifiers) can use organic compounds as a supplementary carbon source, the effect of organic compounds on anammox bacteria was

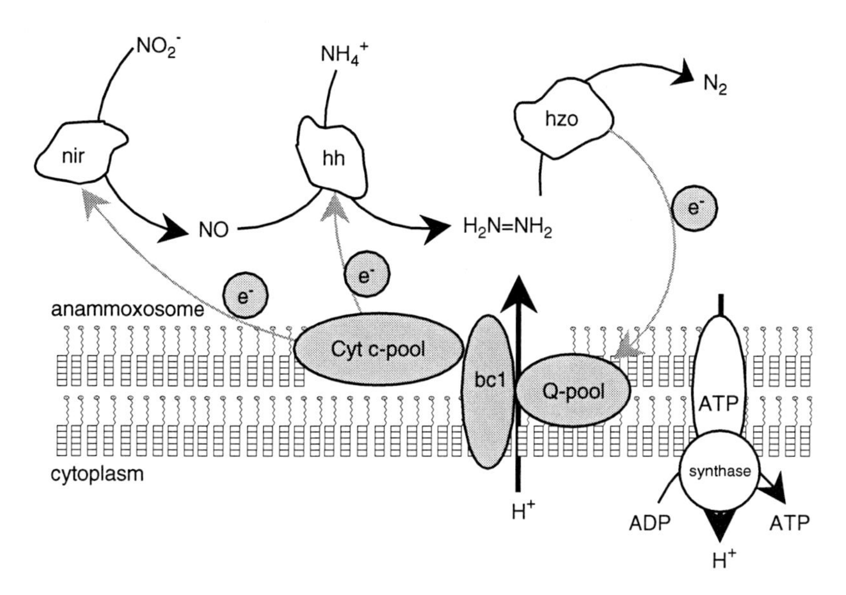

Figure 16-4 Model of the biochemical pathway of anammox. Grey symbols indicate electron transfer. A proton motive force is generated over the anammoxosome membrane. nir, nitrite reductase; hh, hydrazine hydrolase; hzo, hydrazine oxidoreductase.

investigated. It was shown that methanol strongly and irreversibly inhibited anammox bacteria. Acetate, propionate and formate were oxidised mainly to CO_2, with NO_3^- and/or NO_2^- as the electron acceptor [23]. The oxidations occurred simultaneously with anaerobic NH_4^+ oxidation. Apparently, anammox bacteria have a more versatile metabolism than previously assumed [10].

Currently, an environmental genomics project is carried out. This project aims at the reconstruction of genomic data directly from the environment, to assemble the genome of the anammox bacterium *K. stuttgartiensis* from a bioreactor. If successful, the genome data will allow us to exploit the biochemistry of the organism's special properties.

16.9 The application of the anammox process

Treatment of industrial wastewater has become common practice much later than treatment of domestic wastewater. Because industrial wastewater is usually much more concentrated (NH_4^+ concentrations of 1–8 g-N/l are common) in many cases the implementation of physical or chemical techniques is economically feasible, especially when NH_4^+ can be recycled. For NH_4^+, (steam) stripping has been the method of choice. Conventional biological treatment of such wastewaters was not competitive, because often no carbon source was available in water to support denitrification. Supplementation of the wastewater with synthetic methanol has been applied, but is relatively expensive.

With the development of the anammox process, a new opportunity has been created for removal of biological N from industrial wastewater [24]. By combining anammox with a preceding partial nitrification step, NH_4^+ could be converted to N_2 directly. Application of the anammox process in wastewater treatment can result in a 90% reduction of operational costs. The problem regarding the supply of an electron donor (to support conventional denitrification) is circumvented. Furthermore, only half the aeration capacity (required for nitrification) is required. The anammox process is particularly suited for high strength industrial wastewaters that lack a carbon source. For the separate engineering of the nitritification step, a modification of the Sharon process is used. In this process, oxidation of NO_2^- to NO_3^- is prevented by applying a hydraulic retention time that is less than the doubling time of the NO_2^- oxidisers (at somewhat elevated temperatures [>25°C] NO_2^- oxidisers grow slower than NH_4^+ oxidisers). Biomass retention is absent in this configuration. Feasibility studies with various types of wastewater have shown that the activity of the anammox bacteria is not negatively affected by the chemical composition of the applied waters. It has now been independently established in several different laboratories and semi-technical plants that anammox bacteria can be enriched from various types of wastewater sludge indicating that anammox bacteria are indigenous in many treatment plants throughout the world. The highest N removal by anammox bacteria of about 9 kg N m^{-3} reactor day^{-1} was achieved in gas lift reactors.

Alternatively, the anammox process can be combined with partial nitrification in one reactor (the CANON process) [25]. In this process, growth of NO_2^- oxidisers has to be prevented. Long-run experimental studies have shown that this is possible if dual competition occurs for both NO_2^- and O_2. In this process N-load and aeration have to be well balanced.

16.10 Integration and perspectives

This chapter has touched on different aspects of the microbiology of anammox bacteria (physiological, ultrastructural, ecological, phylogenetic). Now where do we go from here? Our understanding of anammox is still far from complete. This section sets out some directions of particular interest, firstly because they integrate different fields of microbiology and secondly because they relate to questions that are of general microbiological significance.

16.10.1 Slow growth, respiration and internal compartments

The combination of slow growth and respiration already makes an interesting case. Mitchells chemiosmotic theory predicts a lower limit for the respiratory rate: as soon as the rate of respiratory proton extrusion drops below the rate of uncoupled proton influx (due to membrane leakiness), energy can no longer be conserved via respiration. Bacteria can limit proton leakiness to a certain extent by adjusting membrane composition and width. Still, calculation shows that in some natural ecosystems activity of the anammox is impossible to explain using standard chemiosmotic theory. These bacteria must have a strategy to overcome this problem, and this strategy might depend on the presence of intracytoplasmic compartments. The anammox compartment does not seem to be designed to boost substrate turnover and to store large amounts of substrate. Still, since the compartment is the location of a key-catabolic enzyme, it does seem to have some role in catabolism. One might argue that the function of the anammoxosome would be to internalise the hydrazine pool and so limit hydrazine losses. In addition, another feature of respiration over internal membranes could be considered: to start-up respiration (to activate the membrane) fewer protons need to be translocated, because (1) the internal membrane surface area is smaller. (2) The internal volume is much smaller than the external medium. This would lead to a closer contact of opposing charges inside the anammoxosome and a larger potential energy per proton translocated. By making use of the internal compartment, the loss of energy is minimised and this would increase the energy efficiency of the bacteria in this dynamic scheme. Internal compartments might thus make extremely slow growth possible.

16.10.2 Growth in clusters

Another problem slowly growing bacteria have to deal with is the loss of catabolic and anabolic intermediates to the external environment due to passive diffusion of these intermediates over the biological membranes. In case of (aerobic and anaerobic) NH_4^+ oxidation, hydroxylamine or hydrazine has been detected experimentally outside the cells. Loss of even small amounts of these compounds might have an impact on the biomass yield, because these bacteria have to make up this loss by investing endogenous electron donors that have been generated by reversed electron transport and CO_2 fixation, both energetically very costly.

Growth in clusters of many cells has been observed for nitrifiers and anammox alike. Clustered growth could serve to decrease the impact of those losses on the biomass yield. In the clusters, each cell would benefit from its neighbours losses. In other words, the surface-to-volume ratio for the cluster is lower than that for a single cell. These considerations show from a different perspective that for nitrifiers a large internal membrane surface area is more advantageous than a large external membrane surface.

There are many environmental conditions that stimulate bacteria to aggregate (protection from grazing protozoa, substrate limitation, etc.), but such explanations are insufficient to describe the typical clustering of cells of NH_4^+ oxidisers. Even compared to neighbouring cells of NO_2^- oxidising cells (that presumably suffer the same conditions in all respects), clusters of NH_4^+ oxidisers are much more compact and conspicuous.

References

[1] E. Broda, Two kinds of lithotrophs missing in nature, Z. Allg. Mikrobiol. 17 (1977) 491–493.

[2] A. Mulder, A.A. van de Graaf, L.A. Robertson, J.G. Kuenen, Anaerobic ammonium oxidation discovered in a denitrifying fluidized bed reactor, FEMS Microbiol. Ecol. 16 (1995) 177–183.

[3] A.A. van de Graaf, P. De Bruijn, L.A. Robertson, M.S.M. Jetten, J.G. Kuenen, Autotrophic growth of anaerobic ammonium-oxidizing micro-organisms in a fluidized bed reactor, Microbiology (UK) 142 (1996) 2187–2196.

[4] A.A. van de Graaf, P. De Bruijn, L.A. Robertson, M.S.M. Jetten, J.G. Kuenen, Metabolic pathway of anaerobic ammonium oxidation on basis of ^{15}N-studies in a fluidized bed reactor, Microbiology (UK) 143 (1997) 2415–2421.

[5] J. Schalk, S. de Vries, J.G. Kuenen, M.S.M. Jetten, Involvement of a novel hydroxylamine oxidoreductase in anaerobic ammonium oxidation, Biochemistry 39 (2000) 5405–5412.

[6] M. Strous, J.J. Heijnen, J.G. Kuenen, M.S.M. Jetten, The sequencing batch reactor as a powerful tool for the study of slowly growing anaerobic ammonium-oxidizing microorganisms, Appl. Microbiol. Biotechnol. 50 (1998) 589–596.

[7] M. Strous, J.G. Kuenen, M.S.M. Jetten, Key physiology of anaerobic ammonium oxidation, Appl. Environ. Microbiol. 65 (1999) 3248–3250.

[8] M. Strous, J.A. Fuerst, E.H.M. Kramer, S. Logemann, G. Muyzer, K.T. Van de Pas-Schoonen, R. Webb, J.G. Kuenen, M.S.M. Jetten, Missing lithotroph identified as new planctomycete, Nature 400 (1999) 446–449.

[9] M.C. Schmid, B. Maas, A. Dapena, K. van de Pas-Schoonen, J. van de Vossenberg, B. Kartal, L. van Niftrik, I. Schmidt, I. Cirpus, J.G. Kuenen, M. Wagner, J.S. Sinninghe Damsté, M. Kuypers, N.P. Revsbech, R. Mendez, M.S.M. Jetten, M. Strous, Biomarkers for in situ detection of anaerobic ammonium-oxidizing (anammox) bacteria, Appl. Environ. Microbiol. 71 (2005) 1677–1684.

[10] M. Strous, E. Pelletier, S. Mangenot, T. Rattei, A. Lehner, M.W. Taylor, M. Horn, H. Daims, D. Bartol-Mavel, P. Wincker, V. Barbe, N. Fonknechten, D. Vallenet, B. Segurens, C. Schenowitz-Truong, C. Médigue, A. Collingro, B. Snel, B.E. Dutilh, H.J.M. Op den Camp, C. van der Drift, I. Cirpus, K.T. van de Pas-Schoonen, H.R. Harhangi, L. van Niftrik, M. Schmid, J. Keltjens, J. van de Vossenberg, B. Kartal, H. Meier, D. Frishman, M.A. Huynen, H.-W. Mewes, J. Weissenbach, M.S.M. Jetten, M. Wagner & D. Le Paslier, Deciphering the evolution and metabolism of an anammox bacterium from a community genome, Nature 440 (2006) 790–794.

[11] M.S.M. Jetten, I. Cirpus, M.B. Kartal, K.T. Van de Pas-Schoonen, A.O. Sliekers, S.C.M. Haaijer, W. Van der Star, M.C. Schmid, J. Van de Vossenberg, I. Schmidt, H.R. Harhangi, M. Van Loosdrecht, J.G. Kuenen, H.J.M. Op den Camp, M. Strous, 1994–2004: 10 years of research on the anaerobic oxidation of ammonium, Biochem. Soc. Trans. 33 (2005) 123–127.

[12] M. Strous, J.G. Kuenen, J.A. Fuerst, M. Wagner, M.S.M. Jetten, The anammox case – A new experimental manifesto for microbiological eco-physiology, Antonie van Leeuwenhoek 81 (2002) 693–702.

[13] J.A. Fuerst, Intracellular compartmentation in planctomycetes, Annu. Rev. Microbiol. 59 (2005) 299–328.

[14] F.O. Glockner, M. Kube, M. Bauer, H. Teeling, T. Lombardot, W. Ludwig, D. Gade, A. Beck, K. Borzym, K. Heitmann, R. Rabus, H. Schlesner, R. Amann, R. Reinhardt, Complete genome sequence of the marine planctomycete *Pirellula* sp. strain 1, Proc. Natl. Acad. Sci. U.S.A. 100 (2003) 8298–8303.

[15] M.R. Lindsay, R.I. Webb, M. Strous, M.S.M. Jetten, M.K. Butler, R.J. Forde, J.A. Fuerst, Cell compartmentalisation in planctomycetes: novel types of structural organisation for the bacterial cell, Arch. Microbiol. 175 (2001) 413–429.

[16] J.S. Sinninghe Damsté, M. Strous, W.I.C. Rijpstra, E.C. Hopmans, J.A. Geenevasen, A.C.T. Van Duin, L.A. Van Niftrik, M.S.M. Jetten, Linearly concatenated cyclobutane lipids form a dense bacterial membrane, Nature 419 (2002) 708–712.

[17] M.M.M. Kuypers, A.O. Sliekers, G. Lavik, M. Schmid, B.B. Jorgensen, J.G. Kuenen, J.S. Sinninghe Damsté, M. Strous, M.S.M. Jetten, Anaerobic ammonium oxidation by anammox bacteria in the Black Sea, Nature 422 (2003) 608–611.

[18] M.M.M. Kuypers, G. Lavik, D. Wobken, M.C. Schmid, B.M. Fuchs, R. Amann, B.B. Jorgensen, M.S.M. Jetten, Massive nitrogen loss from the Benguela upwelling system through anaerobic ammonium oxidation, Proc. Natl. Acad. Sci. U.S.A. 102 (2005) 6478–6483.

[19] T. Dalsgaard, B. Thamdrup, D.E. Canfield, Anaerobic ammonium oxidation (anammox) in the marine environment, Res. Microbiol. 156 (2005) 457–464.

[20] S. Schouten, M. Strous, M.M.M. Kuypers, W.I.C. Rijpstra, M. Baas, C.J. Schubert, M.S.M. Jetten, J.S. Sinninghe Damsté, Stable carbon isotopic fractionations associated with inorganic carbon fixation by anaerobic ammonium oxidizing bacteria, Appl. Environ. Microbiol. 40 (2004) 3785–3788.

[21] T. Dalsgaard, D.E. Canfield, J. Petersen, B. Thamdrup, J. Acuna-Gonzalez, N_2 production by the anammox reaction in the anoxic water column of Golfo Dulce, Costa Rica, Nature 422 (2003) 606–608.

[22] I. Cirpus, M. de Been, H.J.M. Op den Camp, M. Strous, D. Le Paslier, J.G. Kuenen, M.S.M. Jetten, A new soluble 10 kDa monoheme cytochrome c-552 from the anammox bacterium *Candidatus* "Kuenenia stuttgartiensis", FEMS Microbiol. Lett. 252 (2005) 273–278.

[23] D. Güven, A. Dapena, B. Kartal, M.C. Schmid, B. Maas, K.T. van de Pas-Schoonen, S. Sozen, R. Mendez, H.J.M. Op den Camp, M.S.M. Jetten, M. Strous, I. Schmidt, Propionate oxidation and methanol inhibition of the anaerobic ammonium oxidizing bacteria, Appl. Environ. Microbiol. 71 (2005) 1066–1071.

[24] H.J.M. Op den Camp, B. Kartal, D. Güven, L.A.M.P. Van Niftrik, S.C.M. Haaijer, W.R.L. van der Star, K.T. van de Pas-Schoonen, A. Cabezas, Z. Ying, M.C. Schmid, M.M.M. Kuypers, J. van de Vossenberg, H.R. Harhangi, C. Picioreanu, M.C.M. van Loosdrecht, J.G. Kuenen, M. Strous, M.S.M. Jetten, Global impact and application of the anaerobic ammonium-oxidizing (anammox) bacteria, Biochem. Soc. Trans. 34 (2006) 174–178.

[25] A.O. Sliekers, N. Derwort, J.L. Campos Gomez, M. Strous, J.G. Kuenen, M.S.M. Jetten, Completely autotrophic nitrogen removal over nitrite in one single reactor, Water Res. 36 (2002) 2475–2482.

Chapter 17

Nitrate Assimilation in Bacteria

Conrado Moreno-Vivián and Enrique Flores

17.1 Introduction

Nitrate assimilation is a main process of the N-cycle and is carried out by higher plants, fungi, algae, and many bacteria, which use NO_3^- as an N source for growth. More than 2×10^{13} kg of N per year is assimilated by this process [1]. NO_3^- is incorporated into the cells by high-affinity transport systems and further reduced to NH_4^+, via NO_2^-, by two sequential reduction reactions involving two and six electrons that are catalyzed, respectively, by the enzymes nitrate reductase (NR) and nitrite reductase (NiR) (Figure 17-1). The resulting NH_4^+ is then incorporated into carbon skeletons mainly by the glutamine synthetase/glutamate synthase pathway. Although NO_3^- is often found in the environment at relatively low concentrations (e.g., in the 0.03–40 μM range in oceanic seawaters), the excessive use of fertilizers in agricultural activities has led to NO_3^- accumulation up to mM concentrations in the groundwater of many areas, which causes environmental and public health concerns. Consumption of drinking water with high NO_3^- may cause metahemoglobinemia and gastric cancer, whereas the main threat to the environment comes from eutrophication of aquatic ecosystems [2].

Biology of the Nitrogen Cycle
Edited by H. Bothe, S.J. Ferguson and W.E. Newton

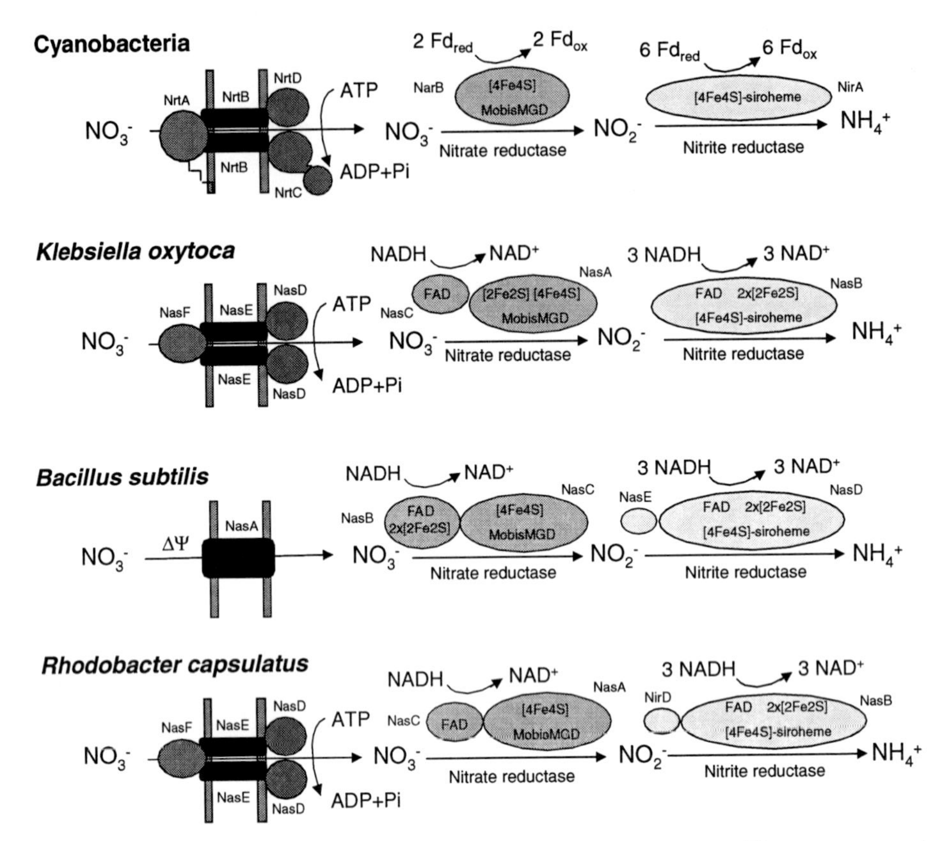

Figure 17-1 Bacterial assimilatory NO_3^- reducing systems. The components of the assimilatory NO_3^- reducing systems of *Synechococcus elongatus* PCC 7942 (as a representative of many cyanobacteria), *Klebsiella oxytoca* strain M5a1, *Bacillus subtilis*, and *Rhodobacter capsulatus* strain E1F1 are shown and have been termed as described in the text (see also Table 17-1). In most cases, the putative cofactors of the structural proteins are predicted from sequence analysis of the corresponding genes, although some of them have been detected by biochemical studies (see the text for details). ATP hydrolysis drives NO_3^- transport by the multicomponent ABC-type NO_3^- transporters, whereas the membrane potential ($\Delta\Psi$) may energize the MFS-family NO_3^- permease. This MFS-type permease is also found in some cyanobacterial marine strains, although *S. elongatus* and most cyanobacteria use a binding protein-dependent transporter.

NO_3^- assimilation has been studied extensively in higher plants, fungi, and algae (see Chapter 18). An analysis of the available prokaryotic genome sequences reveals that the capability to assimilate NO_3^- is widely distributed among bacteria. However, physiological and biochemical studies on NO_3^- assimilation have been performed only in a few bacterial species. Among these are some cyanobacteria, the heterotrophic bacteria *Klebsiella oxytoca*, *Bacillus subtilis* and *Azotobacter vinelandii*, and the anoxygenic phototrophic bacterium *Rhodobacter capsulatus*. These studies reveal significant differences in the bacterial NO_3^--assimilatory systems. In cyanobacteria, the assimilatory NO_3^--reduction is functionally linked to the photosynthetic process and both NR and NiR use photosynthetically reduced ferredoxin (Fd) or flavodoxin (Fld) as the physiological electron donor [3], whereas the reductases from *R. capsulatus* and most heterotrophic bacteria are dependent on NADH as reductant [2, 4, 5]. In addition, the high-affinity NO_3^- and NO_2^- uptake involved in this assimilatory process can take place through either an ATP-dependent ABC-type transporter or a permease from the major facilitator superfamily (MFS) of transport proteins [6]. Finally, bacterial assimilatory NO_3^--reduction is usually regulated by NO_3^- induction (pathway-specific control) and NH_4^+ repression (global N control), but different regulatory patterns, mechanisms, and proteins are involved depending on the organism. The aim of this review is to summarize the organization of the genes coding for proteins of NO_3^- assimilation, the characteristics of the NO_3^- and NO_2^- transport and reduction systems, and the regulation of the NO_3^- assimilation pathway in different bacteria.

The genes coding for the regulatory and structural proteins required for uptake and reduction of both NO_3^- and NO_2^- are, in most cases, clustered. The organization of these gene clusters vary, depending on the organism, mainly due to the presence of different NO_3^- and NO_2^- transport systems (ABC-type transporter or MFS-type permease), different NO_3^- and NO_2^- reductases (ferredoxin- or NADH-dependent enzymes), and distinct regulatory proteins (Figure 17-2).

The gene structure *nirA* (NiR)-*nrtABCD* (NO_3^-/ NO_2^- transporter)-*narB* (NR) found in *Synechococcus elongatus* PCC 7942 is conserved in various cyanobacteria, although in *Synechocystis* PCC 6803 the *nirA* gene is not part of the main gene cluster and in some *Synechococcus* marine strains the NO_3^- assimilation structural genes are clustered with cofactor biosynthesis genes [3]. In *K. oxytoca*, the genes coding for the ABC-type NO_3^--transporter (*nasFED*), the electron transfer subunit of NR (*nasC*), the NiR (*nasB*), and the catalytic subunit of NR (*nasA*) are organized in an operon that is regulated by a gene (*nasR*) located upstream from *nasF* [4, 7, 8]. In *B. subtilis*, a gene cluster has been identified that includes a putative MFS-type permease gene (*nasA*), a gene encoding a flavoprotein (*nasB*) that may act as electron donor to the NR and/or NiR, the genes

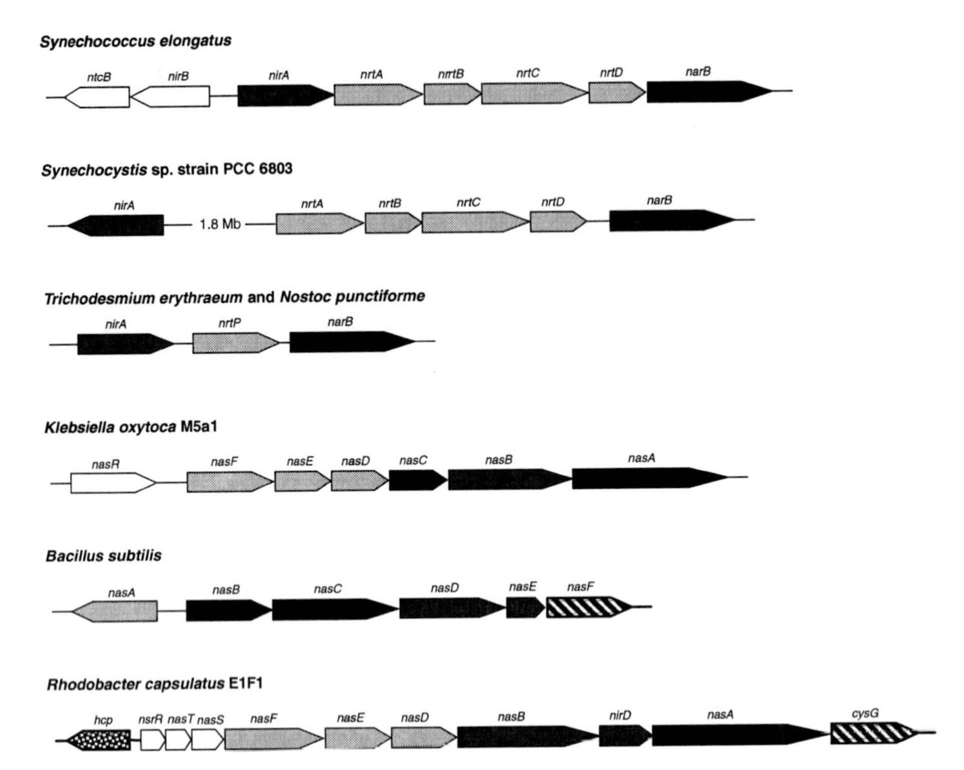

Figure 17-2 Organization of the NO_3^- assimilation gene clusters in different bacteria. The genetic nomenclature used is the same as that in the text and Table 17-1. The NO_3^- transport genes are shown in light gray, the genes coding for NR subunits are indicated in black, the genes encoding the NiR subunits are marked in dark gray, the putative regulatory genes are drawn in white, the genes required for siroheme cofactor biosynthesis are striped, and the *hcp* gene of *R. capsulatus* coding for a putative hydroxylamine reductase is dotted.

coding for the catalytic subunits of the NR (*nasC*) and NiR (*nasD*), a gene coding for a putative small subunit of the NiR (*nasE*), and the *nasF* gene involved in the biosynthesis of siroheme, the NiR cofactor [9]. The *nasBCDEF* genes constitute an operon [9]. In *A. vinelandii*, the genes coding for the NR (*nasB*) and NiR (*nasA*) have been identified and partially sequenced [10]. In addition, two putative genes, *nasS* and *nasT*, required for NO_3^- induction of the *nasAB* operon have been described [11]. Finally, in the anoxygenic phototrophic bacterium *R. capsulatus* strain E1F1, the NO_3^- assimilation (*nas*) gene cluster includes three regulatory genes (*nsrR* and *nasTS*), the *nasFED* genes coding for an ABC-type NO_3^- transporter, the genes encoding the large catalytic (*nasB*) and the small (*nirD*) subunits of NiR, the gene encoding the catalytic subunit of NR (*nasA*), and a siroheme cofactor biosynthesis gene (*cysG*). Interestingly, the *nas* gene region of *R. capsulatus* also includes the *hcp* gene, which codes for a hybrid cluster protein with hydroxylamine reductase activity [12, 13].

The genetic nomenclature for the different NO_3^- assimilation genes in bacteria is confusing due to the use of different designations for homologous genes in different bacteria. Some gene names, like *nar*, *nir*, or *nap*, are also used for respiratory NO_3^- and NO_2^- reductases or periplasmic NRs in bacteria. In this review, we propose a possible unified gene designation based on the cyanobacterial nomenclature for the NO_3^-/ NO_2^- transport genes and the *Klebsiella* nomenclature for the structural genes of NO_3^- and NO_2^- reductases (Table 17-1).

17.2 The NO_3^- and NO_2^- uptake systems

For assimilation purposes, NO_3^- and NO_2^- can be taken up by an ABC-type transporter or by an MFS permease in different bacteria. These proteins are located in the cytoplasmic membrane of either Gram-positive or Gram-negative bacteria, and no specific information on porins required for NO_3^- translocation through the outer membrane of Gram-negative bacteria is available.

17.2.1 NO_3^-– NO_2^- ABC-type transporters

ABC-type uptake transporters that can accept NO_3^- and NO_2^- are found in numerous bacteria. These proteins constitute family 3.A.1.16.1 of the transporter classification of Saier [14]. The basic genetic composition for these transporters is found in some bacteria like *K. oxytoca* and *R. capsulatus* (see above), in which three clustered genes, *nasFED*, encode the transporter. NasF is a periplasmic substrate-binding protein that, according to the sequence of its signal peptide, can be exported to the periplasm by the twin arginine translocation (Tat) system, NasE is a hydrophobic protein likely bearing six transmembrane helices that would act as a homodimer, and NasD is the ATPase of the system and would also act as a homodimer [15].

The cyanobacterial high-affinity NO_3^-–NO_2^- ABC-type uptake transporters are encoded by four genes, *nrtABCD*, and show some interesting variations over the basic ABC-type transporter structure. The periplasmic substrate-binding protein NrtA bears an N-terminal sequence for signal-peptide cleavage by signal peptidase II, which attaches a lipid moiety to the Cys residue that constitutes the N-terminal residue of the processed protein [16]. Thus, NrtA is a lipoprotein that is anchored to the periplasmic side of the cytoplasmic membrane. The integral membrane complex would comprise an NrtB homodimer, but in contrast to the NasE protein discussed

Table 17-1 Nomenclature of the main genes involved in bacterial NO_3^- assimilation.

Protein function	Gene nomenclature					
	Cyanobacteria	*Klebsiella*	*Bacillus*	*Azotobacter*	*Rhodobacter*	Proposed name
ABC-type NO_3^- transporter						
Periplasmic substrate-binding protein	*nrtA*	*nasF*	–	Ni	*nasF*	*nrtA*
Membrane protein	*nrtB*	*nasE*	–	Ni	*nasE*	*nrtB*
Large cytoplasmic ATPase	*nrtC*	–	–	–	–	*nrtC*
Small cytoplasmic ATPase	*nrtD*	*nasD*	–	Ni	*nasD*	*nrtD*
MFS-type ion-driven permease	*nrtP* or *napA*	–	*nasA*	–	–	*nrtP*
NR						
Large catalytic subunit	*narB*	*nasA*	*nasC*	*nasB*	*nasA*	*nasA*
Electron transfer subunit	–	*nasC*	*nasB*	–	Ni	*nasC*
NiR						
Large catalytic subunit	*nirA*	*nasB*	*nasD*	*nasA*	*nasB*	*nasB*
Small subunit	–	–	*nasE*	Ni	*nirD*	*nasD*
Main regulatory proteins						
CAP family transcription factor (global N control)	*ntcA*	–	–	–	–	*ntcA*
LysR family transcription factor (pathway-specific)	*ntcB*	–	–	–	–	*ntcB*

Table 17-1 (Continued)

Protein function	Gene nomenclature					
	Cyanobacteria	*Klebsiella*	*Bacillus*	*Azotobacter*	*Rhodobacter*	Proposed name
AmiR family transcription antiterminator (pathway-specific)	–	*nasR*	–	*nasT*	*nasT*	*nasR* or *nasT*
NO_3^- sensor to regulate antiterminator (pathway-specific)	–	-	-	*nasS*	*nasS*	*nasS*
Rrf2 family negative regulator (pathway-specific)	–	–	–	–	*nsrR*	*nsrR*
Transcription factor (global N control)	–	–	*tnrA*	–	–	*tnrA*

–, homologous gene not present in the organism; Ni, homologous gene probably present in the organism, but not yet isolated or identified.

above, NrtB appears to bear only five transmembrane helices [3]. Finally, the cyanobacterial ABC-type NO_3^-–NO_2^- transporters comprise a heterodimeric ATPase, which is constituted by NrtC and NrtD. Whereas NrtD is a standard ABC-type transporter ATPase subunit, NrtC bears two domains: an N-terminal domain (amino acid residues 1–260, approximately) homologous to NrtD, and a C-terminal domain (amino acid residues 261–600, approximately) homologous to the substrate-binding protein NrtA. This C-terminal extension of NrtC may have a regulatory function [17].

17.2.2 NO_3^-–NO_2^- MFS-type permeases

MFS permeases involved in the assimilation of NO_3^- have been identified in *B. subtilis* and two cyanobacteria. Mutation of the *B. subtilis nasA* gene [9] and the *Synechococcus* PCC 7002 *nrtP* gene [18] impairs growth using NO_3^- as the N source. The *nrtP* mutant has also been shown to be impaired in NO_3^- and NO_2^- uptake. A gene from *Trichodesmium* WH9601 termed *napA* has been shown to complement the *B. subtilis nasA* mutant [19]. The proteins encoded by these genes belong to family 2.A.1.8 of the transporter classification [14], which also includes proteins from other bacteria, algae, and fungi. The well-known NO_3^- transporter NRT2 from higher plants is also homologous to proteins of this family [20]. Some of these proteins, like NarK of *E. coli* or *Paracoccus pantotrophus*, are involved in NO_3^-/ NO_2^- exchange rather than simply in the uptake of NO_3^- or NO_2^-, but the fueling mechanism, whether uniport or H^+ or Na^+ symport, has not been determined for any of the characterized bacterial NO_3^- or NO_2^- MFS importers.

17.3 Assimilatory NRs

The bacterial assimilatory NRs (Nas) are cytoplasmic molybdoenzymes that catalyze the two-electron reduction of NO_3^- to NO_2^- (Figure 17-1). These enzymes are structurally and functionally different from the dissimilatory periplasmic NRs (Nap) and the respiratory membrane-bound NRs (Nar) present in many bacteria (see Chapter 2). Prokaryotic and eukaryotic NRs share no significant sequence similarity and have little in common beyond their physiological function.

According to the specificity for the electron donor, two main types of bacterial NRs can be distinguished: the Fd- or Fld-dependent NR, which is typically found in cyanobacteria [1, 3] but is also present in *A. vinelandii* [21] and the archaeon *Haloferax mediterranei* [22] and the NADH-dependent NR, present in most heterotrophic bacteria [4] and in the phototrophic bacterium *R. capsulatus* [23].

Phylogenetic relationships deduced from comparisons of the available NR sequences reveal that assimilatory NRs have the least conserved sequences of all the bacterial NRs [24]. Assimilatory (Nas) and periplasmic (Nap) NRs form two subgroups of a clade clearly different from the membrane-bound respiratory enzyme (Nar) clade and the eukaryotic NR clade. A specific clustering based on preferred electron donor, NADH or Fd, in the bacterial NRs is not observed [24].

17.3.1 Fd- or Fld-dependent NRs

Cyanobacterial NR is partially associated to the thylakoid membrane and uses photosynthetically reduced Fd or Fld as physiological electron donor [3]. The enzyme is a monomer of about 80 kDa, encoded by the *narB* gene, that contains a Mo-*bis*-molybdopterin guanine dinucleotide cofactor (Mo-*bis*-MGD) and one [4Fe–4S] cluster [25, 26]. The metal centers in the *S. elongatus* enzyme have midpoint redox potentials of −190 mV for $[4Fe–4S]^{1+/2+}$ and −150 mV for $Mo^{6+/5+}$ [26]. Cyanobacterial NRs have been reported to show K_m values for NO_3^- between 1 and 10 mM, but a recent report on the *S. elongatus* enzyme indicates a value of 50–80 μM [26].

Electrons from the [2Fe–2S] cluster-containing ferredoxin (Fd) or the flavin-containing flavodoxin (Fld), which are very low negative redox potential electron donors, seem to be transferred to the Fe–S cluster and from this center to the Mo-cofactor for the reduction of the substrate [26]:

$$\mathrm{Fd}\{[2\mathrm{Fe}-2\mathrm{S}]\}\,\mathrm{or}\,\mathrm{Fld}\{\mathrm{FMN}\} \rightarrow \underbrace{\{[4\mathrm{Fe}-4\mathrm{S}] \rightarrow \mathrm{Mo}\text{-}bis\text{-}\mathrm{MGD}\}}_{\mathrm{NO_3^-\ reductase\,(NarB)}} \rightarrow \mathrm{NO_3^-}$$

The purified NR from *A. vinelandii* uses Fld 1 as electron donor, but not NAD(P)H [21]. The enzyme is a monomer (encoded by *nasB* [10]) with a molecular mass of 105 kDa and contains one Mo, four Fe, and four acid-labile S atoms per molecule according to the presence of a [4Fe–4S] cluster and a Mo-*bis*-MGD cofactor [27]. As in the case of the cyanobacterial NR, the enzyme of *A. vinelandii* does not contain flavin. Apparent K_m values for NO_3^- of 70 and 220 μM are found with reduced methyl viologen or bromophenol blue as artificial electron donor [27]. The pathway of electron flow from reduced Fld 1 to NO_3^- in the *A. vinelandii* NR may be similar to that in the cyanobacterial enzyme:

$$\mathrm{Fld1}\{\mathrm{FMN}\} \rightarrow \underbrace{\{[4\mathrm{Fe}-4\mathrm{S}] \rightarrow \mathrm{Mo}\text{-}bis\text{-}\mathrm{MGD}\}}_{\mathrm{NO_3^-\ reductase\,(NasB)}} \rightarrow \mathrm{NO_3^-}$$

The NR of the haloarchaeon *H. mediterraneii* is a dimer of 105- and 50-kDa subunits, uses reduced Fd as reductant, but not NAD(P)H, and exhibits an apparent K_m for NO_3^- of 0.95 mM [22]. The enzyme seems to contain an Fe–S center but lacks FAD and requires a pH value of 9, elevated temperatures (up to 80°C), and high salt concentrations (3.1 M NaCl) for optimal activity [22].

17.3.2 NADH-dependent NRs

The properties of the NADH-dependent NRs of *K. oxytoca* and *B. subtilis* have been deduced from DNA sequence analysis, despite the absence of biochemical studies [4, 7, 9]. In *K. oxytoca*, the enzyme is predicted to be a dimer composed of a large catalytic subunit of 92 kDa and a small electron transfer subunit of 43 kDa. The catalytic subunit (NasA) probably binds a Mo-*bis*-MGD cofactor, one [4Fe–4S] cluster in the N-terminal domain and one [2Fe–2S] cluster in a C-terminal extension that is not present in the homologous proteins of other bacteria. This C-terminal domain has sequence similarity to NifU and other [2Fe–2S] cluster-containing proteins. The small subunit (NasC) is predicted to be an NADH oxidoreductase (diaphorase) that bears FAD and mediates electron transfer from NADH to the catalytic subunit NasA [4, 7]:

$$\underbrace{\mathrm{NADH} \rightarrow \{\mathrm{FAD}\}}_{\mathrm{Diaphorase\,subunit\,(NasC)}} \rightarrow \underbrace{\{[2\mathrm{Fe}-2\mathrm{S}] \rightarrow [4\mathrm{Fe}-4\mathrm{S}] \rightarrow \mathrm{Mo}\text{-}bis\text{-}\mathrm{MGD}\}}_{\mathrm{Catalytic\,subunit\,(NasA)}} \rightarrow \mathrm{NO_3^-}$$

DNA sequence and genetic analysis of *B. subtilis* also indicate the presence of a dimeric NADH-dependent NR [9]. In this organism, the large catalytic subunit (termed NasC) seems to contain the Mo-*bis*-MGD cofactor and the N-terminal [4Fe–4S] cluster, but not the NifU-like [2Fe–2S] cluster present in the *Klebsiella* NasA protein. Instead, two tandem NifU-like motifs are present in the small electron transfer subunit (called NasB in this bacterium), which also contains the FAD- and NADH-binding domains [9]. The pathway of electrons from NADH to NO_3^- through the *B. subtilis* NR may be as follows:

$$\underbrace{NADH \rightarrow \{FAD \rightarrow 2 \times [2Fe-2S]\}}_{\text{Diaphorase subunit (NasB)}} \rightarrow \underbrace{\{[4Fe-4S] \rightarrow \text{Mo-}bis\text{-MGD}\}}_{\text{Catalytic subunit (NasC)}} \rightarrow NO_3^-$$

In the phototrophic bacterium *R. capsulatus* strain E1F1, both biochemical and DNA sequence analysis of the NADH-dependent NR have been carried out [12, 13, 23]. The purified enzyme is a heterodimer composed of a 95-kDa catalytic subunit carrying a Mo-*bis*-MGD cofactor and one N-terminal [4Fe–4S] cluster and a 45-kDa FAD-containing NADH oxidoreductase (diaphorase), which mediates electron transfer from NADH to the catalytic subunit. The enzyme exhibits an apparent K_m value for NO_3^- of 0.1 mM and is inhibited by NADH under aerobic conditions, probably by formation of superoxide anion at the diaphorase flavin center [23]. Recently, a 17-kb *nas* gene region that includes the *nasA* gene encoding the catalytic subunit of the NR has been sequenced, but the gene coding for the small diaphorase subunit is not present in the isolated *nas* gene cluster and should be located elsewhere in the genome of this bacterium [12, 13]. Sequence analysis reveals that, as occurs in *B. subtilis* [9], the predicted catalytic subunit NasA lacks the C-terminal NifU-like [2Fe–2S] cluster that is present in the *Klebsiella* NasA protein [13].

17.4 Assimilatory NiRs

Bacterial assimilatory NiRs are cytoplasmic enzymes that catalyze the six-electron reduction of NO_2^- to NH_4^+ (Figure 17-1). Two types of NiRs, according to the electron donor specificity, have been described: the Fd-dependent NiRs, characteristic of both eukaryotic and prokaryotic photosynthetic organisms like the cyanobacteria, algae, and plants, and the NAD(P)H-dependent NiRs, which are present in fungi and most heterotrophic bacteria. The available information from DNA sequence analysis reveals that both types of NiRs share significant structural features, and also that a high sequence similarity with the siroheme-containing sulphite reductase exists [4]. Owing to these similarities, it has been proposed that eukaryotic NiRs were originally derived from cyanobacterial endosymbionts during chloroplast evolution [28].

17.4.1 *Fd-dependent NiRs*

Cyanobacterial NiRs are monomers (encoded by the *nirA* gene) with a molecular mass of about 55 kDa, which contain as prosthetic groups a [4Fe–4S] cluster and siroheme. The physiological electron donor is the photosynthetically reduced Fd or Fld [3]. Purified NiRs from different cyanobacteria show biochemical properties similar to those of plants and algae, with apparent K_m values of 5–20 μM for Fd and 40–50 μM for NO_2^- [3]. Two intermediates, NO and NH_2OH, may be formed in the catalytic cycle of the enzyme, as described for plant NiRs [29], and the probable pathway for the electron flow from reduced Fd to NO_2^- may be schematized as follows:

$$\text{Fd}\{[2\text{Fe}-2\text{S}]\} \rightarrow \{[4\text{Fe}-4\text{S}] \rightarrow \text{siroheme}\} \rightarrow \text{NO}_2^-$$
$$\text{NO}_2^- \text{ reductase (NirA)}$$

However, the NiR from *Plectonema boryanum* contains a C-terminal domain extension, similar to plant-type [2Fe–2S]-containing Fds, probably involved in the intramolecular electron flow to the [4Fe–4S] cluster [30].

A Fd-dependent NiR has also been purified from the haloarchaeon *H. mediterranei* [31]. The enzyme is a 66-kDa monomer that contains siroheme and a [4Fe–4S] center, exhibits an apparent K_m for NO_2^- of 8 mM, and shows maximal activity at 60°C with 3.3 M NaCl [31].

17.4.2 *NADH-dependent NiRs*

Most information about NADH-NiRs of heterotrophic bacteria, such as *K. oxytoca* and *B. subtilis*, has been deduced by DNA sequence analysis or the biochemical studies carried out with the NAD(P)H-dependent enzyme of the fungus *N. crassa* and the NADH-NiR of *E. coli*. The bacterial NADH-NiRs also share a high C-terminal sequence similarity with SO_3^{2-} reductases. It is worth noting that the *E. coli* NiR is not a proper assimilatory enzyme, although uses NADH as electron donor and shows the biochemical characteristics of the assimilatory siroheme-containing NiRs. However, the enzyme is only induced under anaerobic conditions and may be involved in detoxification of the NO_2^- formed during NO_3^- respiration rather than in aerobic NO_3^- assimilation [4, 32]. It is also interesting to note that the NADH-NiR from *E. coli* consists of two polypeptides: a large catalytic subunit (NirB) and a small subunit (NirD) that is highly similar to the carboxyl termini of the monomeric NiRs of *Klebsiella* (NasB protein) and fungi (NiiA protein). The NiR from *B. subtilis* is also composed of two subunits, NasDE, homologous to the *E. coli* NirBD proteins [4, 9], and genes coding for NirD-like proteins are also present in the available genome sequences of many heterotrophic bacteria.

The NADH-NiRs contain noncovalently bound FAD, a [4Fe–4S] cluster, and siroheme as prosthetic groups. The FAD- and NADH-binding sites are located in an N-terminal extension that is not present in the Fd-dependent NiR. Two additional [2Fe–2S] centers could also be present in the proteins of *K. oxytoca*, *B. subtilis*, and *E. coli*, as suggested by the presence of tandem NifU-like motifs in the deduced sequence of these proteins [4]. Therefore, the following linear scheme of electron flow from NADH to NO_2^- is predicted for these enzymes:

$$\text{NADH} \rightarrow \{\text{FAD} \rightarrow [\text{2Fe}-\text{2S}]? \rightarrow [\text{4Fe}-\text{4S}] \rightarrow \text{siroheme}\} \rightarrow NO_2^-$$
$$NO_2^- \text{ reductase (catalytic subunit)}$$

The NO_3^- assimilation *nas* gene region of *R. capsulatus* strain E1F1 contains two putative NiR genes: the *nasB* gene product shows high similarity to the large catalytic subunit of the NADH-NiR of *K. oxytoca*, *E. coli* and other bacteria, whereas the *nirD* gene product is homologous to the *E. coli* NirD protein [12, 13]. The purification and biochemical characterization of the NiR of this phototrophic bacterium revealed it as an NADH-dependent enzyme bearing FAD, siroheme, and labile Fe and S similar to the NiRs from nonphotosynthetic organisms (M.F. Olmo-Mira et al., unpublished).

17.5 Regulation of NO_3^- assimilation

NO_3^- assimilation in bacteria is usually regulated by NO_3^- and/or NO_2^- induction (pathway-specific control) and NH_4^+ repression (general N control). However, the regulatory proteins and mechanisms involved in both types of control are different depending on the organism. In cyanobacteria, the global N control is exerted by NtcA, a transcription factor of the CAP family, whereas in the pathway-specific control may participate NtcB, a regulator of the LysR family of bacterial transcription factors, and other regulators. On the other hand, in most heterotrophic bacteria the general N control takes place by the Ntr system. In *K. oxytoca*, the specific NO_3^-/ NO_2^- induction occurs by a transcription antitermination mechanism in which a single antiterminator protein (NasR) participates, whereas in *A. vinelandii* this mechanism requires two components, a NO_3^- sensor (NasS) and an antiterminator protein (NasT) [4]. Significant differences also exist in the regulation of NO_3^- assimilation in *B. subtilis* and *R. capsulatus*.

17.5.1 Control of NO_3^- assimilation in cyanobacteria

Expression of NO_3^- assimilation genes is subjected to repression by NH_4^+ (the so-called N control) in all cyanobacteria investigated to date, whereas induction by NO_3^- is observed in the N_2-fixing, heterocyst-forming cyanobacteria (such as *Anabaena* PCC 7120), but not in unicellular, non-N_2-fixing cyanobacteria like *S. elongatus*. The expression of NO_3^- assimilation genes requires the NtcA transcription factor, which mediates N control [3]. NtcA activates transcription when the cellular C to N balance is high, which in cyanobacteria is signaled by increased 2-oxoglutarate levels, and 2-oxoglutarate has been shown to increase the binding affinity of NtcA for some NtcA-dependent promoters [33, 34]. NtcA binds to a DNA binding site with consensus sequence $GTAN_8TAC$, which is located about 22 nucleotides upstream of the promoter −10 hexamer TAN_3T in most NtcA-dependent promoters [35], like those of the *nir* operon in *Anabaena* PCC 7120 or *S. elongatus* [3].

In cyanobacteria, the C to N balance is determined by the level of available CO_2 and by the N source, usually NH_4^+, NO_3^- or N_2, with the C to N balance being lowest in the presence of NH_4^+ for air levels of CO_2. In non-N_2-fixing cyanobacteria, in the absence of a source of combined N (conditions under which an extreme N deficiency may be reached), the P_{II} protein (*glnB* gene product) is required to attain high levels of expression of NtcA-dependent genes like the NO_3^- assimilation genes [36, 37]. The P_{II} protein is an important 2-oxoglutarate sensor [38], but the molecular mechanism through which it influences the NtcA-dependent transcription is unknown (see Chapter 9).

The NtcB transcription factor positively influences the expression of NO_3^- assimilation genes at different stringency levels in different cyanobacteria [3]. NtcB is strictly required for expression of the *nir* operon in *Anabaena* PCC 7120, it is not required but has a clear positive effect on transcription of NO_3^- assimilation genes in *Synechocystis* PCC 6803, and it has a demonstrable effect under conditions of inhibition of the assimilation of NH_4^+ in *S. elongatus* (reviewed in [3]). In *Anabaena* PCC 7120, NtcB binds to the *nir* operon promoter just upstream from NtcA and both transcription factors can bind simultaneously to the promoter [39]. In this strain, requirement of NtcB is independent of the presence of NO_3^-. On the other hand, in the case of the *Synechococcus* and *Synechocystis* strains, NtcB appears to mediate a NO_2^- induction effect on the expression of NO_3^- assimilation genes [40].

At least two other regulators influence NO_3^- assimilation in cyanobacteria. CnaT, which is homologous to proteins with glycosyl transferase activity, is required for *nir* operon expression in *Anabaena* PCC 7120 [41]. NirB is needed to attain high levels of NiR in *S. elongatus* [42]. The molecular mechanisms through which these proteins affect NO_3^- assimilation gene expression or enzyme activity remain to be elucidated.

To summarize, expression of the capability to assimilate NO_3^- in cyanobacteria is subjected to a complex regulation. A strict requirement for activation of gene expression by NtcA (which mediates N control) is generally found, whereas a variably dependence on NO_3^-, NtcB, and possibly other regulators (like CnaT or NirB) is observed in different cyanobacterial strains. Additionally, the NO_3^- assimilation system is subjected to post-translational regulation. When NO_3^--grown cyanobacterial cells are supplemented with NH_4^+, they quickly lose the capacity to take up NO_3^- that is restored once NH_4^+ is exhausted from the medium [43]. This effect is mediated by the P_{II} protein, which in its nonphosphorylated form [38] appears to inhibit the NO_3^-- NO_2^- transporter [44]. The NrtA-like C-terminal extension of NrtC (see above) might be a target for P_{II} [17].

17.5.2 Control of NO_3^- assimilation in heterotrophic and anoxygenic phototrophic bacteria

The regulation of the *B. subtilis* NO_3^- assimilation genes is still emerging and is rather complex. This bacterium has two different NRs (the cytoplasmic assimilatory NasBC enzyme and the membrane-bound respiratory NarGHI complex), but only one NiR (the soluble NADH-dependent enzyme encoded by the *nasDE* genes described above). Therefore, this NiR is required for both anaerobic respiratory and aerobic assimilatory processes. Because the NR *nasBC* genes and the NiR *nasDEF* genes constitute an operon (Figure 17-2), transcription of the *nasDEF* genes can take place from the *nas* promoter in front of *nasB* or from an internal promoter between *nasC* and *nasD* [45]. Transcription from both promoters is activated under aerobiosis and N limitation by the general N regulator TnrA, a transcription factor that also activates the expression of other N assimilation operons in *B. subtilis* (it should be noted that this bacterium has no known orthologue of the Ntr system). Under anaerobic conditions, expression of the *nasDEF* genes from the internal promoter is induced in the presence of nitrite. The ResDE two-component regulatory system is required for anaerobic induction of the *nasDEF* genes and the expression of the *narGHI* genes of respiratory NR [45].

Expression of the *nas* operon in *K. oxytoca* is subjected to dual control by NH_4^+ and NO_3^- or nitrite. The general N control is mediated by the Ntr system, which regulates multiple N assimilation operons in enterobacteria, and the pathway-specific NO_3^- or NO_2^- induction is elicited by the transcription antitermination protein NasR [4, 8]. The *nasR* gene is located upstream from the *nasFEDCBA* operon (Figure 17-2) and codes for a protein that shows similarity with the AmiR protein of *Pseudomonas aeruginosa*, an activator of the amidase catabolism operon that acts by a transcription antitermination mechanism. Both proteins, NasR and AmiR, possess the so-called ANTAR motif for directly binding to leader sequences in the mRNA [4, 8]. Whereas the AmiR protein requires the AmiC protein for the negative regulation of its transcription antitermination function, the *Klebsiella* NasR protein does not involve an AmiC-like partner for the response to NO_3^- and NO_2^-. Thus, in response to N limitation the *Klebsiella* Ntr system activates the expression of the *nasF* operon, but transcription terminates in the *nasF* leader sequence. When NO_3^- or NO_2^- is present, the

NasR protein binds to the *nasF* leader sequence in the mRNA and exerts an antitermination effect, allowing the full expression of the *nasF* operon [4, 8].

In *A. vinelandii* the expression of the *nasAB* operon encoding the assimilatory NiR and NR is also subjected to a dual control, the general N regulation by the Ntr system and the specific NO_3^- or NO_2^- induction that operates also by a transcription antitermination mechanism [4, 11]. Two regulatory genes, *nasS* and *nasT*, are required for the NO_3^- induction of the *nasAB* genes. The *nasT* gene product is homologous to the AmiR and NasR proteins and possesses the ANTAR domain for binding to the mRNA. The *nasS* gene product has significant similarity with the NrtA or NasF proteins, the periplasmic substrate-binding components of the ABC-type NO_3^- transporters, although lacks the signal leader sequence for the translocation to the periplasm. NasS is required for negative regulation [11] and could play a similar role to the AmiC protein. Therefore, it has been suggested that NasS is a NO_3^- sensor that, in absence of NO_3^-, inhibits the transcription antitermination effect of NasT. When NO_3^- is present, the NasS protein is unable to block NasT, and this transcription antiterminator binds to the leader mRNA sequence to allow the complete transcription of the *nasAB* genes [4]. It is worth noting that genes homologous to *nasT* and *nasS* are present in the genomes of many bacteria, suggesting that this regulatory mechanism could be frequent in bacteria. However, in many cases the putative *nasS* genes are described as additional copies of the *ntrA* / *nasF* NO_3^- transport gene. As a general rule, the products of the *ntrA* / *nasF* genes involved in NO_3^- transport contain a signal peptide for translocation to the periplasm, whereas the putative regulatory NasS NO_3^- sensor proteins lack this pre-sequence.

The NO_3^- assimilation system of *R. capsulatus* strain E1F1 is induced by NO_3^- or NO_2^- and repressed in the presence of NH_4^+, which also inhibits – rapidly and completely – the transport of NO_3^-. Three genes are probably involved in the regulation of expression of the NO_3^- assimilation *nas* gene region: the *nsrR* gene, which encodes a putative nitrite-sensitive transcription repressor of the Rrf2 family, and the *nasTS* genes that could mediate a transcription antitermination mechanism similar to that described for *A. vinelandii* [12, 13]. However, the *R. capsulatus nasS* gene has a premature stop codon and a complete NasS protein can be synthesized only if a translational −1 frameshift takes place in an mRNA sequence coding for five almost consecutive arginine residues (an RRRRKR sequence). This RNA sequence is located upstream from a hairpin secondary structure that includes the premature UGA stop codon. The −1 frameshift could be induced under N limitation, which probably causes a drop in the arginine pool that favors the ribosome to pause and slippage in this sequence [13]. This would allow the synthesis of a whole NasS protein (NO_3^- sensor) under N limitation. In the absence of NO_3^-, the transcription antiterminator NasT is blocked by NasS and the *nas* genes are not expressed. In the presence of

NO_3^-, the NasS protein is unable to inhibit NasT and this protein acts as an antiterminator to allow *nas* gene expression. In addition, the *R. capsulatus* NsrR protein could repress NO_3^- assimilation in the absence of nitrite, but in the presence of this compound, NsrR is inactivated and the *nas* genes are expressed. Therefore, *nas* gene expression is induced by NO_3^- or NO_2^- only in the absence of NH_4^+, without the involvement of the Ntr system [13]. However, the occurrence of this N-dependent -1 frameshift and its possible role in the control of *R. capsulatus nas* gene expression need to be confirmed experimentally.

17.6 Concluding remarks

In bacteria, the NO_3^- assimilation system exhibits a universal architectural design with some variations in specific bacterial groups or strains. Basically, bacteria assimilate NO_3^- with the concourse of an uptake transporter, which is necessary to translocate low concentrations of NO_3^- through the cytoplasmic membrane and concentrate it in the cytoplasm, and a reduction system that renders NH_4^+. This system consists of two enzymes, NR and NiR. Whereas two different types of membrane transporters, an ABC-type and an MFS-type, can do the uptake work, the basic structures for reduction of NO_3^-, a molybdenum cofactor in the form Mo-*bis*-MGD, and nitrite, siroheme, are universal in bacterial NR and NiR, respectively. Although the immediate intramolecular electron donor is a [4Fe–4S] cluster in both cases, the physiological donor can be a Fd or a Fld, which can transfer electrons directly to the reductases, or NADH, which requires an additional intermediate enzyme, a diaphorase, to transfer the electrons to the reductases.

Also in the regulation of gene expression a general feature is evident: the NO_3^- assimilation system is repressed by growth of bacteria in the presence of NH_4^+. The general N-control system of each particular bacterium (for instance, the Ntr regulatory system or, in cyanobacteria, NtcA) usually mediates this regulation. On the other hand, pathway-specific regulation is the most variable aspect of bacterial NO_3^- assimilation. Induction by NO_3^- or NO_2^- operates in many bacteria, but it cannot be considered universal. Additionally, different mechanisms appear to be involved in this regulation, among which an antitermination regulation of transcription has been well characterized in *K. oxytoca*. Pathway-specific regulation is clearly an aspect of bacterial NO_3^- assimilation that deserves further investigation.

References

[1] M.G. Guerrero, J.M. Vega, M. Losada, The assimilatory nitrate-reducing system and its regulation, Ann. Rev. Plant Physiol. 32 (1981) 169–204.

[2] C. Moreno-Vivián, P. Cabello, M. Martínez-Luque, R. Blasco, F. Castillo, Prokaryotic nitrate reduction: molecular properties and functional distinction among bacterial nitrate reductases, J. Bacteriol. 181 (1999) 6573–6584.
[3] E. Flores, J.E. Frías, L.M. Rubio, A. Herrero, Photosynthetic nitrate assimilation in cyanobacteria, Photosynth. Res. 83 (2005) 117–133.
[4] J.T. Lin, V. Stewart, Nitrate assimilation by bacteria, Adv. Microb. Physiol. 39 (1998) 1–30.
[5] D.J. Richardson, B.C. Berks, D.A. Russell, S. Spiro, C.J. Taylor, Functional, biochemical and genetic diversity of prokaryotic nitrate reductases, Cell Mol. Life Sci. 58 (2001) 165–178.
[6] J.W.B. Moir, N.J. Wood, Nitrate and nitrite transport in bacteria, Cell Mol. Life Sci. 58 (2001) 215–224.
[7] J.T. Lin, B.S. Goldman, V. Stewart, The *nasFEDCBA* operon for nitrate and nitrite assimilation in *Klebsiella pneumoniae* M5a1, J. Bacteriol. 176 (1994) 2551–2559.
[8] J.T. Lin, V. Stewart, Nitrate and nitrite-mediated transcription antitermination control of *nasF* (nitrate assimilation) operon expression in *Klebsiella pneumoniae* M5a1, J. Mol. Biol. 256 (1996) 423–435.
[9] K.I. Ogawa, E. Akagawa, K. Yamane, Z.W. Sun, M. LaCelle, P. Zuber, M.M. Nakano, The *nasB* operon and the *nasA* gene are required for nitrate/nitrite assimilation in *Bacillus subtilis*, J. Bacteriol. 177 (1995) 1409–1413.
[10] F. Ramos, G. Blanco, J.C. Gutiérrez, F. Luque, M. Tortolero, Identification of an operon involved in the assimilatory nitrate-reducing system of *Azotobacter vinelandii*, Mol. Microbiol. 8 (1993) 1145–1153.
[11] J.C. Gutiérrez, F. Ramos, L. Ortner, M. Tortolero, *nasST*, two genes involved in the induction of the assimilatory nitrite-nitrate reductase operon (*nasAB*) of *Azotobacter vinelandii*, Mol. Microbiol. 18 (1995) 579–591.
[12] P. Cabello, C. Pino, M.F. Olmo-Mira, F. Castillo, M.D. Roldán, C. Moreno-Vivián, Hydroxylamine assimilation by *Rhodobacter capsulatus* E1F1. Requirement of the *hcp* gene (hybrid cluster protein) located in the nitrate assimilation *nas* gene region for hydroxylamine reduction, J. Biol. Chem. 279 (2004) 45485–45494.
[13] C. Pino, M.F. Olmo-Mira, P. Cabello, M. Martínez-Luque, F. Castillo, M.D. Roldán, C. Moreno-Vivián, The assimilatory nitrate reduction (Nas) system of the phototrophic bacterium *Rhodobacter capsulatus* E1F1, Biochem. Soc. Trans. 34 (2006) 127–129.
[14] W. Busch, M.H. Saier, Jr., The transporter classification (TC) system, 2002, Crit. Rev. Biochem. Mol. Biol. 37 (2002) 287–337.
[15] Q. Wu, V. Stewart, NasFED proteins mediate assimilatory nitrate and nitrite transport in *Klebsiella oxytoca* (*pneumoniae*) M5al, J. Bacteriol. 180 (1998) 1311–1322.

[16] S.I. Maeda, T. Omata, Substrate-binding lipoprotein of the cyanobacterium *Synechococcus* sp. strain PCC 7942 involved in the transport of nitrate and nitrite, J. Biol. Chem. 272 (1997) 3036–3041.

[17] M. Kobayashi, R. Rodríguez, C. Lara, T. Omata, Involvement of the C-terminal domain of an ATP-binding subunit in the regulation of the ABC-type nitrate/nitrite transporter of the cyanobacterium *Synechococcus* sp. strain PCC 7942, J. Biol. Chem. 272 (1997) 27197–27201.

[18] T. Sakamoto, K. Inoue-Sakamoto, D.A. Bryant, A novel nitrate/nitrite permease in the marine cyanobacterium *Synechococcus* sp. strain PCC 7002, J. Bacteriol. 181 (1999) 7363–7372.

[19] Q. Wang, H. Li, A.F. Post, Nitrate assimilation genes of the marine diazotrophic, filamentous cyanobacterium *Trichodesmium* sp. strain WH9601, J. Bacteriol. 182 (2000) 1764–1767.

[20] B.G. Forde, Nitrate transporters in plants: structure, function and regulation, Biochim. Biophys. Acta 1465 (2000) 219–235.

[21] R. Gangeswaran, R.A. Eady, Flavodoxin 1 of *Azotobacter vinelandii*: characterization and role in electron donation to purified nitrate reductase, Biochem. J. 317 (1996) 103–108.

[22] R.M. Martínez-Espinosa, F.C. Marhuenda-Egea, M.J. Bonete, Nitrate reductase from the haloarchaeon *Haloferax mediterraneii*: purification and characterization, FEMS Microbiol. Lett. 204 (2001) 381–385.

[23] R. Blasco, F. Castillo, M.M. Martínez-Luque, The nitrate reductase from the phototrophic bacterium *Rhodobacter capsulatus* E1F1 is a flavoprotein, FEBS Lett. 414 (1997) 45–49.

[24] J.F. Stolz, P. Basu, Evolution of nitrate reductase: molecular and structural variations on a common function, ChemBioChem 3 (2002) 198–206.

[25] L.M. Rubio, E. Flores, A. Herrero, Purification, cofactor analysis and site-directed mutagenesis of *Synechococcus* ferredoxin-nitrate reductase, Photosynth. Res. 72 (2002) 13–26.

[26] B.J.N. Jepson, L.J. Anderson, L.M. Rubio, C.J. Taylor, C.S. Butler, E. Flores, A. Herrero, J.N. Butt, D.J. Richardson, Tuning a nitrate reductase for function. The first spectropotentiometric characterization of a bacterial nitrate reductase reveals novel redox properties, J. Biol. Chem. 279 (2004) 32212–32218.

[27] R. Gangeswaran, D.J. Lowe, R.A. Eady, Purification and characterization of the nitrate reductase of *Azotobacter vinelandii*, Biochem. J. 289 (1993) 335–342.

[28] I. Luque, E. Flores, A. Herrero, Nitrite reductase gene from *Synechococcus* sp. PCC 7942: homology between cyanobacterial and higher-plant nitrite reductases, Plant Mol. Biol. 21 (1993) 1201–1205.

[29] S. Kuznetsova, D.B. Knaff, M. Hirasawa, B. Lagoutte, P. Setif, Mechanism of spinach chloroplast ferredoxin-dependent nitrite

reductase: spectroscopic evidence for intermediate states, Biochemistry 43 (2004) 510–517.
[30] I. Suzuki, H. Kikuchi, S. Nakanishi, Y. Fujita, T. Sugiyama, T. Omata, A novel nitrite reductase gene from the cyanobacterium *Plectonema boryanum*, J. Bacteriol. 177 (1995) 6137–6143.
[31] R.M. Martínez-Espinosa, F.C. Marhuenda-Egea, M.J. Bonete, Purification and characterization of a possible nitrite reductase from the halophile archaeon *Haloferax mediterraneii*, FEMS Microbiol. Lett. 196 (2001) 113–118.
[32] J. Cole, Nitrate reduction to ammonia by enteric bacteria: redundancy or a strategy for survival during oxygen starvations? FEMS Microbiol. Lett. 136 (1996) 1–11.
[33] R. Tanigawa, M. Shirokane, S.I. Maeda, T. Omata, K. Tanaka, H. Takahashi, Transcriptional activation of NtcA-dependent promoters of *Synechococcus* sp. PCC 7942 by 2-oxoglutarate *in vitro*, Proc. Natl. Acad. Sci. USA 99 (2002) 4251–4255.
[34] M.F. Vázquez-Bermúdez, A. Herrero, E. Flores, 2-Oxoglutarate increases the binding affinity of the NtcA (nitrogen control) transcription factor for the *Synechococcus glnA* promoter, FEBS Lett. 512 (2002) 71–74.
[35] A. Herrero, A.M. Muro-Pastor, E. Flores, Nitrogen control in cyanobacteria, J. Bacteriol. 183 (2001) 411–425.
[36] M.F. Aldehni, J. Sauer, C. Spielhaupter, R. Schmid, K. Forchhammer, Signal transduction protein P_{II} is required for NtcA-regulated gene expression during nitrogen deprivation in the cyanobacterium *Synechococcus elongatus* strain PCC 7942, J. Bacteriol. 185 (2003) 2582–2591.
[37] J. Paz-Yepes, E. Flores, A. Herrero, Transcriptional effects of the signal transduction protein P_{II} (*glnB* gene product) on NtcA-dependent genes in *Synechococcus* sp. PCC 7942, FEBS Lett. 543 (2003) 42–46.
[38] K. Forchhammer, Global carbon/nitrogen control by P_{II} signal transduction in cyanobacteria: from signals to targets, FEMS Microbiol. Rev. 28 (2004) 319–333.
[39] J.E. Frías, E. Flores, A. Herrero, Activation of the *Anabaena nir* operon promoter requires both NtcA (CAP family) and NtcB (LysR family) transcription factors, Mol. Microbiol. 38 (2000) 613–625.
[40] M. Aichi, N. Takatani, T. Omata, Role of NtcB in activation of nitrate assimilation genes in the cyanobacterium *Synechocystis* sp. strain PCC 6803, J. Bacteriol. 183 (2001) 5840–584.
[41] J.E. Frías, A. Herrero, E. Flores, Open reading frame *all0601* from *Anabaena* sp. strain PCC 7120 represents a novel gene, *cnaT*, required for expression of the nitrate assimilation *nir* operon, J. Bacteriol. 185 (2003) 5037–5044.

[42] I. Suzuki, N. Horie, T. Sugiyama, T. Omata, Identification and characterization of two nitrogen-regulated genes of the cyanobacterium *Synechococcus* sp. strain PCC7942 required for maximum efficiency of N assimilation, J. Bacteriol. 177 (1995) 290–296.
[43] E. Flores, M.G. Guerrero, M. Losada, Short-term ammonium inhibition of nitrate utilization by *Anacystis nidulans* and other cyanobacteria, Arch. Microbiol. 128 (1980) 137–144.
[44] H.M. Lee, E. Flores, K. Forchhammer, A., Herrero, N. Tandeau de Marsac, Phosphorylation of the signal transducer P_{II} protein and an additional effector are required for the P_{II}-mediated regulation of nitrate and nitrite uptake in the cyanobacterium *Synechococcus* sp. PCC 7942, Eur. J. Biochem. 267 (2000) 591–600.
[45] M.M. Nakano, T. Hoffmann, Y. Zhu, D. Jahn, Nitrogen and oxygen regulation of *Bacillus subtilis nasDEF* encoding NADH-dependent nitrite reductase by TnrA and ResDE, J. Bacteriol. 180 (1998) 5344–5350.

Chapter 18

Nitrate Assimilation in Plants

Rudolf Tischner and Werner Kaiser

18.1 Introduction

Plants require greater amounts of fixed-nitrogen (N) than any other mineral nutrient. Due to large N losses by leaching and volatilization, soil N generally limits plant growth in both natural and agricultural ecosystems. To cope with the low N supply, plants developed sensitive and selective uptake systems and the flexibility to grow on different N sources. They also developed a regulatory network to tightly coordinate N assimilation with C- and S-assimilation for optimal growth. The major N sources include: (i) NH_4^+, which is used by several forest species, especially in low-pH soils, and by paddy rice; (ii) amino acids, which are consumed by plants in areas where nitrification and mineralization are limited by the climate (e.g., in arctic regions); and (iii) NO_3^-, which is the most commonly utilized N compound.

Nitrate taken up by the roots is either reduced or stored in the vacuoles or is translocated to the shoot for subsequent reduction and vacuolar storage; it is also used for osmoregulation. Reduction of NO_3^- to NO_2^- is catalyzed in the cytosol by assimilatory nitrate reductase (NR). NO_2^-

Biology of the Nitrogen Cycle
Edited by H. Bothe, S.J. Ferguson and W.E. Newton

enters the plastid (or chloroplast in the shoot) to be reduced to NH_4^+ by assimilatory nitrite reductase (NiR). NH_4^+ is then assimilated by the GS/GOGAT (glutamine synthetase/glutamate-oxoglutarate aminotransferase) pathway into glutamine/glutamate, which serve as substrates for transamination reactions to produce all of the other proteinaceous amino acids. This chapter focuses on the physiological and molecular aspects of NO_3^- uptake and reduction. For other reviews, consult the list of references [1, 2].

18.2 Physiology and molecular biology of nitrate uptake

18.2.1 Nitrate sensing

18.2.1.1 General

Nitrate acts as both a nutrient and a signal for the initiation of various processes [2]. It triggers the induction of NO_3^--assimilating enzymes, but also shifts carbohydrate metabolism from starch synthesis to increased sucrose synthesis. Sucrose metabolism results in the production of organic acids, like oxoglutarate, which acts as an acceptor for reduced N in the GS/GOGAT pathway, or like malate, which acts as a counter-ion in NO_3^- uptake and reduction. Experiments on NO_3^- sensing are rare, but the effects of very low concentrations of NO_3^- both on the expression of NR and on NO_3^--uptake systems have been studied [3]. NR expression was evaluated by the use of nitrate pulses on N-free cultivated barley [4], where 120 nmol NO_3^- per g fresh weight of root was sufficient to induce NR mRNA synthesis.

In *E. coli*, the P_{II} protein is part of the signaling chain that controls the expression of the NO_3^--assimilation pathway. The P_{II} protein might sense either glutamine (through the action of an uridydyl-transferase/uridydyl-removing enzyme) or 2-oxoglutarate (by the P_{II} protein directly). In *Synechococcus* sp. PC 7942, the nonphosphorylated form of P_{II} protein is involved in the short-term inhibition of NO_3^- uptake by NH_4^+ and also in the coordination of C and N metabolism [5]. Similarly, in *Arabidopsis* and castor bean, a P_{II}-like protein (encoded by the gene *GLB1*) might be involved in sensing the C/N relationship [6, 7]. The expression of this P_{II}-like protein, which is located in the chloroplast, is affected by light, reduced N-compounds, and sucrose and so direct control of NO_3^- sensing by this protein seems unlikely. However, the protein may be part of a signaling chain leading to feedback control of NO_3^- uptake.

18.2.1.2 Nitrate is involved in gene regulation

In *Arabidopsis*, a "NO_3^- box" [8] is present in the promoter region of many NO_3^--regulated genes, including the *NRT2* genes, which encode high-affinity NO_3^- transporters [9]. The product of the "NO_3^- box" is believed to bind to the promoters of the NR genes *NIA1* and *NIA2* because a mutation in the "NO_3^- box" abolished induction of these promoters. Furthermore, the barley-gene promoter, p*ANR1*, is also NO_3^- inducible and belongs to the MADS family of transcription factors, which are involved in nutrient-induced changes in root architecture [10]. Its expression is specific for both roots and NO_3^-, and so is consistent with this gene being a prerequisite for the modulation of lateral root proliferation by NO_3^-.

18.2.1.3 Nitrate also appears to be involved in cytokinin production

In maize leaves, a response regulator homologue (ZmCip1) [11], which appears involved in the N-signaling pathway, is induced by added cytokinin. When NO_3^- was supplied to roots, but not when supplied to excised leaves, the accumulation of isopentyladenosine, a cytokinin precursor [11], increased; the amounts of both the ZmCip1 transcript and its corresponding protein also increased. This result suggests a NO_3^--sensing mechanism in the roots that leads to cytokinin synthesis and its subsequent transport to the shoot. In *Arabidopsis*, addition of either cytokinin to the leaves or NO_3^- to the medium increased the expression of the *ARR3-ARR7* genes [12], which are likely involved in inorganic-N signal transduction.

18.2.1.4 But how might NO_3^- be sensed?

One suggestion is that the membrane-bound form of NR, which is located on the outer side of the plasma membrane [13], is a NO_3^- sensor. This idea was later extended to show how the signal might be transmitted, when a novel plasma membrane-bound enzyme was shown to produce NO from external NO_3^- [14]. Nitric oxide, which is readily membrane permeable, is believed to be the trigger that induces the expression of NO_3^--uptake systems. This suggestion is consistent with the known induction of expression of the *Chlorella* NO_3^--transporter gene (*ChNTR2.1*) by the NO donor, sodium nitroprusside [15].

In summary, few genes, which code for either a sensor protein or transcription factors, have been detected so far. Further research is required to identify the individual steps of the NO_3^- signaling chain. Determining the location of the P_{II} analogue and the expression of the corresponding gene appears essential in this context.

18.2.2 Nitrate uptake

Plants, which have never been in contact with NO_3^- before, take up NO_3^- immediately upon exposure to it. NO_3^- uptake increases even more when plants are continuously exposed to NO_3^-. These results indicate the existence of two separate NO_3^--uptake systems, termed the constitutive high-affinity transport system (cHATS) and the inducible high-affinity transport (iHATS) system. The residual HATS activity found in an

Arabidopsis mutant lacking *AtNRT2.1* and *AtNRT2.2* (both coding for iHATS) confirmed the existence of cHATS [16]. The difference between the inducible system and the constitutive system lies in their V_{max}, which depends on the absolute number of NO_3^--carriers. In addition to the two HATS, a low-affinity transport system (LATS) has also been suggested [17]. This LATS does not show saturation kinetics, but is likely also to involve H^+-cotransport with NO_3^- as with HATS. Recently, LATS has been suggested to make a minor contribution (due to its low affinity) to overall net NO_3^- uptake at low NO_3^- concentrations. Thus, at low NO_3^- concentrations both systems are active, with HATS being by far the most active, whereas at high NO_3^- concentrations, the contribution of LATS is high because HATS is already saturated. When a T-DNA mutant of *Arabidopsis*, where the *AtNRT2.1* gene was completely deleted and the 3′-region of *AtNRT2.2* partially deleted [18], was complemented with *NRT2* from *Nicotiana plumbaginifolia*, NO_3^- influx was restored, indicating that these genes definitely code for proteins with a role in HATS.

NO_3^--uptake capacity is not equally distributed along the root axis and is not identical in roots of different age or ontogeny [19]. A high uptake rate, together with the highest NR activity, occurs at the root tip [19]. Older root parts are more active in NO_3^- uptake but NR activity is low, possibly indicating a higher rate of nitrate translocation to the shoot from these root parts. This suggests that HATS is likely located close to the root tip, whereas LATS would be present in older root parts. A different HATS may occur in root tips [21] as compared to mature roots, because *NRT2.1*, which codes for iHATS, was not expressed in younger root parts. Although NO_3^--uptake rates depend on the NO_3^- concentrations in the surrounding medium, xylem loading depends on the shoot's demand and on reduced N-compounds in the phloem.

NO_3^- uptake is subject to feedback regulation mainly by glutamine [22], which was found to inhibit HATS, but not LATS, in *Lolium perenne* [23]. Although the inhibition of NO_3^- uptake was attributed to NH_4^+ [17], glutamine as the inhibitor could not be excluded. However, inhibition of glutamine synthetase by methionine sulphoximine did not prevent the inhibitory effect of ammonium [17, 24]. Further, the expression of one of the *Arabidopsis* genes (*NRT2.1*; coding for iHATS) was downregulated by the inhibition of either glutamate-synthase or glutamine-synthetase activity [25]. The first result points to glutamine, but the second to NH_4^+, as repressor of *NRT2.1*. With *Arabidopsis* plants starved for NO_3^- but continuously supplied with NH_4^+ for 3 days, the induction of net NO_3^- uptake occurred after refeeding with 250 μM NO_3^-. The same result occurred when glutamine was fed during NO_3^- starvation (R. Tischner, unpublished). The effects of reduced N-compounds on the induction of NO_3^--uptake systems are complicated by the fact that several genes, e.g., at least seven in *Arabidopsis*, code for NO_3^- transporters and the expression of these genes may respond differentially to added nitrogen compounds, the developmental stage, and environmental conditions.

In fungi, NO_3^- uptake is affected by the presence of NR [26]. In plant cells, strong evidence exists for glutamine and other amino compounds [27, 28] being responsible for decreasing the NO_3^--uptake rate but not the accumulation of a specific amino acid [29, 30]. The supply of either glutamine or aspartic acid to beech via the phloem [27] resulted in NO_3^- accumulation in the fine roots and also decreased NO_3^- uptake, probably reflecting an adaptation to the lower N-demand. Also, NO_3^- itself may inhibit its own net uptake if supplied either in high concentrations [17] or for long time. This "substrate inhibition" occurs when nitrate uptake matches the growth rate, and the storage pools are overloaded. The sink-strength, determined by the C/N relationship, affects NO_3^- uptake. In Chenopodiaceae, the vacuoles may contain up to 100 mM NO_3^-, but with a cytosolic concentration of only 5 mM, without a negative effect on the NO_3^--uptake rate [31]. A similar absence of feedback inhibition due to a high NO_3^- content was also reported for mutants or transgenic plants with altered NR expression [32].

18.2.3 Mechanism of NO_3^- uptake

Nitrate uptake usually involves H^+/nitrate cotransport as first demonstrated for *Lemna* [33] and subsequently for other organisms [34–36]. The commonly observed alkalization of the medium during NO_3^- uptake is explained by a 2:1 rather than a 1:1 stoichiometry of $H^+NO_3^-$ during cotransport. However, variations of this basic type of cotransport exist. For cyanobacteria [37] and for a marine diatom *Coscinodiscus wailesii* [38], Na^+/ NO_3^- cotransport is likely [37]. In *Zostera marina*, a Na^+/ NO_3^- transport system may also include P_i uptake [39]. NO_3^- uptake is always stimulated by cations ($K^+ \gg Mg^{2+} > Na^+ > Ca^{2+}$); they affect V_{max}, but not K_m, of the uptake system [40]. A greater depolarization at the plasma membrane and smaller changes in cytosolic pH occurred when K^+ accompanied NO_3^- uptake [41]. These results may be due to effects on the plasma membrane-bound ATPase. Lysine residues have been found to play a crucial role in the activity of the tonoplast electro-neutral NO_3^-/H^+ antiporter [42].

18.2.4 Molecular studies on NO_3^--uptake systems

NO_3^--transport proteins from both eukaryotic and prokaryotic organisms belong to three clearly different transporter superfamilies: (i) the major facilitator superfamily (MFS); (ii) the ATP-binding cassette superfamily; and (iii) the proton-dependent oligopeptide transporter (POT) superfamily. The first NO_3^--transporter gene, termed *CRNA*, was found in *Aspergillus nidulans* [43]. The *CRNA* locus is part of a gene cluster that includes the NR and NiR structural genes.

In *Chlamydomonas*, three genes have been cloned (*NAR2*, *NRT2;1*, and *NRT2;2*), all of which are involved in NO_3^- uptake as demonstrated by complementation of mutants with defects in this activity [44]. Complementation was achieved, however, only if *NAR2* was used in combination with either one of the other two genes, but never with any of the three genes alone nor with the combination of *NRT2;1* and *NRT2;2*. In contrast, the NO_3^--transporter gene from *Chlorella* (*ChNRT2.1*) was functionally expressed in oocytes without a requirement for a second gene [15] and only one *NRT2* gene was found in Southern blots of genomic DNA.

Both the *NRT2;1* and *NRT2;2* genes code for the typical carrier structure with 12 putative transmembrane domains. Both the intracellular C-terminal and the hydrophilic domains, which occur between the transmembrane domains 6 and 7 of members of the MFS superfamily, may be essential for function and substrate recognition [45]. However, the *NAR2* gene codes for a protein with only one transmembrane domain. These three genes in *Chlamydomonas* occur in a gene cluster that also includes *NIT1*, the gene for NR, and a gene *NAR1* of unknown function. Details of the genes encoding both NO_3^-- and NO_2^--transport proteins are available [46, 47].

The first NO_3^--transporter gene identified in higher plants was the *Arabidopsis* gene *CHL1* (now *AtNRT1;1*). Its gene product, which is involved in NO_3^- transport at low NO_3^- concentrations, has 12 membrane-spanning helices and is a typical membrane-transport protein belonging to the POT family. Expression of *CHL1* in *Xenopus* oocytes facilitated NO_3^- uptake, but the CHL1 product may also transport other anions. The location of *CHL1* in epidermal cells near the root tip and in the cortex, including endodermal cells, of mature root sections [48] confirmed its importance in NO_3^- uptake. The CHL1 protein may be a component of both the low- and high-affinity uptake systems [49], probably coding for a dual-affinity NO_3^- transporter. *CHL1* mutants show reduced growth, root elongation, and delayed flowering. Auxin may play a role in both regulation of expression and the targeting of CHL1 [50]. The expression of *CHL1* in guard cells suggests a function for NO_3^- accumulation in stomata opening [51].

Using degenerate PCR primers and starting with mRNA from NO_3^--induced barley roots as the target, a 130-bp PCR product, named *pBCRNA*, was produced [45]. Screening of a barley cDNA library led to the isolation of two clones, *pBCH1* and *pBCH2* (now *HvNRT2;1* and *HvNRT2;2*). Both genes code for proteins that belong to the MFS superfamily of transporters with 12 predicted membrane-spanning domains and both are expressed with NO_3^-, but not with NH_4^+. From Southern blot analysis, 7–10 related genes could be expected in barley roots. A functional barley NO_3^- transporter, HvNRT2;1, could only be expressed in oocytes after co-injection of a *NAR2*-related barley gene [52]. NAR2 is essential for a functional NO_3^--uptake system, but its precise role is unknown [45]. A full-length clone, named *NpNRT2.1*, was obtained from NO_3^--induced *N. plumbaginifolia* [53] after RT-PCR. The sequence obtained was homologous to *CRNA*, *CrNRT2;1*, and *CrNRT2;2*. From Southern blots, only 1–2 members of the NRT2 family can be expected in *N. plumbaginifolia*.

Arabidopsis is now known to possess both types of nitrate transporters (HATS and LATS), both of which are encoded by multigene families [54]. These genes are named *NTR1* and *NRT2*; the former group encodes LATS (note that *NRT1.1*, formerly *CHL1*, is an exception), whereas the latter codes for HATS. The *NRT2*-gene family now consists of seven genes and the *NRT1*-gene family consists of four genes [22]. The major challenge is to clarify the function of each gene and to understand how the interactions among them are orchestrated. One interaction has already been demonstrated; when *NRT2;1* was de-repressed in an *CHL1-5* mutant, it was no longer sensitive to high NH_4NO_3 concentrations [55]. This result points to a regulatory role for *CHL1* in *NRT2;1* expression in the presence of NH_4^+. It further suggests that the coordination of *CHL1* and *NRT2;1* expressions enables plants to overcome the disadvantage of pure NH_4^+ nutrition and to ensure NO_3^- uptake in the presence of NH_4^+.

Other research showed that a mutant, which is defective in both *NRT2.1* and NRT2.2 (*AtNRT2-A*) coding for iHATS [56], had decreased NO_3^- uptake and accumulation but increased glucose, fructose, sucrose, and starch content. In addition, the *NRT2.4* and *NRT2.5* were overexpressed in this mutant indicating that gene expression was controlled from inside of this gene family. The *NRT2*-gene family falls into three groups [57]. Of these, one group consists of the *NRT2.1/2.2/2.4* genes, which are preferentially expressed in roots and respond to external NO_3^- supply, whereas another group contains *NRT2.5/2.7*, which are similar to the genes from *Chlamydomonas* and *E. coli*, mainly expressed in the shoot, and may be involved in leaf NO_3^- uptake.

On the basis of an analysis of the *Arabidopsis NRT*-gene families [58] with respect to NO_3^- sensitivity and root/shoot distribution, four genes, *NRT1.1/2.1/2.2/2.4*, are mainly induced in the roots (*NRT1.1* also in the shoot), whereas *NRT1.3/1.4/2.3* are induced in both plant organs although with different distribution patterns. Further, *NRT2.1/2.2/2.4* appears to participate in iHATS formation, *NRT1.1* mainly represents LATS, and the *NRT1.3/1.4/2.3* genes may represent cLATS. Electrophysiological studies employing oocytes have provided additional evidence for a LATS. *AtNRT1.4* may have a specific role in petiole NO_3^- homeostasis [59]. Thus, although some NO_3^- transporters have been characterized, there remains a lack of information on the regulation of gene expression and on the location, structure, and processing of the native protein. Further research should also focus both on NO_3^- efflux systems (possibly based on separate genes) and on NO_3^- transporters involved in vacuole and xylem loading.

18.3 Assimilatory NO_3^- reduction in higher plants

18.3.1 Reactions catalyzed by assimilatory NR

NO_3^- reduction in higher plants, as catalyzed by NR, is a two-electron transfer from NAD(P)H to NO_3^- to yield NO_2^-. For NADH:NR, the K_m value for NADH is 1–7 μM, for NO_3^- is 20–40 μM, and the k_{cat} is 200 s^{-1} [60]. In addition to this basic reaction, higher plant (and algal) NR also catalyzes one-electron transfer either to O_2 (equation 1) or to its normal product NO_2^- (equation 2), resulting in the formation of the free radicals, superoxide anion, or nitric oxide (NO), respectively [61–63].

$$NADH + H^+ + 2O_2 \rightarrow NAD^+ + O_2^{\cdot -} + H_2O \quad (1)$$

$$NADH + H^+ + NO_2^- \rightarrow NAD^+ + NO^{\cdot} + H_2O \quad (2)$$

The rates of the two latter reactions appear to be marginal (<1%) compared to the "main reaction" [63, 64]. Nevertheless, NO production may be physiologically relevant because of its role as an almost universal signaling molecule in plants [65].

18.3.2 In higher plants, different organs contribute differentially to NO_3^- reduction

One part of assimilatory NO_3^- reduction takes place in the roots and another part in the shoot. This basic distribution of the overall process between below- and above-ground organs is highly variable and depends, among other factors, on NO_3^- supply and on the plant species. Generally, NR in higher plants is inducible (iNR), and NO_3^- is one obligatory inducer. NO_3^- from the soil is first "sensed" by roots, where the whole NO_3^-/ NO_2^--reduction machinery is induced first. If NO_3^- uptake exceeds assimilation by the roots, the excess will be transported to the shoot and leaves to induce NR and NiR there. In addition to NO_3^-, induction of NR is affected by other signals (light, carbohydrate, and phytohormones; see below).

18.3.3 Structure and types of NR in higher plants

The structure of NR was recently reviewed in detail [60]. Briefly, the holoenzyme is a dimer of two identical polypeptide chains, each of about 900 amino acids and containing the cofactors, FAD, heme-Fe, and Mo-pterin, all of which are bound into structurally independent domains. Each 100-kDa monomer has two active sites. One is located at the C-terminus (the FAD-domain) and accepts electrons from NAD(P)H and the other at the N-terminus (the Mo-pterin domain) for the reduction of bound NO_3^- to NO_2^-. Both are connected to a central heme (cyt-b) domain. Higher plant NR has three other sequence regions with no similarity to other proteins. These are an N-terminal extension preceding the Mo-pterin domain and two "hinge" domains. Hinge 1 connects the Mo-pterin and the heme domains, whereas hinge 2 connects the FAD/NADH domain with the other side of the heme. The two hinge regions are suggested to be structurally flexible [60]. Hinge 1 contains the serine residue, which is posttranslationally modified by reversible phosphorylation (see below).

NR, using NADH as electron donor (EC 1.6.1.1), is the dominant form in higher plants. A NADH/NADPH dual-specific form (EC 1.6.1.2) exists in monocots and also in some dicots, like soybean. Mono-specific NADPH:NR (EC 1.6.1.3) occurs in mosses and fungi. All NR forms are located in the cytosol of plant cells. In addition to cytosolic NR (cNR), plants possess a second NR (PM-NR), which is bound to the plasma membrane of root and shoot cells [66] and which catalyzes extracellular NO_3^- reduction. The root and leaf PM-NR differ in molecular weight (63 kDa in roots vs. 98 kDa in leaves). The root PM-NR lacks the covalently bound FAD domain and uses succinate as reductant, whereas the leaf PM-NR is primarily NADH-dependent. Levels of both proteins are responsive to NO_3^-. The root PM-NR level is highest at low NO_3^- supply, whereas the leaf PM-NR level is increased at very high NO_3^- contents. The physiological roles of the two PM-NR remain to be elucidated.

18.3.4 Regulation of NR expression and activity

18.3.4.1 Transcriptional regulation

Most higher plants contain two or more genes that code for NR. *Arabidopsis* possesses two different genes that code for NADH:NR (*NIA1* and *NIA2*), and NO_3^- is the key regulator for both (see above). Plants appear to have a highly sensitive NO_3^--detection system, of which a PM-bound NR is likely an important part (see above); the NR (and NiR) genes in roots and leaves can be induced by NO_3^- concentrations as low as 1 μM and within 15 min [67]. Recent microarray analysis suggested that many other genes are strongly NO_3^- responsive [68]. Genes for Mo-pterin synthesis are, however, constitutively expressed [69] probably because, besides NR, three other enzymes not directly involved in N-metabolism require Mo-pterin. These are xanthine dehydrogenase, aldehyde oxidase (involved in ABA synthesis), and sulphite oxidase. However, no nucleotide sequence to which a NO_3^--stimulated trans-acting factor could bind has yet been identified within a promoter of NR of higher plants. Studies of transgenic tobacco, using *NIA1* and *NIA2* probes from *Arabidopsis*, revealed NO_3^--responsive sequences in the 5′-region of both genes. NO_3^- induction of transcription appears to depend on a 12-bp conserved sequence with a core consensus of AGTCA or ACTCA (the " NO_3^- element") [70].

In addition to NO_3^-, sugars or organic acids positively affect NR expression, whereas glutamine exerts negative feedback [71]. The known positive effect of light on NR induction is usually attributed to an improved carbohydrate supply; however, a more direct light effect is mediated by the phytochrome system [72]. The only known specific transcription factor that mediates light (or sugar) effects on the expression of the *NIA* genes is the HY5 protein. The HY5 locus encodes a bZIP transcription factor that is important for expression of several genes related both to photosynthesis and to de-etiolation. In *Arabidopsis*, HY5 binds to light-responsive elements in promoters and this binding is affected by phytochromes. Details of light regulation of NR expression in higher plants can be found in Ref. [73]. Obviously, to be effective, transcriptional regulation (by whatever signal) requires that the NR protein can be rapidly degraded. Controlled proteolysis is, indeed, part of a posttranslational regulation that is unique to higher plant NR.

18.3.4.2 Posttranslational regulation

On top of the transcriptional regulation described above, higher plants are able to "fine-tune" NR activity via posttranslational phosphorylation/dephosphorylation [67]. In leaves, either light or high CO_2 rapidly activate NR [74, 75] and the response to CO_2 suggests that carbohydrates serve as positive triggers. In roots, either hypoxia or anoxia activates NR [76], and the activation of NR by anoxia can be mimicked by cellular acidification [74]. A number of different effectors are involved in the posttranslational regulation of NR. NR is more active and less phosphorylated in the light than in the dark [76, 77], and a specific Ser residue (Ser-534 in *Arabidopsis* NR2 and Ser-543 in spinach) in hinge 1 was identified as the only regulatory phosphorylation site on the NR protein. Protein kinases (PK) known to be involved in NR phosphorylation [78] mostly belong to the CDPK-type (*c*alcium-*d*ependent protein *k*inases), but one is a SNF1 (*s*ucrose-*n*on-*f*ermenting *1*-related kinase). Protein phosphatases (PP) that dephosphorylate NR are of type 2A [77; see Figure 18-1].

Phosphorylation alone is now known not to be sufficient to modulate NR activity. An "inhibitor protein," which belongs to the 14-3-3 family, must bind to phosphorylated NR (P-NR) to cause inactivation [79]. In *Arabidopsis*, the 14-3-3 family consists of 14 isoforms. Divalent metal ions bound to 14-3-3 are required for complex formation with NR [80, 81], which uses the binding motive, R-X-X-pS/pT-X-P. The 14-3-3 proteins are homodimers with two binding sites, both of which can be filled simultaneously [82]. However, when binding to NR, these sites can only be used one at a time, but two independent 14-3-3-molecules can bind to the dimeric NR. Mutations affecting the regulatory phosphorylation site result in constitutive activation of NR and in NO_2^- accumulation and increased NO emissions from the plants [83, 84], all of which confirm the importance of the posttranslational regulation mechanism. Thus, the activity of PK and PP creates a dynamic equilibrium between NR (active), P-NR (active), and (14-3-3)-P-NR (inactive), which can be rapidly shifted in response to external conditions. The ratios of NR, P-NR, and (14-3-3)-P-NR in illuminated spinach leaves have

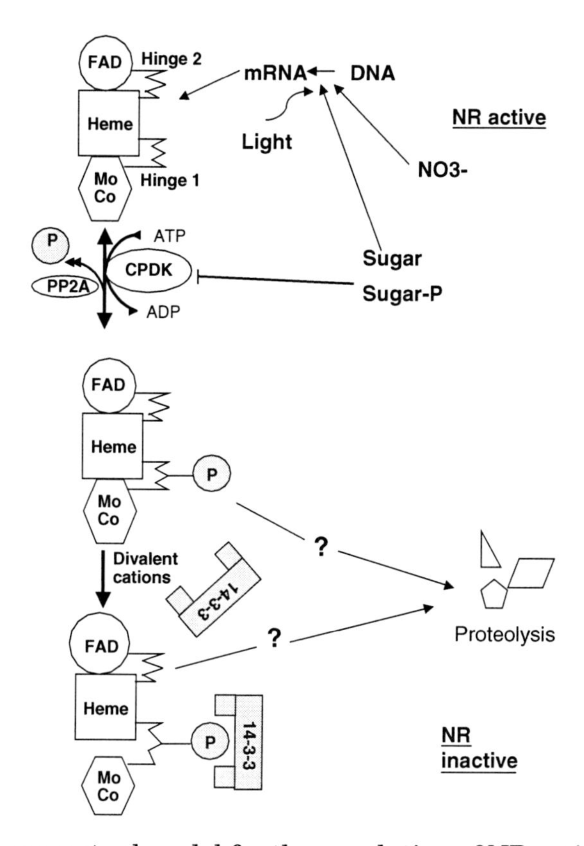

Figure 18-1 A conceptual model for the regulation of NR activity in plants. For more details and abbreviations, see text.

been estimated using site-specific antibodies and revealed that about one-third of the total NR may exist as free P-NR not bound to 14-3-3. Upon darkening, the P-NR increased and much more 14-3-3 appeared to be bound. Thus, not only the phosphorylation state of NR, but also the binding affinity of the 14-3-3 proteins to P-NR may vary [85].

It is not yet clear how the participating PK and PP enzymes are regulated in response to the above-mentioned triggers. However, hexose-phosphates are inhibitors of the CDPK (see Figure 18-1) and, therefore, cytosolic hexose-phosphate levels could be the link connecting the plants carbohydrate status with NO_3^- reduction.

How do phosphorylation and 14-3-3 binding affect NR activity? Phosphorylation and 14-3-3 binding block only one of the partial activities (the methyl viologen-dependent NR activity) of NR. Binding of 14-3-3 to the phosphorylated hinge 1 may cause a local conformational change, which inhibits electron transfer from the heme-Fe of cyt b to the Mo-pterin (see Figure 18-1). In line with this suggestion, activation or inactivation of

NR did not change the kinetic constants at all. Circumstantial evidence suggests that 14-3-3-binding also triggers NR proteolysis [85, 86], thereby controlling not only activity of the existing NR protein, but also the amount of NR protein. However, this conclusion has recently been questioned based on experiments with transgenic tobacco [83].

Altogether, it appears that, apart from the NO_3^- induction of NR, transcriptional and posttranslational regulation of NR (including NR proteolysis) are co-controlled, to a large extent, by the same triggers, namely photosynthesis, carbohydrate status, O_2 availability, and possibly pH, all of which enable plants to respond to small and rapid changes in environmental conditions. This signaling network helps to optimize investments of C, N, and energy and avoids accumulation of toxic compounds, like NO_2^-, NO, peroxynitrite, and NH_4^+.

18.3.5 Reduction of NO_2^- to NH_4^+ and glutamine/glutamate synthesis

NO_2^- produced by NR in the cytosol does not accumulate in the plant under physiological conditions, except in water-logged fields when O_2 supply is low and NO_2^- is excreted by the roots. NO_2^- is transported into the plastids via a specific membrane channel or carrier and further reduced to NH_4^+ according to equation (3).

$$\text{glutamate} + NH_4^+ + \text{ATP} \rightarrow \text{glutamine} + \text{ADP} + \text{Pi} \tag{3}$$

Equation (3) requires six electrons and is catalyzed by NiR; no intermediates are released. In chloroplasts in light, electrons are supplied by reduced ferredoxin. In darkened chloroplasts or in plastids of nongreen organs, electrons are supplied via NADPH, provided by the oxidative pentose phosphate cycle.

NiR is a 68-kDa protein that contains a siro-heme and a iron–sulphur cluster (4Fe–4S). Siro-heme is a cyclic tetrapyrol that contains a central Fe atom. Its acetate and propionate residues make it clearly distinct from other heme molecules. NiR has a high affinity for NO_2^-, thus ensuring that toxic NO_2^- does not accumulate in the plant. Electrons are transferred from ferredoxin to the 4Fe–4S cluster, then to siroheme, and finally to NO_2^- to form NH_4^+. In higher plants, NH_4^+ also does not accumulate, although it may reach higher concentrations than NO_2^-, because glutamine synthetase (GS-2), which catalyzes incorporation of NH_4^+ into glutamine using glutamate as the NH_4^+ acceptor, has a high affinity (K_m = 5 μM) for NH_4^+. GS-2 is a heterooctamer with two subunit types (39 and 45 kDa) and is, therefore, distinct from bacterial GS. The reaction requires ATP (see equation 3), and so is indirectly connected to photosynthesis. In plants, GS activity is probably redox-regulated by light. For algae, a thioredoxin-dependent reduction/oxidation has been found. A second GS in plants

(GS-1) is located in the cytosol. This enzyme functions during senescence, when NH_4^+ is released largely from catabolic processes, and eventually also when NH_4^+ is released from the mitochondria during photorespiration [87].

In a subsequent reaction catalyed by glutamate synthase (equation 4), NH_4^+ from the amido-group of glutamine is transferred to oxo-glutarate to form two molecules of glutamate.

$$\text{Glutamine} + \text{Oxo-glutarate} + \text{NADH} + \text{H}^+ \rightarrow 2\,\text{Glutamate} + \text{NAD}^+ \quad (4)$$

This enzyme functions as GOGAT and requires reduced ferredoxin as electron donor. Isoforms exist in leucoplasts and in the cytosol; the former accepts electrons from NADPH, the latter from NADH. The function of these isoforms is still unclear. This reaction again demonstrates the close connection between N and C metabolism.

References

[1] H.M. Lam, K.T. Coschigano, I. Oliveira, R. Melo-Oliveira, G.M. Coruzzi, The molecular-genetics of nitrogen assimilation into amino acids in higher plants, Ann. Rev. Plant Physiol. Plant Mol. Biol. 47 (1966) 569–593.

[2] N.M. Crawford, Nitrate: nutrient and signal for plant growth, Plant Cell 7 (1995) 859–868.

[3] M.E. Samuelson, E. Ohlen, M. Lind, C.M. Larsson, Nitrate regulation of nitrate uptake and nitrate reductase expression in barley grown at different nitrate: ammonium ratios at constant relative nitrogen addition rate, Physiol. Plant. 94 (1995) 254–260.

[4] R. Tischner, B. Waldeck, S.S. Goyal, W.D. Rains, Effect of nitrate pulses on the nitrate-uptake rate, synthesis of mRNA coding for nitrate reductase, and nitrate-reductase activity in the roots of barley seedlings, Planta 189 (1993) 533–537.

[5] H.M. Lee, E. Flores, A. Herrero, J. Houmard, N. Tandeau de Marsac, A role for the signal transduction protein P_{II} in the control of nitrate/nitrite uptake in a cyanobacterium, FEBS Lett. 427 (1998) 291–295.

[6] M.H. Hsieh, H.M. Lam, F.J. Van de Loo, G. Coruzzi, A PII-like protein in Arabidopsis: putative role in nitrogen sensing, Proc. Natl. Acad. Sci. USA 95 (1998) 13965–13970.

[7] R.C. Chiang, R. Cavicchioli, P. Gunsalus, 'Locked-on' and 'locked-off' signal transduction mutations in the periplasmic domain of the *Escherichia coli*. NarQ and NarX sensors affect nitrate- and nitrite-dependent regulation by NarL and NarP, Mol. Microbiol. 24 (1997) 1049–1060.

[8] C.F. Hwang, Y. Lin, T. D'Souza, C.L. Cheng, Sequences necessary for nitrate-dependent transcription of *Arabidopsis* nitrate reductase genes, Plant Physiol. 113 (1997) 853–862.
[9] J.J. Vidmar, D.G. Zhuo, M.Y. Siddiqi, A.D.M. Glass, Isolation and characterization of *HvNRT2.3* and *HvNRT2.4*, cDNAs encoding high-affinity nitrate transporters from roots of barley, Plant Physiol. 122 (2000) 783–792.
[10] H. Zhang, B.G. Forde, An *Arabidopsis* MADS box gene that controls nutrient-induced changes in root architecture, Science 279 (1998) 407–409.
[11] H. Sakakibara, M. Suzuki, K. Takei Kentaro, A. Deji, M. Taniguchi, T. Sugiyama, A response regulator homologue possibly involved in nitrogen signal transduction mediated by cytokinin in maize, Plant J. 14 (1998) 337–344.
[12] M. Taniguchi, T. Kiba, H. Sakakidara, C. Ueguchi, T. Mizuno, T. Sugiyama, Expression of *Arabidopsis* response regulator homologs is induced by cytokinins and nitrate, FEBS Lett. 429 (1998) 259–262.
[13] M. Kunze, J. Riedel, U. Lange, R. Hurwitz, R. Tischner, Evidence for the presence of GPI-anchored PM-NR in leaves of *Beta vulgaris* and for PM-NR in barley leaves, Plant Physiol. Biochem. 35 (1997) 507–512.
[14] C. Stöhr, F. Strube, G. Marx, W.R. Ullrich, P. Rockel, A plasma membrane-bound enzyme of tobacco roots catalyses the formation of nitric oxide from nitrite, Planta 212 (2001) 835–841.
[15] M. Koltermann, A. Moroni, S. Gazzarini, D. Nowara, R. Tischner, Cloning, functional expression and expression studies of the nitrate transporter gene from *Chlorella sorokiniana* (strain 211-8k), Plant Mol. Biol. 52 (2003) 855–864.
[16] M. Cerezo, P. Tillard, S. Filleur, S. Munos, F. Daniel-Vedele, A. Gojon, Major alterations of the regulation of root NO_3^- uptake are associated with the mutation of *Nrt2.1* and *Nrt2.2* genes in *Arabidopsis*, Plant Physiol. 127 (2001) 262–271.
[17] B.J. King, M.Y. Siddiqi, T.J. Ruth, R.L. Warner, A.D.M. Glass, Feedback regulation of nitrate influx in barley roots by nitrate, nitrite, and ammonium, Plant Physiol. 102 (1993) 1279–1286.
[18] S. Filleur, M.F. Dorbe, M. Cerezo, M. Orsel, F. Granier, A. Gojon, F. Daniel-Vedele, An *Arabidopsis* T-DNA mutant affected in *Nrt2* genes is impaired in nitrate uptake, FEBS Lett. 489 (2001) 220–224.
[19] C. Cruz, S.H. Lips, M.A. Martins-Loucao, Uptake regions of inorganic nitrogen in roots of carob seedlings, Physiol. Plant. 95 (1995) 167–175.
[20] S. Siebrecht, G. Mäck, R. Tischner, Function and contribution of the root tip in the induction of NO_3^- uptake along the barley root axis, J. Exp. Bot. 46 (1995) 1669–1676.

[21] P. Nazoa, J.J. Vidmar, T.J. Tranbarger, K. Mouline, I. Damiani, P. Tillard, D. Zhuo, A.D.M. Glass, B. Touraine, Regulation of the nitrate transporter gene *AtNRT2.1* in *Arabidopsis thaliana*: responses to nitrate, amino acids and developmental stage, Plant Mol. Biol. 52 (2003) 689–703.
[22] A.D.M. Glass, J.E. Shaff, L.V. Kochian, Studies of the uptake of nitrate in barley: IV. Electrophysiology, Plant Physiol. 99 (1992) 456–463.
[23] B. Thornton, Inhibition of nitrate influx by glutamine in *Lolium perenne* depends upon the contribution of the HATS to the total influx, J. Exp. Bot. 55 (2004) 761–769.
[24] J.N. Feng, R.J. Volk, W.A. Jackson, Inward and outward transport of ammonium in roots of maize and sorgum: contrasting effects of methionine sulfoximine, J. Exp. Bot. 45 (2004) 429–439.
[25] D. Zhuo, M. Okamoto, J.J. Vidmar, A.D.M. Glass, Regulation of a putative high-affinity nitrate transporter (*Nrt2;1At*) in roots of *Arabidopsis thaliana*, Plant J. 17 (1999) 563–568.
[26] S.E. Unkles, R. Wang, Y. Wang, A.D. Glass, N.M. Crawford, J.R. Kinghorn, Nitrate reductase activity is required for nitrate uptake into fungal but not plant cells, J. Biol. Chem. 279 (2004) 28182–28186.
[27] A. Gessler, M. Schultz, S. Schrempp, H. Rennenberg, Interaction of phloem-translocated amino compounds with nitrate net uptake by roots of beech (*Fagus silvatica*) seedlings, J. Exp. Bot. 49 (1998) 1529–1537.
[28] A. Gojon, L. Dapoigny, L. Lejay, P. Tillard, T.W. Rufty, Effects of genetic modification of nitrate reductase expression on $^{15}NO_3^-$ uptake and reduction in *Nicotiana* plants, Plant Cell Environ. 21 (1998) 43–53.
[29] A.J. Barneix, H.F. Causin, The central role of amino acids on nitrogen utilization and plant growth, J. Plant Physiol. 149 (1996) 358–362.
[30] C. Dzuibany, S. Haupt, H. Fock, K. Biehler, A. Migge, T.W. Becker, Regulation of nitrate reductase transcript levels by glutamine accumulating in the leaves of a ferredoxin-dependent glutamate synthase-deficient *gluS* mutant of *Arabidopsis thaliana*, and by glutamine provided via the roots, Planta 206 (1998) 515–522.
[31] A.J. Miller, S.J. Smith, Nitrate transport and compartmentation in cereal root cells, J. Exp. Bot. 47 (1996) 843–854.
[32] W.R. Scheible, M. Lauerer, E.D. Schulze, M. Caboche, M. Stitt, Accumulation of nitrate in the shoot acts as a signal to regulate shoot-root allocation in tobacco, Plant J. 11 (1997) 671–691.
[33] W.R. Ullrich, Nitrate and ammonium uptake in green (algae) and higher plants: Mechanism and Relationship with nitrate metabolism in: W.R. Ullrich, P.J. Aparicio, P.J. Syrett, F. Castillo (Eds.) Inorganic Nitrogen Metabolism, Springer, Berlin, 1987, pp. 32–38.

[34] A.A. Meharg, M.R. Blatt, NO_3^- transport across the plasma membrane of *Arabidopsis thaliana* root hairs: kinetic control by pH and membrane voltage, J. Memb. Biol. 145 (1995) 49–66.
[35] G.H. Hendriksen, D.R. Raman, L.P. Walker, R.M. Spanswick, Measurement of net fluxes of ammonium and nitrate at the surface of barley roots using ion-selective microelectrodes: II. Patterns of uptake along the root axis and evaluation of the microelectrode flux estimation technique, Plant Physiol. 99 (1992) 734–747.
[36] I. Mistrik, Nitrate uptake and assimilation by individual roots of *Zea mays*, Biol. Bratislava 52 (1997) 567–571.
[37] R. Rodrigez, C. Lara, M.G. Guerrero, Nitrate transport in the cyanobacterium *Anacystis nidulans* R2. Kinetics and energetic aspects, Biochem. J. 282 (1992) 639–643.
[38] C.M. Boyd, D. Gradmann, Three types of membrane excitations in the marine diatom *Coscinodiscus wailesii*, J. Memb. Biol. 175 (2000) 149–160.
[39] L. Rubio, A. Linares-Rueda, M.J. Garcia-Sanchez, J.A. Fernandes, Physiological evidence for a sodium-dependent high-affinity phosphate and nitrate transport at the plasma membrane of leaf and root cells of *Zostera marina* L., J. Exp. Bot. 56 (2005) 613–622
[40] N.V. Ivashikina, M. Feyziev-Ya, Regulation of nitrate uptake in maize seedlings by accompanying cations, Plant Sci. 131 (1998) 25–34.
[41] L. Espen, F.F. Nocito, M. Cocucci, Effect of NO_3^- transport and reduction on intracellular pH: an in vivo NMR study in maize roots, J. Exp. Bot. 55 (2004) 2053–2061.
[42] K.M. Kabala, G. Lobus, M. Janicka-Russak, Nitrate transport across the tonoplast of *Cucumis sativus* L. root cells, J. Plant. Physiol. 160 (2003) 523–530.
[43] S.E. Unkles, K.L. Hawker, C. Grieve, E.I. Campbell, P. Montague, J.R. Kinghorn, *crnA* encodes a nitrate transporter in *Aspergillus nidulans*, Proc. Natl. Acad. Sci. USA 88 (1991) 204–208.
[44] A. Quesada, A. Galván, E. Fernández, Identification of nitrate transporter genes in *Chlamydomonas reinhardtii*, Plant J. 5 (1994) 407–419.
[45] L.J. Trueman, I. Onyeocha, B.G. Forde, Recent advances in the molecular biology of a family of eukaryotic high affinity nitrate transporters, Plant Physiol. Biochem. 34 (1996) 621–627.
[46] A. Quesada, J. Hidalgo, E. Fernandez, Three *Nrt2* genes are differentially regulated in *Chlamydomonas reinhardtii*, Mol. Gen. Genet. 258 (1998) 373–377.
[47] Y.F. Tsay, J.I. Schroeder, K.A. Feldmann, N.M. Crawford, The herbicide sensitivity gene *CHL1* of *Arabidopsis* encodes a nitrate-inducible nitrate transporter, Cell 72 (1993) 1–20.
[48] N.C. Huang, C.S. Chiang, N.M. Crawford, Y.F. Tsay, *CHL1* encodes a component of the low-affinity nitrate uptake system in *Arabidopsis*

and shows cell type-specific expression in roots, Plant Cell 8 (1996) 2183–2191.

[49] R. Wang, D. Liu, N.M. Crawford, The *Arabidopsis* CHL1 protein plays a major role in high-affinity nitrate uptake, Proc. Natl. Acad. Sci. USA 95 (1998) 15134–15139.

[50] F.Q. Guo, R. Wang, N.M. Crawford, The *Arabidopsis* dual-affinity nitrate transporter gene *AtNRT1.1* (*CHL1*) is regulated by auxin in both shoots and roots, J. Exp. Bot. 53 (2002) 835–844.

[51] F.Q. Guo, J. Young, N.M. Crawford, The nitrate transporter *AtNRT1.1* (*CHL1*) functions in stomatal opening and contributes to drought susceptibility in *Arabidopsis*, Plant Cell 15 (2003) 107–117.

[52] Y. Tong, J.J. Zhou, A.J. Miller, A two-component high-affinity nitrate uptake system in barley, Plant J. 41 (2004) 442–450.

[53] A. Quesada, A. Krapp, L.J. Trueman, F. Daniel-Vedele, E. Fernandez, B.G. Forde, M. Caboche, PCR-identification of a *Nicotiana plumbaginifolia* cDNA homologous to the high-affinity nitrate transporters of the *crnA* family, Plant Mol. Biol. 34 (1997) 265–274.

[54] M. Orsel, S. Filleur, V. Fraisier, F. Daniel-Vedele, Nitrate transport in plants: which gene and which control? J. Exp. Bot. 53 (2002) 825–833.

[55] S. Munos, C. Cazettes, C. Fizames, F. Gaymard, P. Tillard, M. Lepetit, L. Lejay, A. Gojon, Transcript profiling in the *chl1-5* mutant of *Arabidopsis* reveals a role of the nitrate transporter NRT1.1 in the regulation of another nitrate transporter, NRT2.1, Plant Cell 16 (2004) 2433–2447.

[56] M. Orsel, K. Eulenburg, A. Krapp, F. Daniel-Vedele, Disruption of the nitrate transporter genes *AtNRT2.1* and *AtNRT2.2* restricts growth at low external nitrate concentration, Planta 219 (2004) 714–721.

[57] M. Orsel, A. Krapp, F. Daniel-Vedele, Analysis of the NRT2 nitrate transporter family in *Arabidopsis*. Structure and gene expression, Plant Physiol. 129 (2002) 886–896.

[58] M. Okamoto, J.J. Vidmar, A.D.M. Glass, Regulation of *NRT1* and *NRT2* gene families of *Arabidopsis thaliana*: responses to nitrate provision, Plant Cell Physiol. 44 (2003) 304–317.

[59] C.C. Chiu, C.S. Lin, A.P. Hsia, R.C. Su, H.L. Lin, Y.E. Tsay, Mutation of a nitrate transporter, AtNRT1:4, results in a reduced petiole nitrate content and altered leaf development, Plant Cell Physiol. 45 (2004) 1139–1148.

[60] W.H. Campbell, Nitrate reductase structure, function and regulation: bridging the gap between biochemistry and physiology, Ann. Rev. Plant Physiol. Mol. Biol. 50 (1999) 277–303.

[61] P. Ruoff, C. Lillo, Molecular oxygen as electron acceptor in the NADH-nitrate reductase system, Biochem. Biophys. Res. Comm. 172 (1990) 1000–1005.

[62] H. Yamasaki, Y. Sakihama, Simultaneous production of nitric oxide and peroxynitrite by plant nitrate reductase: in vitro evidence for the

NR-dependent formation of active nitrogen species, FEBS Lett. 468 (2000) 89–92.

[63] P. Rockel, F. Strube, A. Rockel, J. Wildt, W.M. Kaiser, Regulation of nitric oxide (NO) production by plant nitrate reductase in vivo and in vitro, J. Exp. Bot. 53 (2002) 103–110.

[64] E. Planchet, K.J. Gupta, M. Sonoda, W.M. Kaiser, Nitric oxide emission from tobacco leaves and cell suspensions: rate-limiting factors and evidence for the involvement of mitochondrial electron transport, Plant J. 41 (2005) 732–743.

[65] S.J. Neill, R. Desikan, J.T. Hancock, Nitric oxide signalling in plants, New Phytol. 159 (2003) 11–35.

[66] C. Meyer, C. Stöhr, Soluble and plasma membrane-bound enzymes involved in nitrate and nitrite metabolism, in: C.H. Foyer, G. Noctor (Eds.) Photosynthetic Nitrogen Assimilation and Associated Carbon and Respiratory Metabolism, Kluwer Academic Publishers, Dordrecht, 2002, pp. 49–62.

[67] W.H. Campbell, Molecular control of nitrate reductase and other enzymes involved in nitrate assimilation, in: C.H. Foyer, G. Noctor (Eds.) Photosynthetic Nitrogen Assimilation and Associated Carbon and Respiratory Metabolism, Kluwer Academic Publishers, Dordrecht, 2002, pp. 35–48.

[68] R. Wang, K. Guegler, S.T. LaBrie, N.M. Crawford, Genomic analysis of a nutrient response in Arabidopsis reveals diverse expression patterns and novel metabolic and potential regulatory genes induced by nitrate, Plant Cell 12 (2000) 1491–1509.

[69] R.R. Mendel, Molybdenum cofactor of higher plants: biosynthesis and molecular biology, Planta 203 (1997) 399–405.

[70] C.F. Hwang, T. Lin, Y. D'Souza, C.L. Cheng, Sequences necessary for nitrate-dependent transcription of *Arabidopsis* nitrate reductase genes, Plant Physiol. 113 (1997) 853–862.

[71] G. Coruzzi, D.R. Bush, Nitrogen and carbon nutrient and metabolite signalling in plants, Plant Physiol. 125 (2001) 61–64.

[72] C. Lillo, K.J. Appenroth, Light regulation of nitrate reductase in higher plants. Which photoreceptors are involved? Plant Biol. 3 (2001) 455–465.

[73] C. Lillo, Light regulation of nitrate uptake, assimilation and metabolism, in: S. Amancio,I. Stulen (Eds.) Nitrogen Acquisition and Assimilation in Higher Plants, Kluwer Academic Publishers, Dordrecht, 2004, pp. 149–184.

[74] W.M. Kaiser, S.C. Huber, Post-translational regulation of nitrate reductase: mechanism, physiological relevance and environmental triggers, J. Exp. Bot. 52 (2001) 1981–1989.

[75] W.M. Kaiser, E. Brendle-Behnisch, Acid-base-modulation of nitrate reductase in leaf tissues, Planta 196 (1995) 1–6.

[76] W.M. Kaiser, E. Planchet, M. Stoimenova, M. Sonoda, Modulation of nitrate reduction-environmental triggers and internal factors involved, in: S. Amancio and I. Stulen (Eds.) Nitrogen Acquisition and Assimilation in Higher Plants, Kluwer Academic Publishers, Dordrecht, 2004, pp. 185–205.

[77] J.L. Huber, S.C. Huber, W.H. Campbell, M.G. Redinbaugh, Reversible light/dark modulation of spinach leaf nitrate reductase activity involves protein phosphorylation, Arch. Biochem. Biophys. 296 (1992) 58–65.

[78] C. MacKintosh, Regulation of spinach-leaf nitrate reductase by reversible phosphorylation, Biochim. Biophys. Acta 1137 (1992) 121–126.

[79] R.W. McMichael, M. Bachmann, S.C. Huber, Spinach leaf sucrose-phosphate synthase and nitrate reductase are phosphorylated/inactivated by multiple protein kinases in vitro, Plant Physiol. 108 (1995) 1077–1082.

[80] J. Glaab, W.M. Kaiser, Inactivation of nitrate reductase involves NR-protein phosphorylation and subsequent 'binding' of an inhibitor protein, Planta 195 (1995) 514–518.

[81] G.S. Athwal, J.L. Huber, S.C. Huber, Phosphorylated nitrate reductase and 14-3-3 proteins: site of interaction, effects of ions, and evidence for an AMP-binding site on 14-3-3 proteins, Plant Physiol. 118 (1998) 1041–1048.

[82] C. Petosa, S.C. Masters, L.A. Bankston, J. Pohl, B. Wang, H. Fu, R.C. Liddington, 14-3-3 binds a phosphorylated Raf peptide and an unphosphorylated peptide via its conserved amphipathic groove, J. Biol. Chem. 273 (1998) 16035–16310.

[83] C. Lillo, U.S. Lea, M.T. Leydecker, C. Meyer, Mutation of the regulatory phosphorylation site of tobacco nitrate reductase results in constitutive activation of the enzyme *in vivo* and nitrite accumulation, Plant J. 35 (2003) 566–573.

[84] U.S. Lea, F. ten Hoopen, F. Provan, W.M. Kaiser, C. Meyer, C. Lillo, Mutation of the regulatory phosphorylation site of tobacco nitrate reductase results in high nitrite excretion and NO emission from leaf and root tissue, Planta 219 (2004) 59–65.

[85] H. Weiner, W.M. Kaiser, 14-3-3 proteins control proteolysis of nitrate reductase in spinach leaves, FEBS Lett. 455 (1999) 75–78.

[86] H. Weiner, W.M. Kaiser, Antibodies to assess phosphorylation of spinach leaf nitrate reductase on serine-543 and its binding to 14-3-3 proteins, J. Exp. Bot. 52 (2001) 1165–1172.

[87] B. Hirel, G.H. Miao, D.P.S. Verma, Metabolic and developmental control of glutamine synthetase genes in legume and non-legume plants, in: D.P.S. Verma (Ed.) Control of Plant Gene Expression, CRC Press, Boca Raton, 1993, pp. 443–458.

Chapter 19

Characterization of Proteolytic Microbes and Their Activities in Soils

Mirna Mrkonjic Fuka, Marion Engel,
Jean-Charles Munch and Michael Schloter

19.1 Introduction

Nitrogen in soil occurs in several organic and inorganic forms. Proteins, nucleic acids, and other heterocyclic N-compounds, such as peptidoglycan and chitin, are the major sources of nitrogen. Degradation of proteins occurs through intra- and extracellular proteolytic enzymes. Whereas intracellular proteases are more related to the regulation of metabolic processes in cells, hydrolysis of large polypeptidic compounds through the extracellular enzymes is of prime importance for the global N-turnover, thus resulting in the formation of low-molecular-weight products that could be transported across the cell membrane.

Classification of proteolytic enzymes are based on the three major criteria: (i) type of reaction catalyzed, (ii) chemical nature of catalytic site, and (iii) evolutionary relationship with reference to structure [1]. In general, proteases are divided into two major groups: (i) endopeptidases that

Biology of the Nitrogen Cycle
Edited by H. Bothe, S.J. Ferguson and W.E. Newton

act on the interior linkages of substrate and (ii) exopeptidases that attack free ends of proteins. On the basis of the functional group present at the active site, proteases could be separated into four prime groups: serine proteases, aspartic proteases, cysteine proteases, and metalloproteases. Proteases are further divided in different families and clans depending on their amino acid sequences and evolutionary relationships. On the basis of the pH values of their optimal activity, they are classified as acidic, neutral, or alkaline proteases.

Proteases are ubiquitous and can be found in a wide diversity of taxa across the plants, animals, and microorganisms. Proteases that origin from microbes are found in almost all proteolytic groups. Predominant fungal extracellular proteases are supposedly cysteine and aspartic proteases, whereas those of bacterial origin are mainly alkaline metalloproteases (Apr), neutral metalloproteases (Npr), and serine peptidases (Sub) [2]. Microbial proteases have a higher substrate affinity compared to animal and plant proteases thus being the main source for N mineralization in soil [3].

19.2 Methods to assess proteolysis

19.2.1 Methods for measuring proteolytic activity

Proteolytic activity can be easily measured in soil. The assays are generally sensitive, short term, and reproducible. A small quantity of soil is incubated for a short period using dye- or fluorescence-labeled proteins as a substrate analogue. Proteolytic activity is usually detected by either (i) following the decrease of initial substrate or more often (ii) measuring the increase of amino acids or peptides released during the incubation period. Detection of released chromogenic or fluorogenic compounds can be measured spectrophotometrically or fluorometrically, respectively [4, 5]. However, measurements of enzyme activities are rather showing a potential and do not reflect the in situ activity. During the measurements, conditions such as pH value, temperature, and substrate concentration are adapted to disclose the optimal activity that is often completely different from those of native systems. Furthermore, the assays do not distinguish between intra- and extracellular proteases. To overcome that problem, soil samples have been treated with toluene or other bacteriostatic agents prior to enzymatic measurements. However, this approach has been often criticized because of the release of intracellular enzymes during the plasmolysis which, together with microbial residues, could enhance overall enzymatic activity [4]. Additionally, you have to take into account that measurements with soil samples do not discriminate between bacterial and fungal proteases. In summary, enzyme assays could give important information about potential activity in soil but do not give any information about the actual process in situ and do not distinguish between microbial species directly involved in measured processes.

19.2.2 Cultivation of proteolytic microbes

Methods to study organisms involved in proteolysis are based on isolation of proteolytic microorganisms on special selective media. To estimate the number of proteolytic bacteria/fungi, plate counting and most probable number (MPN) techniques have been applied. However, traditional cultivation-dependent techniques are time consuming and do not represent in situ situation because cultivation conditions could be favorable just for a small part of the investigated microbial species.

19.3 Tools to study the gene- and transcript pool of proteolytic organisms

Genomic techniques are based on the isolation of DNA/RNA without the need to isolate or culture a species, although very promising molecular techniques are not without their own limitations. Insufficient efficiency of cell lysis, impropriety extraction of nucleic acids, coextraction of inhibitory substances such as humic acids, selective amplification in the presence of nucleic acids from mixed communities, formation of chimeric or heteroduplex DNA molecules during the PCR as well as relatively high manipulative costs may result in biases. Despite this, molecular-based techniques still provide the best approach for characterizing complex community structures.

19.3.1 Polymerase chain reaction (PCR) and reverse transcription-PCR (RT-PCR)

PCR primers and probes for the detection of genes encoding for extracellular bacterial proteases were first designed as given in Ref. [4]. To estimate the number of genes in environmental samples, PCR-based quantitative techniques such as MPN-PCR and real-time PCR have been employed. The MPN-PCR method focuses on the quantification of gene fragments at the end of amplification and requires post-PCR steps to ensure visualization of the generated products. Real-time PCR techniques allow detection of the product once linear amplification is achieved and do not require post-PCR procedure. These techniques are more rapid, accurate, and sensitive than MPN-PCR. Real-time PCR analysis of different soil samples showed subtilisin gene (*sub*) densities between 8.4×10^5 and 3.1×10^7, whereas neutral metalloprotease gene (*npr*) copy number per gram dry soil ranged from 4.2×10^5 to 1.2×10^7 [6].

Numerous fingerprinting approaches have been developed to study microbial diversity. However, the most promising method to fingerprint proteolytic communities from the soil samples is the terminal restriction fragment length polymorphism technique (t-RFLP) so far. T-RFLP analysis is based on the restriction endonuclease digestion of fluorescently end-labeled PCR products. The digested products are separated by capillary gel electrophoresis and detected on an automated sequence analyzer. The technique provides diverse profiles (fingerprints) dependent on the species composition of the communities of the samples. Other commonly used methods to monitor highly diverse soil microbial communities are DGGE and SSCP [7, 8]. To gain detailed information about the composition of the proteolytic microbes, PCR or RT-PCR products have to be cloned and sequenced. However, not necessarily the fragments used for quantitative PCR (qPCR) or fingerprinting are well suited for phylogenetic analysis, as the length of the generated sequences is too short. Furthermore, this approach is not suitable for high-throughput analysis. At least 50 clones have to be analyzed per sample to get information on the most abundant species.

The presence of the target gene in complex communities can be simply demonstrated by hybridization of the DNA/cDNA recovered from soil to the specific labeled probe. The technique has been successfully used not only to detect subtilisin (*sub*), neutral metalloprotease (*npr*), and alkaline metalloprotease (*apr*) gene fragments from the soil samples [9], but also to detect transcripts of *sub* and *npr* genes in the rhizospheres [10].

19.3.2 Metagenomics

Metagenomics represent a new approach in a genomic analysis. This method accesses the potential reservoir of novel genes in soil. To explore this reservoir, DNA from an environmental sample is extracted, cloned into an appropriate vector, and transformed into competent *E. coli* cells. The resulting transformants in metagenomic libraries are screened for novel physiological, metabolic, and genetic features. Although time-consuming and labor-intensive, metagenomic is the most powerful environmental approach that offers possibilities to discover novel genes and novel biomolecules through the expression of genes from uncultivated and unknown bacteria in recipient host cell. Theoretically, a metagenomic database should contain DNA sequences for all the genes in the microbial community [11]. However, often those genes cannot be expressed, folded, or excreted correctly in the corresponding host system.

19.4 Investigation of proteolysis in terrestrial ecosystems

19.4.1 Temporal variation

A study of the temporal variations of proteolytic activities in an arable field during the vegetation period showed that the activity was highest in spring after the application of mineral N-fertilizer [18]. Application of mineral fertilizer accelerated decomposition rate [12] and increased soil organic matter (SOM) [13]. The abundance of genes encoding for subtilisin (*sub*) and neutral metallopeptidase (*npr*) as well as activity of proteolytic bacteria was measured in soil samples, taken in April, July, and October from four soil types at three different depths [13]. Higher potential proteolytic activities were detected in April than in July. This could be attributed to the application of higher amounts of fertilizer in spring and the decrease in size of proteolytic communities as well as a reduction of moisture in summer. A reduction of soil moisture decreased protease activity by 15–66% depending on annual period and soil depth [14]. Potential proteolytic activity increased again in autumn after harvesting. Previous studies suggested that an addition of high-energy material increases overall enzyme activity [15]. Increased proteolytic activity in autumn may be explained by an elevated N demand caused by enhanced microbial growth [16].

The copy number of *npr* was highest in spring, whereas that of *sub* showed maximum in autumn. Factors like substrate composition obviously influence the abundance of soil bacteria during the seasons and thus promote the growth of specific groups of microorganisms with different enzyme patterns. Therefore, it has been suggested that different properties of available substrates in spring and autumn could be the major cause for differences in observed patterns of *npr* and *sub* copy numbers during the year [6].

19.4.2 Site variation

The variation of proteolytic bacteria at different soil sites was examined, and the lowest proteolytic activity was detected in an arable field, slightly higher in the beach forest, and highest in the grassland rhizosphere [5]. Different activities between sites may result from diverse soil properties, such as soil texture and nutrient contents as well as changing numbers of soil proteolytic bacteria. Proteolytic activity was dominated in the clay fractions where also most of the amino acids are located [12]. A high correlation could be shown between clay contents and proteolytic activity [11], and higher number of proteolytic bacteria and their activity were found in soil samples containing a higher proportion of clay and silt particles [13]. The reason might be the stabilizing effect of clay particles and humic colloids for free enzymes in soil.

The effect of depth on the properties of proteolytic communities has been rarely studied. Significant differences between top- and subsoil in MPN count of proteolytic bacteria in March and no differences in October at the same arable field [5]. Only a slight decrease of the abundance of proteolytic bacteria with increasing depth was observed in other terrestrial ecosystems [17]. Significant variations of the number and activity of proteolytic bacteria were, however, seen within soil profiles [13]. During the whole vegetation period and at all sites, potential proteolytic activity and the number of proteolytic bacteria significantly decreased with increasing depth following the decrease in nutrient contents. With increasing depth, organic matter contents decline and its composition and structure changes [18]. Thus the substrate availability was likely the main factor governing proteolytic properties among the soil profiles.

In conclusion, although huge efforts have been made in the last five decades to understand proteolytic processes in soils, more information, mainly coming from molecular approaches, is needed for a better understanding of the proteolytic community structure and function in any ecosystem.

References

[1] M.B. Rao, A.M. Tanksale, M.S. Ghatge, V.V. Deshpande, Molecular and biotechnological aspects of microbial proteases, Microbiol. Mol. Biol. Rev. 62 (1998) 597–635.

[2] H.M. Kalisz, Microbial proteinases, Adv. Biochem. Eng. Biotechnol. 36 (1988) 165–171.

[3] R. Ji, A. Brune, Digestion of peptidic residues in humic substances by an alkali-stable and humic-acid-tolerant proteolytic activity in the gut of soil-feeding termites, Soil Biol. Biochem. 37 (2005) 1648–1655.

[4] J.N. Ladd, J.H.A. Butler, Short-term assays of soil proteolytic enzyme activities using proteins and dipeptide derivatives as substrates, Soil Biol. Biochem. 4 (1972) 19–30.

[5] H.-G. Hoppe, S.-J. Kim, K. Gocke, Microbial decomposition in aquatic environments: combined process of extracellular enzyme activity and substrate uptake, Appl. Environ. Microbiol. 54 (1988) 784–790.

[6] M. Mrkonjic Fuka, M. Engel, A. Gattinger, A. Embacher, U. Bausenwein, M. Sommer, J.C. Munch, M. Schloter, Spatial and temporal variability of proteolytic genes and their activities in an arable field, Appl. Environ. Microbiol., 2006, in press.

[7] F. Schwieger, C.C. Tebbe, A new approach to utilize PCR-single-strand-conformation polymorphism for 16S rRNA gene-based microbial community analysis, Appl. Environ. Microbiol. 64 (1998) 4870–4876.

[8] S. Sharma, M.K. Aneja, J. Mayer, J.C. Munch, M. Schloter, Diversity of transcripts of nitrite reductase genes [nirK and nirS] in rhizospheres of grain legumes, Appl. Environ. Microbiol. 71 (2005) 2001–2007.

[9] H.-J. Bach, A. Hartmann, M. Schloter, J.C. Munch, PCR primers and functional probes for amplification and detection of bacterial genes for extracellular peptidases in single strains and in soil, J. Microbiol. Methods 44 (2001) 173–182.

[10] S. Sharma, M.K. Aneja, J. Mayer, M. Schloter, J.C. Munch, RNA fingerprinting of microbial community in the rhizosphere soil of grain legumes, FEMS Microbiol. Lett. 240 (2004) 181–186.
[11] J. Handelsman, Microbiol. Metagenomics: application of genomics to uncultured microorganisms, Mol. Biol. Rev. 68 (2004) 669–685.
[12] S.Y. Newell, T.L. Arsuffi, L.A. Palm. Misting and nitrogen fertilization of shoots of a saltmarsh grass: effects upon fungal decay of leaf blades. Oecologia 108 (1996) 495–502.
[13] J. Galantini, R. Rosell, Long-term fertilization effects on soil organic matter quality and dynamics under different production systems in semiarid Pampean soils, Soil Tillage Res, 87 (2006) 72–79.
[14] J. Sardans, J. Penuelas, Drought decreases soil enzyme activity in a Mediterranean Quercus ilex L. forest, Soil Biol. Biochem. 37 (2005) 455–446.
[15] F. Asmar, F. Eiland, N.E. Nielsen, Interrelationship between extracellular-enzyme activity, ATP content, total counts of bacteria and CO2 evolution, Biol. Fertil. Soils 14 (1992) 288–292.
[16] L. Alden, F. Demoling, E. Baath, Rapid method of determining factors limiting bacterial growth in soil. Appl. Environ. Microbiol. 67 (2001) 1830–1838.
[17] H.-J. Bach, J.C. Munch, Identification of bacterial sources of soil peptidases, PCR primers and functional probes for amplification and detection of bacterial genes for extracellular peptidases in single strains and in soil, Biol. Fert. Soils 31 (2000) 219–224.
[18] A. Agnelli, L. Celi, G. Corti, A. Degl'Innocenti, F. Ugolini, The changes with depth of humic and fluvic acids extracted from the fine earth and rock fragments of a forest soil, Soil Sci. Soc. Am. J. 167 (2002) 524–538.

Part IV

Applications of Reactions of the Nitrogen Cycle, with Emphasis on Denitrification

Chapter 20

Molecular Tools to Assess the Diversity and Density of Denitrifying Bacteria in Their Habitats

Sara Hallin, Gesche Braker and Laurent Philippot

20.1 Introduction

Denitrifiers are commonly found in many natural environments such as soil, marine and freshwater sediment, as well as in wastewater treatment systems. Studies of the ecology of denitrifiers started with their cultivation from diverse environments. In one of the first and most comprehensive studies exploring soil denitrifier communities, Gamble et al. [1] isolated 1,500 bacteria and 146 of those were capable of complete denitrification. In this and other cultivation-based studies of this functional community, the genera *Pseudomonas*, *Ralstonia*, *Alcaligenes*, *Paracoccus*, *Rhodobacter*, *Rubrivivax*, *Thauera*, *Burkholderia*, *Bacillus*, and *Streptomyces* have been pointed out as the dominant denitrifiers in various environments. Marine denitrifiers were dominated by *Shewanella baltica* [2] and *Marinobacter* spp. [3]. However, cultivation is known to be highly selective for certain

Biology of the Nitrogen Cycle
Edited by H. Bothe, S.J. Ferguson and W.E. Newton

organisms and the lack of appropriate tools to study these bacteria in the environment have limited our knowledge of denitrifier ecology.

Today, molecular tools are being developed or refined to assess both diversity and numbers of denitrifying populations in different ecosystems. The major concern is the role of denitrifier diversity for ecosystem functioning and especially for N fluxes and N_2O emissions. The diversity of denitrifying bacteria in different environments seems to be immense, but basic questions, such as redundancy of populations or consequences of population shifts within the denitrifying community and how this is linked to denitrifying activity, remain unanswered. Especially our understanding of the active population's role and their response to environmental factors is still limited by current methods.

20.2 Molecular markers for denitrifying bacteria

The ability to denitrify is sporadically distributed both within and between different genera and cannot be associated with any specific taxonomic group. Therefore, existing techniques to study the ecology of denitrifiers are based on the use of the functional genes in the denitrification pathway or their transcripts as molecular markers of this community [4].

DNA extraction followed by PCR amplification of denitrification genes has been the most common way to start-off the analysis of denitrifier communities. For this, reliable primers are a prerequisite. The first attempts were aimed at amplification of partial *nirK*, *nirS*, and *nosZ* genes [5–7], and alternatives or modifications followed [8–11]. Also for *narG* and *napA*, primers have been developed and especially the former has been targeted frequently in the environment [12–14]. Recently, the gene *norB*, encoding nitric oxide reductase, was used as a marker for denitrifying bacteria in sediments [15]. The number of sequences deposited in the databases is increasing exponentially, and so is the sequence variation of these genes, which calls for continual refinement of the primers. Genome sequencing and metagenomic projects might even provide new denitrification gene sequences, which could aid in designing more broad range primers.

20.3 Genetic fingerprinting of denitrifier communities

Several different techniques are available to resolve PCR-amplified denitrification genes. Most information is obtained by cloning and sequencing the PCR amplicons, but a more rapid analysis is achieved using fingerprinting techniques. These separate PCR amplicons of the same size that are obtained from mixed templates based on their nucleotide-sequence polymorphism. Terminal restriction fragment length polymorphism (T-RFLP) and denaturing gradient gel electrophoresis (DGGE) are techniques that are widely used to analyze microbial communities from the environment and these methods have also been applied to study denitrifier communities. As all PCR-based analyses, the fingerprinting techniques are subjected to well-known biases introduced by, e.g. DNA extraction procedures, primer selection, and PCR conditions.

20.3.1 PCR-RFLP

For denitrifiers, the PCR-RFLP fingerprint technique has mainly been applied to study the diversity of the *narG* gene encoding the membrane-bound nitrate reductase [13], but was recently also used for the N_2O-reductase gene [16]. This technique uses the variation in the position of restriction sites among the PCR amplicons to unravel sequence polymorphism. The resulting restriction fragments are then sorted by size on a polyacrylamide gel.

Similarly to T-RFLP, the choice of restriction enzyme is crucial. Ideally, to get a visual signal for each fragment, the enzyme should cut in variable regions to yield fragments larger than 50 bp. In silico analysis showed that *Alu*I was a good candidate for *narG* and several studies have confirmed this choice [13, 17–19]. In contrast to DGGE, this approach requires PCR products longer than 500 bp to facilitate both the choice of the restriction enzyme and the resolution of the gel analysis. PCR-RFLP is the most basic fingerprint technique and does not require any expensive equipment or specific skills. It is only useful to monitor changes in the composition of denitrifier communities, since a single population can be represented by one to several bands in the RFLP gel. Therefore, data on diversity (richness) cannot be deduced using this approach.

20.3.2 T-RFLP

Several studies used T-RFLP based on *nir* and *nosZ* genes as functional marker genes to explore denitrifier communities in different habitats [20–23]. T-RFLP is based on the difference in size of terminal restriction fragments from PCR-labeled amplicons. During PCR one or both primers, carrying a 5′-fluorescent dye, is used to amplify a fluorescent-labeled fragment, which is subsequently digested with one or several restriction enzymes in parallel reactions. Resulting fragments are separated on an automated sequencer that detects only terminal restriction fragments (T-RFs) with the fluorescent label, thus reducing the complexity of the profiles compared to RFLP analyses. By comparing the peaks to an internal size-standard, the lengths of the T-RFs are calculated. T-RFLP is technically limited neither by the use of degenerate primers in the PCR nor by the size of the amplicons, e.g. 1.1 kb *nosZ* PCR products were successfully analyzed [20].

Data are normalized to the lowest total fluorescence of all profiles to be compared and the relative abundance of each T-RF is determined as the ratio of fluorescence from that T-RF to the total fluorescence of the sample. T-RFLP is semi-quantitative and allows the processing and comparison of many samples. The number and relative abundance of T-RFs is a suitable means to determine the diversity and evenness of a community [23, 24]. Statistical analyses of denitrifier T-RFLP profiles can be based on both presence/absence and relative abundance of T-RFs, respectively.

Depending on the choice of restriction enzyme, the T-RFs from denitrification genes rather represent operational taxonomic units (OTUs) than specific sequence clusters or phylogenetic groups of organisms. The enzyme yielding the highest number of distinct T-RFs from a sample is often considered to provide the highest level of resolution [20, 21, 25]. Another option is to in silico determine which enzyme gives rise to specific T-RFs for the prevalent groups of denitrifiers from sequence data [26] to be able to later reveal the sequence groups from a database by using several enzymes [27, 28]. To assign T-RFs to sequences in the databases, two approaches are currently used. One is cloning and sequence analysis together with experimental determination of T-RFs from cloned PCR products [22, 25, 29, 30] and the other is to use in silico tools to assign database sequences to T-RFs and vice versa (TRFCUT [26], PAT [27], TreFID, [28]). Generally, PAT and TreFID [27, 28] allow to assess the composition of denitrifier communities, but their comprehensiveness is dependent on the number of sequence entries currently in the database. Moreover, a significant and non-linear discrepancy (up to 11 bp) is frequently observed between T-RFs that were determined experimentally and in silico. This is presumably due to differences in the GC-content and sequence length of the target gene [31], which requires an additional experimental evaluation.

20.3.3 DGGE

The use of DGGE to fingerprint denitrifier communities in the environment is rather new, although the technique has been exploited since around 1990. DGGE of *nirS*, *nirK*, and *nosZ* fragments was evaluated to analyze denitrifier communities from different environments [11] and DGGE of *nosZ* was later employed to compare soil samples [32]. Goregues et al. [3] also elaborated a method to separate *nirS* fragments by DGGE with the aim to identify sequence variations in marine denitrifying isolates.

DGGE separates gene fragments of the same size on a polyacrylamide gel cast with a gradient of increasing concentration of the denaturants formamide and urea. For optimal resolution, DGGE for each gene must be optimized regarding both gradient concentration and running time. On denaturation, double-stranded DNA partially melts in domains that are sequence specific and then adopts a secondary structure, which slows down the migration rate of the fragment through the gel. The resulting banding patterns thus depend on fragment mobility. DGGE is simplified for fragments having a single melting domain in the region of interest, but functional genes such as the *nirS*, *nirK*, and *nosZ* have several melting domains [11]. This is known to hamper band resolution and typically result in cloudy bands [33]. To avoid complete denaturing of the PCR-amplified fragments and to minimize the effects of multiple melting domains, a GC-clamp is added to one of the primers. With short, GC-clamped fragments, DGGE of *nirK* and *nosZ* genes was proven to be a good tool for fingerprinting denitrifier communities in environmental samples, but further optimization is required for *nirS* [11].

For successful DGGE analysis, the optimal fragment size is about 500 bp. This limits the amount of sequence information and restricts the possibilities of finding appropriate PCR primers. DNA fragments with different sequences can sometimes have similar mobility characteristics and more than one *nirK* or *nosZ* sequence was detected in some bands when the method was evaluated [11]. These were in most cases closely related and thereby difficult to separate accurately. Therefore, conclusions on denitrifier diversity based on the DGGE patterns alone can be ambiguous. DGGE is sometimes also used as a semi-quantitative method in the same way as T-RFLP, but the intensity of the bands does not necessarily correspond to the relative abundance of a sequence type. One advantage as compared to T-RFLP is that DGGE not only provides a fingerprint of the communities, but also allows sequencing of the bands appearing on the gel after excising them. In this way, screening of hundreds of clones can be avoided, at least if information on the predominant populations in the community is requested.

20.4 Quantification of denitrifier communities

On the basis of the analysis of the number of bands or peaks and of their relative intensity, the fingerprint techniques described in the previous section can give estimates of both richness and evenness. However, estimation of the total number of denitrifiers in a given environment has been often neglected even though denitrifying activity is likely to be positively correlated to it. During the last years, different approaches based on either PCR or hybridization have been developed to quantify denitrifiers without any cultivation step.

20.4.1 Competitive PCR (cPCR) and real-time PCR

cPCR is based on the simultaneous amplification of the target DNA and a control DNA with a known concentration, the so-called competitor. The target and competitor DNA competes for the primers during amplification and, by assuming the same amplification efficiency, the mass ratio between the two amplicons is used to determine the initial amount of target DNA. This ratio is estimated by agarose gel analysis of PCRs of multiple dilutions of the competitor with the target DNA. The method was first applied for denitrifiers to quantify *nirS* genes in marine samples [8]. When compared to culture-dependent most-probable-number (MPN) estimates, up to a thousand times more cytochrome cd_1 nitrite reductase denitrifiers were detected by cPCR. More recently, cPCR was used to quantify both *nirS* and *nirK* at different depths in a biofilm reactor [34]. The *nirK* gene was also quantified in soil and stream-sediment samples with numbers in the range of 10^8–10^9 copies g soil^{-1} dry weight [35].

In contrast to cPCR, real-time PCR does not need a gel migration step and multiple dilutions for each sample and is thereby more suitable for high-throughput analysis. Real-time PCR is based on either the Taq-man technology or the SYBR green detection system. SYBR green is a fluorescent dye that non-specifically binds to double-stranded DNA. The Taq-man assay is more specific since it involves both primers and a fluorescently labeled probe and this could make the method more difficult to apply for functional genes with high sequence polymorphism. Both methods rely on the direct proportionality between the intensity of the fluorescent signal measured during the exponential phase of the PCR reaction and the initial amount of target DNA. The copy number of initial target DNA is thereby determined by comparison to a standard curve constructed using target DNA of a known concentration.

A Taq-man real-time PCR assay was developed to estimate the abundance of the *nirS* gene in marine samples [36]. It was shown that primers and probe were specific for *nirS* from *Pseudomonas stutzeri* and they cannot be used to quantify the abundance of the total NirS-type denitrifier community. A more general assay employs the SYBR green system to quantify *nirK* genes in soil samples. The NirK encoding genes were in the range of 10^5–10^8 copies g soil^{-1} dry weight, which was about 100–1,000 times lower than the 16S rDNA copy number [37].

20.4.2 DNA arrays

A simple way to compare abundance of denitrifiers in the environment is by quantifying Southern hybridization signals using denitrifier genes as probes [38]. However, this technique cannot estimate actual numbers of bacteria. The recent microarray techniques are promising for quantification of denitrifiers but the use is still limited by the low sensitivity. Wu et al. [39], who developed the first gene array for assessing diversity and relative abundance of *nirS* and *nirK*, reported that sensitivity obtained with the glass-based microarray hybridization may be 1,000- to 10,000-fold less than with PCR amplification [39]. Even though sensitivity was improved with 70-mer oligonucleotides for *nirK* and *nirS* quantification [40], the detection limit was approximately 10^7 copies. Similarly, other microarrays [41, 42] need improvement of the detection limit before becoming useful. Nonetheless, these oligonucleotide microarray prototypes demonstrated the potential of this approach for studying both diversity and abundance.

20.5 Examples of denitrifier communities in habitats

20.5.1 Soil ecosystems

Use of molecular tools to analyze the denitrifier community in soil has mainly been done to study the impact of fertilization on these communities. In a laboratory experiment, T-RFLP analysis of the *nirK* gene was used to explore the effect of ammonium addition at different concentrations on soil denitrifiers after one month [29]. This resulted in a shift in the structure of the denitrifier community at medium and high ammonia concentration, which probably was attributed to an increased supply of nitrate through nitrification of the added ammonium. Recent field studies focused on comparing effects of mineral versus organic fertilizer on the denitrifying community. Thus, the effect of a 6-year application of mineral fertilizer or cattle manure has been studied using T-RFLP analysis of *nirK* genes [30]. A strong seasonal shift in the structure of the *nirK* community was observed, whereas fertilizer type, rather than amount, appeared to affect the communities to a minor extent. Comparison of seven-year application of composted pig manure or ammonium nitrate also showed significant differences in *narG* and *nosZ* fingerprints and clone libraries obtained from the samples [43]. Interestingly, while potential denitrifying activity was higher under organic than under mineral fertilization, the $N_2O/(N_2O+N_2)$ ratio was lower under organic fertilization. The most impressive effect of the fertilization regimes on the denitrifying community has been observed in a 50 years long-term field experiment, where plots were fertilized with cattle manure, sewage sludge, $Ca(NO_3)_2$, and $(NH_4)_2SO_4$ [32]. As expected, plots treated with the organic fertilizers manure and sewage sludge had the highest potential denitrification rate compared to the other treatments. In contrast, fingerprint analyses of the *nosZ* and *narG* genes showed big differences of the corresponding communities in the plots treated with $(NH_4)_2SO_4$ and sewage sludge compared to the plots treated with the other types of fertilizers and the non-fertilized plots. A significantly lower pH was observed in the plots treated with $(NH_4)_2SO_4$ and sewage sludge suggesting that this long-term fertilization effect on the structure of the denitrifying community was, at least partially, attributed to an indirect effect by soil acidification. A similar effect of pH was also recorded in a study on the long-term effect of elevated pCO_2 on the structure of the *narG* community [44]. Altogether, these experiments demonstrate that composition of the denitrifier community can be affected by fertilizer addition and that soil pH is probably an important driver of the structure of this functional community.

Molecular studies on denitrifier communities in soils were also performed to investigate the diversity in various soil types. Analyses of *nirK*, *nirS*, and *nosZ* gene clone libraries in two acid forest soils revealed a high diversity of the *nosZ* gene but only few sequences were obtained with *nirK* specific primers [45]. The abundance of denitrifiers in layers of one of these forest soils by denitrification gene probing decreased with soil depth [38]. Since nitrate concentration was high in all the layers, it has been suggested that nitrate did not influence denitrifier distribution in this soil. Diversity of the denitrifier communities was also compared between forest and meadow soils at two sampling sites. Analysis of *nosZ* T-RFLP profiles showed that denitrifier community composition differed significantly according to vegetation type and site with the majority of cloned PCR-products clustering with *nosZ* from *α-Proteobacteria* species [22]. Measurements of denitrifying enzyme activity showed mainly a strong effect of the vegetation type. More recently, comparison of the diversity of the denitrifier communities was performed between agricultural and riparian soils and creek sediment. While communities in agricultural and riparian soils were more similar, the creek sediment community was unique [23]. These results, together with comparison of *narG* RFLP fingerprints from seven agricultural soils [46], have shown that each soil has a nearly unique denitrifier community, due to inherent soil properties and specific environmental conditions at the site.

20.5.2 Marine environments

Marine denitrifier communities were explored from sediments [20, 21, 47–49] and the water column at oxic–anoxic interfaces [24, 25, 50]. Except for the studies by Scala and Kerkhof [5, 20, 47], analyses of marine denitrifier communities focused mainly on *nirS*-type denitrifiers. While some of the published primers detected *nirS* genes from marine sediments with sequence identity levels as low as 45% [48], PCR amplification of *nirK* genes using the known primer pairs generally seems to be restricted to more conserved sequences. Whether this is an effect introduced by the primers or of ecological relevance is not resolved yet. However, there is evidence that environmental factors may have different impacts on *nirK*- and *nirS*-type denitrifier communities from sediments of the eastern tropical South Pacific [49]. *nirS*-type denitrifiers were diverse and readily found from layers of the water column of the Arabian Sea with high NO_2^- levels implying a high denitrification activity. In contrast, they were less efficiently detected from layers with high levels of hydrogen sulfide and oxygen [50]. Temporal variation in *nirS*-type denitrifier occurrence was also found for the oxic zone but not for the sulfidic zone within the water column of the Baltic Sea [24].

The gene *nosZ* was used as a functional marker gene to explore temporal and spatial variations of sediment denitrifier communities along scales of centimeters, meters, and kilometers [20]. Variations ranked in the order cm < m < km and were possibly caused by disturbances of the sediment through fauna activities, by its surface topology, and by differing environmental factors at different geographic locations, respectively. These results point toward a high horizontal patchiness of denitrifier communities in the sediment indicating that microbial communities may vary substantially along very small scales. Major differences in the structure of sediment denitrifier communities were also observed between locations that were separated by hundreds of kilometers in the Pacific, which suggests an independent development of denitrifier communities separated by large geographic distances [21, 49]. However, similar environmental conditions along biogeochemical gradients (e.g. C-concentration and quality, N-oxide and O_2-availability, presence of H_2S) at large geographic distances seem to shape similar but not closely related denitrifier communities. These factors were revealed in more recent studies trying to link *nirS*-type denitrifier community structure to the prevalent environmental factors observed within the oxygen minimum zones of the water column of the Arabian Sea, the Eastern South Pacific, and the Baltic Sea [24, 25, 50]. The respective *nirS* clone libraries indicated that denitrification genes from distinct locations were not closely related to but clustered near clones derived from locations with similar biogeochemical properties. In contrast, the surprising stability of marine denitrifier communities that varied only by the relative abundance of *nirS*-T-RFs in sediments of the Pacific Northwest implies that the steep vertical gradients of oxidants (O_2, NO_3^-) observed reflect differential activities of microorganisms with similar genetic and physiological potential rather than distinct communities [21].

20.5.3 Wastewater treatment processes and eutrophic river sediments

The composition of denitrifying communities in municipal and industrial wastewater treatment systems (WWTS), such as activated sludge processes, biofilm process, and constructed treatment wetlands, is largely unknown although considerable effort has been made to improve the technology for efficient and economic removal of N. Moreover, the few studies carried out have mainly focused on laboratory-scale reactors of various process designs with synthetic wastewater making it difficult to draw any general conclusions and to evaluate whether up-scaling of the results is possible. Data reported from clone library analyses of rRNA genes and fluorescence in situ hybridizations showed that members of α-, β-, γ-, and δ-Proteobacteria are present in sludge and biofilm WWTS designed for N removal treating either municipal or industrial wastewater [34, 51–54]. Which of these populations denitrify is impossible to infer since bacteria closely related to a denitrifier do not necessarily have this trait.

Surprisingly, only a few functional gene approaches have been published. Sequencing of the dominant *nirK, nirS*, and *nosZ* genes in activated sludge from two different full-scale treatment plants revealed that some were distantly related to those from *Rhizobium* spp., *Azospirillum* spp., *Rhodobacter sphaeroides*, and *Paracoccus denitrificans*, although the majority clustered with other environmental clones [11]. In contrast, *nosZ* genes related to those found in commonly cultured denitrifiers were enriched in a closed WWTS built for re-cycling of water during long-term space missions [55]. In a saline metallurgic WWTS the diversity of *nirS* was greater than of *nirK*, but they were both much lower than what has been reported from soil [56]. This was also the case in municipal-activated sludge processes (Hallin, unpublished data). In a biofilm reactor, the highest numbers of *nirK* and *nirS* genes were observed in the middle and outer region, respectively, of an aerated biofilm, demonstrating a differential stratification of the copper- and cytochrome cd_1-nitrite reductase [34]. The difference in diversity and density between *nirS* and *nirK* reported in these studies could be a bias introduced by the use of selective primers, but this could also suggest that there is a niche differentiation between *nirK* and *nirS* denitrifiers in these systems.

Denitrification in river sediments or biofilms contributes to control of eutrophication in N-polluted areas. Even though these environments often have strong gradients of nutrients, which make them interesting for gaining more knowledge on the impact of environmental factors on denitrifier ecology in general, they are unfortunately less explored than soil and marine ecosystems. Denitrification activity and *nirK* community composition in river biofilms vary both with nutrient load and time [57] and the diversity of expressed *nirS* genes was high in nitrogen-rich estuarine sediments [58].

20.6 Improving our ways to study denitrifiers in habitats

20.6.1 How to connect functional gene diversity and phylogenetic diversity?

In the studies described above and in other studies as well, it was found that most of the denitrification genes from environmental clone libraries are not related to those from known denitrifiers, suggesting a great, undiscovered diversity of this functional community. Estimations of the diversity of denitrification genes from environmental clone libraries using statistical approaches clearly show that we are still only aware of a very small fraction of this diversity. In a close future, metagenomic approaches with direct cloning and sequencing of large DNA fragments extracted from environmental samples without any PCR step may be very helpful to improve the inventory of denitrification gene diversity.

One shortcoming with only using functional genes or enzymes for surveys of denitrifying communities is the unresolved connection of functional gene/enzyme diversity to phylogenetic diversity. Therefore, directed cultivation approaches that link these genes to the respective organism are needed. A few studies with a rather limited number of denitrifying isolates have tried to correlate the *nir* or *nos* genes to the phylogenetic affiliation of the same isolates and the most likely is that there is no correlation [3, 59]. A likely explanation for this is that not only evolution but also horizontal gene transfer drives the composition of the denitrifying community.

The fact that denitrification can be found in distantly related bacteria with completely different physiological features raises the question as to what ecological implications these other traits may have on the denitrification activity of these bacteria. Presently, stable isotope probing (SIP) is gaining interest to elucidate the active populations contributing to a particular metabolic pathway. With SIP combined with PCR, it is possible to analyze bacterial populations incorporating ^{13}C into their DNA or RNA when growing on a labeled substrate under denitrifying conditions. The technique was used to identify that members of the family Methylophilales were enriched in a laboratory-scale reactor operated under nitrate-reducing conditions with methanol as the sole carbon source [60]. By subsequently targeting these bacteria in the reactor using fluorescence in situ hybridization in combination with microautoradiography, the authors elegantly confirmed that the targeted cells were the dominant methanol-utilizing bacteria. The major drawback with this approach is the lack of foolproof evidence that the targeted cells are actually denitrifiers. To avoid this problem, but loosing the connection to phylogeny, a modified SIP method in combination with *nosZ* amplification was recently used to detect benzoate-degrading denitrifiers in sediment enrichments [61]. Although these recently developed approaches are very useful, there is yet no set of tools available that readily detects denitrifiers and at the same time gives information on the phylogeny and physiology of these organisms.

20.6.2 *Is there a link between diversity and activity of denitrifiers?*

The composition of the denitrifying community was suggested to be important for the in situ denitrification activity and N_2O emissions [62, 63]. Nevertheless, others have shown that potential denitrification activity and community structure appeared uncoupled in agroecosystem [23, 32]. The discrepancy between activity and diversity estimated from DNA extracts highlights two major limits of most molecular studies of denitrifiers. First, while activity of microorganisms is strongly correlated to their numbers, ecological studies have rather focused on denitrifier community composition and neglected to estimate the density of this functional community.

The application of recently developed molecular tools to quantify denitrifiers in the environment should circumvent this limit in the near future. Second, the DNA-based approaches only target bacteria potentially capable to denitrify when appropriate conditions are faced, but do not provide information on their actual activity.

The challenge for microbial ecologists is now to identify active members of this functional community. To do so, better tools must be developed to target bacteria driving the denitrification process in the environment. mRNAs are candidate molecules to target active denitrifiers, but direct mRNA extraction from the environmental samples is problematic. However, the first results to detect denitrifiers that express denitrification genes were achieved by reverse transcription amplification of nitrite reductase and nitrous oxide reductase genes from RNA extracted from sediment [58], pond bacterioplancton [64], and soil [65]. To more directly relate community structure and activity, proteins are better candidates than mRNA because they are at the end of the regulatory cascade. Metz et al. [66], who used antibodies to target NirK and combined this technique with either 16S rRNA oligonucleotide probes or a flow cytometry system to sort out antibody-labeled cells, obtained promising results. Such polyphasic studies, which have the potential to provide insight into both structural and functional aspects of the denitrifier community, will probably be the key to elucidate the functional link between diversity, density, and activity of denitrifying community.

References

[1] T.N. Gamble, M.R. Betlach, J.M. Tiedje, Numerically dominant denitrifying bacteria from world soils, Appl. Environ. Microbiol. 33 (1977) 926–939.

[2] I. Brettar, E.R.B. Moore, M.G. Höfle, Phylogeny and abundance of novel denitrifying bacteria isolated from the water column of the central Baltic Sea, Microb. Ecol. 42 (2001) 295–305.

[3] C.M. Goregues, V. Michotey, P. Bonin, Molecular, biochemical, and physiological approaches for understanding the ecology of denitrification, Microb. Ecol. 49 (2005) 198–208.

[4] L. Philippot, S. Hallin, Molecular analyses of soil denitrifying bacteria, in: J.E. Cooper and J. R. Rao (Eds.) Molecular Approaches to Soil, Rhizosphere and Plant Microorganism Analysis, CABI Publishing, Wallingford, UK. (2006) 146–165.

[5] D.J. Scala, L.J. Kerkhof, Nitrous oxide reductase (*nos*Z) gene-specific PCR primers for detection of denitrifiers and three *nos*Z genes from marine sediments, FEMS Microbiol. Lett. 162 (1998) 61–68.

[6] G. Braker, A. Fesefeldt, K.-P.Witzel, Development of PCR primer systems for amplification of nitrite reductase genes (*nirK* and *nirS*) to detect denitrifying bacteria in environmental samples, Appl. Environ. Microbiol. 64 (1998) 3769–3775.
[7] S. Hallin, P.-E. Lindgren, PCR detection of genes encoding nitrite reductases in denitrifying bacteria, Appl. Environ. Microbiol. 65 (1999) 1652–1657.
[8] V. Michotey, V. Méjean, P. Bonin, Comparison of methods for quantification of cytochrome cd_1-denitrifying bacteria in environmental marine samples, Appl. Environ. Microbiol. 66 (2001) 1564–1571.
[9] K. Kloos, A. Mergel, C. Rösch , H. Bothe, Denitrification within the genus *Azospirillum* and other associative bacteria, Aust. J. Plant Physiol. 28 (2001) 991–998.
[10] T. Yan, M.W. Fields, L. Wu, Y. Zu, J.M. Tiedje, J. Zhou, Molecular diversity and characterization of nitrite reductase gene fragments (*nirK* and *nir*S) from nitrate- and uranium-contaminated groundwater, Environ. Microbiol. 5 (2003) 13–24.
[11] I.N. Throbäck, K. Enwall, A. Jarvis, S. Hallin, Reassessing PCR primers targeting *nirS*, *nirK* and *nosZ* genes for community surveys of denitrifying bacteria with DGGE, FEMS Microbiol. Ecol. 49 (2004) 401–417.
[12] D.A. Flanagan, L.G. Gregory, J.P. Carter, A. Karakas-Sen, J.P. Richardson, S. Spiro, Detection of genes for periplasmic nitrate reductase in nitrate respiring bacteria and in community DNA, FEMS Microbiol. Lett. 177 (1999) 263–270.
[13] L. Philippot, S. Piutti, F. Martin-Laurent, S. Hallet, J.C. Germon, Molecular analysis of the nitrate-reducing community from unplanted and maize-planted soils, Appl. Environ. Microbiol. 68 (2002) 6121–6128.
[14] L.G. Gregory, A. Karakas-Sen, D.J. Richardson, S. Spiro, Detection of genes for membrane-bound nitrate reductase in nitrate-respiring bacteria and in community DNA, FEMS Microbiol. Lett. 183 (2000) 275–279.
[15] G. Braker, J.M. Tiedje, Nitric oxide reductase (*norB*) genes from pure cultures and environmental samples, Appl. Environ. Microbiol. 69 (2003) 3476–3483.
[16] B. Stres, I. Mahne, G. Avgustin, J.M. Tiedje, Nitrous oxide reductase (*nosZ*) gene fragments differ between native and cultivated Michigan soils, Appl. Environ. Microbiol. 70 (2004) 301–309.
[17] D. Chèneby, S. Hallet, M. Mondon, F. Martin-Laurent, J.C. Germon, L. Philippot, Genetic characterization of the nitrate reducing community based on *narG* nucleotide sequence analysis, Microb. Ecol. 46 (2003) 113–121.

[18] E. Mounier, S. Hallet, D. Cheneby, E. Benizri, Y. Gruet, C. Nguyen, S. Piutti, C. Robin, S. Slezack-Deschaumes, F. Martin-Laurent, J.C. Germon, L. Philippot, Influence of maize mucilage on the diversity and activity of the denitrifying community, Environ. Microbiol. 6 (2004) 301–312.
[19] A.K. Patra, L. Abbadie, A. Clays-Josserand, V. Degrange, S.J. Grayston, P. Loiseau, F. Louault, S. Mahmood, S. Nazareth, L. Philippot, F. Poly, A. Richaume, X.L. Roux, Effect of grazing on microbial functional groups involved in soil N dynamics, Ecol. Monogr. 75 (2005) 65–80.
[20] D.J. Scala, L.J. Kerkhof, Horizontal heterogeneity of denitrifying bacterial communities in marine sediments by terminal restriction fragment length polymorphism analysis, Appl. Environ. Microbiol. 66 (2000) 1980–1986.
[21] G. Braker, H.L. Ayala-del-Rio, A.H. Devol, A. Fesefeldt, J.M Tiedje, Community structure of denitrifiers, Bacteria, and Archaea along redox gradients in Pacific Northwest marine sediments by terminal restriction fragment length polymorphism analysis of amplified nitrite reductase (*nirS*) and 16S rRNA genes, Appl. Environ. Microbiol. 67 (2001) 1893–1901.
[22] J.J. Rich, R.S. Heichen, P.J. Bottomley, K. Cromack Jr., D.D. Myrold, Community composition and functioning of denitrifiying bacteria from adjacent meadow and forest soil, Appl. Environ. Microbiol. 69 (2003) 5974–5982.
[23] J.J. Rich, D.D. Myrold, Community composition and activities of denitrifying bacteria from adjacent agricultural soil, riparian soil, and creek sediment in Oregon, USA, Soil Biol. Biochem. 36 (2004) 1431–1441.
[24] M. Hannig, G. Braker, J.W. Dippner, K. Jürgens, Linking denitrifier community structure and prevalent biogeochemical parameters in the pelagial of the central Baltic proper (Baltic Sea), FEMS Microbiol. Ecol. 57 (2006) 260–271.
[25] M. Castro-González, G. Braker, L. Farías, O. Ulloa, Communities of *nirS*-type denitrifiers in the water column of the oxygen minimum zone in the eastern South Pacific, Environ. Microbiol. 7 (2005) 1298–1306.
[26] P. Ricke, S. Kolb, G. Braker, Application of a newly developed ARB software-integrated tool for in silico terminal restriction fragment length polymorphism analysis reveals the dominance of a novel *pmoA* cluster in a forest soil, Appl. Environ. Microbiol. 71 (2005) 1671–1673.
[27] A.D. Kent, D.J. Smith, B.J. Benson, E.W. Triplett, Web-based phylogenetic assignment tool for analysis of terminal restriction fragment length polymorphism profiles of microbial communities, Appl. Environ. Microbiol. 69 (2003) 6768–6776.

[28] C. Rösch, H. Bothe, Improved assessment of denitrifying, N_2-fixing, and total-community bacteria by terminal restriction fragment length polymorphism analysis using multiple restriction enzymes, Appl. Environ. Microbiol. 71 (2005) 2026–2035.
[29] S. Avrahami, R. Conrad, G. Braker, Effect of soil ammonium concentration on N_2O release and on the community structure of ammonia oxidizers and denitrifiers, Appl. Environ. Microbiol. 68 (2002) 5685–5692.
[30] M. Wolsing, A. Priemé, Observation of high seasonal variation in community structure of denitrifying bacteria in arable soil receiving artificial fertilizer and cattle manure by determining T-RFLP of *nir* gene fragments, FEMS Microbiol. Ecol. 48 (2004) 261–271.
[31] C.W. Kaplan, C.L. Kitts, Variation between observed and true Terminal Restriction Fragment length is dependent on true TRF length and purine content, J. Microbiol. Meth. 54 (2003) 121–125.
[32] K. Enwall, L. Philippot, S. Hallin, Activity and composition of the denitrifying community respond differently to long-term fertilization, Appl. Environ. Microbiol. 71 (2005) 8335–8343.
[33] V. Kisand, J. Wikner, Limited resolution of 16S rDNA DGGE caused by melting properties and closely related DNA sequences, J. Microbiol. Meth. 54 (2003) 183–191.
[34] A.C. Cole, M.J. Semmens, T.M. LaPara, Stratification of activity and bacterial community structure in biofilms, grown on membranes transferring oxygen, Appl. Environ. Microbiol. 70 (2004) 1982–1989.
[35] X.Y. Qiu, R.A. Hurt, L.Y. Wu, C.H. Chen, J.M. Tiedje, Z. Zhou, Detection and quantification of copper-denitrifying bacteria by quantitative competitive PCR, J. Microbiol. Meth. 59 (2004) 199–210.
[36] V. Grüntzig, S.C. Nold, J. Zhou, J.M. Tiedje, *Pseudomonas stutzeri* nitrite reductase gene abundance in environmental samples measured by real-time PCR, Appl. Environ. Microbiol. 67 (2001) 760–768.
[37] S. Henry, E. Baudoin, J.C. Lopez-Gutierrez, F. Martin-Laurent, A. Brauman, L. Philippot, Quantification of denitrifying bacteria in soils by *nirK* gene targeted real-time PCR, J. Microbiol. Meth. 59 (2004) 327–335; Erratum in: J Microbiol. Meth. 61 (2005) 289–290.
[38] A. Mergel, O. Schmitz, T. Mallmann, H. Bothe, Relative abundance of denitrifying and dinitrogen-fixing bacteria in layers of a forest soil, FEMS Microbiol. Ecol. 36 (2001) 33–42.
[39] L. Wu, D.K. Thompson, X. Liu, M.W. Fields, C.E. Bagwell, J.M. Tiedje, J. Zhou, Development and hybridization of microarray-based whole-genome hybridization for detection of microorganisms within the context of environmental applications, Environ. Sci. Technol. 38 (2004) 6775–6782.
[40] G. Taroncher-Oldenburg, E.M. Griner, C.A. Francis, B.B. Ward, Oligonucleotide microarray for the study of functional gene diversity

in the nitrogen cycle in the environment, Appl. Environ. Microbiol. 69 (2003) 1159–1171.

[41] J.-C. Cho, J.M. Tiedje, Quantitative detection of microbial genes by using DNA microarrays, Appl. Environ. Microbiol. 68 (2002) 1425–1430.

[42] S.M. Tiquia, L. Wu, S.C. Chong, S. Passovets, D. Xu, T. Xu, J. Zhou, Evaluation of 50-mer oligonucleotide arrays for detecting microbial populations in environmental samples, Biotechniques 36 (2005) 664–670.

[43] C. Dambreville, S. Hallet, C. Nguyen, C. Morvan, J.C. Germon, L. Philippot, Structure and activity of the denitrifying community in a maize-cropped field fertilized with composted pig manure or ammonium nitrate, FEMS Microbiol. Ecol. 56 (2006) 119–131.

[44] K. Deiglmayr, L. Philippot, U.A. Hartwig, E. Kandeler, Structure and activity of the nitrate-reducing community in the rhizosphere of *Lolium perenne* and *Trifolium repens* under long-term elevated atmospheric pCO_2, FEMS Microbiol. Ecol. 49 (2004) 445–454.

[45] C. Rösch, A. Mergel, H. Bothe, Biodiversity of denitrifying and dinitrogen-fixing bacteria in an acid forest soil, Appl. Environ. Microbiol. 68 (2002) 3818–3829.

[46] L. Philippot, Tracking nitrate reducers and denitrifiers in the environment, Biochem. Soc. Trans. 33 (2005) 200–204.

[47] D.J. Scala, L.J. Kerkhof, Diversity of nitrous oxide reductase (*nos*Z) genes in continental shelf sediments, Appl. Environ. Microbiol. 65 (1999) 1681–1687.

[48] G. Braker, J. Zhou, L. Wu, A.H. Devol, J.M. Tiedje, Nitrite reductase genes (*nirK* and *nirS*) as functional markers to investigate diversity of denitrifying bacteria in Pacific Northwest marine sediment communities, Appl. Environ. Microbiol. 66 (2000) 2096–2104.

[49] X. Liu, S. Tiquia, G. Holguin, L. Wu, S. Nold, A. Devol, K. Luo, A. Palumbo, J. Tiedje, J. Zhou, Molecular diversity of denitrifying genes in continental margin sediments within the oxygen-deficient zone off the Pacific coast of Mexico, Appl. Environ. Microbiol. 69 (2003) 3549–3560.

[50] D.A. Jayakumar,C.A. Francis, S.W.A. Naqvi, B.B. Ward, Diversity of nitrite reductase genes (*nirS*) in the denitrifying water column of the coastal Arabian Sea, Aquat. Microb. Ecol. 34 (2004) 69–78.

[51] J. Snaidr, R. Amann, I. Huber, W. Ludwig, K.H. Schleifer, Phylogenetic analysis and in situ identification of bacteria in activated sludge, Appl. Environ. Microbiol. 63 (1997) 2884–2896.

[52] S. Juretschko, A. Loy, A. Lehner, M. Wagner, The microbial community composition of a nitrifying-denitrifying activated sludge from an industrial sewage treatment plant analyzed by the full-cycle rRNA approach, System. Appl. Microbiol. 25 (2002) 84–99.

[53] N. Lee, H. Aspegren, M. Henze, P.H. Nielsen, M. Wagner, Population dynamics in wastewater treatment plants with enhanced biological phosphorus removal operated with and without nitrogen removal, Water Sci. Technol. 46 (2002) 163–170.
[54] M. Eschenhagen, M. Schuppler, I. Roske, Molecular characterization of the microbial community structure in two activated sludge systems for the advanced treatment of domestic effluents, Water Res. 37 (2003) 3224–3232.
[55] Y. Sakano, K.D. Pickering, P. E. Strom, L.J. Kerkhof, Spatial distribution of total, ammonia-oxidizing, and denitrifying bacteria in biological wastewater treatment reactors for bioregenerative life support, Appl. Environ. Microbiol. 68 (2002) 2285–2293.
[56] S. Yoshie, N. Noda, S. Tsuneda, A. Hirata, Y. Inamori, Salinity decreases nitrite reductase gene diversity in denitrifying bacteria of wastewater treatment systems, Appl. Environ. Microbiol. 70 (2004) 3152–3157.
[57] M.R. Chenier, D. Beaumier, R. Roy, B.T. Driscoll, J.R. Lawrence, C.W. Greer, Impact of seasonal variations and nutrient inputs on nitrogen cycling and degradation of hexadecane by replicated river biofilms, Appl. Environ. Microbiol. 69 (2003) 5170–5177.
[58] B. Nogales, K.N. Timmis, D.B. Nedwell, A.M. Osborn, Detection and diversity of expressed denitrification genes in estuarine sediments after Reverse Transcription-PCR amplification from mRNA, Appl. Environ. Microbiol. 68 (2002) 5017–5025.
[59] L.G. Gregory, P.L. Bond, D.J.Richardson, S. Spiro, Characterization of a nitrate respiring bacterial community using the nitrate reductase gene (*narG*) as a functional marker, Microbiology 149 (2003) 229–237.
[60] M.P. Ginige, P. Hugenholtz, H. Daims, M. Wagner, J. Keller, L.L. Blackall, Use of stable-isotope probing, full-cycle rRNA analysis, and fluorescence in situ hybridization-microautoradiography to study a methanol-fed denitrifying microbial community, Appl. Environ. Microbiol. 70 (2004) 588–596.
[61] E. Gallagher, L. McGuinness, C. Phelps, L.Y. Young, L.J. Kerkhof, ^{13}C-carrier DNA shortens the incubation time needed to detect benzoate-utilizing denitrifying bacteria by stable-isotope probing, Appl. Environ. Microbiol. 71 (2005) 5192–5196.
[62] M.A. Cavigelli, G.P. Robertson, Role of denitrifier diversity in rates of nitrous oxide consumption in a terrestrial ecosystem, Soil Biol. Biochem. 33 (2001) 297–310.
[63] L. Holtan-Hartwig, P. Dörsch, L.R. Bakken, Comparison of denitrifying communities in organic soils: kinetics of NO_3^- and N_2O reduction, Soil Biol. Biochem. 32 (2002) 833–843.
[64] M.G. Weinbauer, I. Fritz, D.F. Wenderoth, M.G. Höfle, Simultaneous extraction from bacterioplankton of total RNA and DNA suitable for

quantitative structure and function analysis, Appl. Environ. Microbiol. 68 (2002) 1082–1087.

[65] S. Sharma, M.K. Aneja, J. Mayer, J.C. Munch, M. Schloter, Diversity of transcripts of nitrite reductase genes (*nirK* and *nirS*) in rhizospheres of grain legumes, Appl. Environ. Microbiol. 71 (2005) 2001–2007.

[66] S. Metz, M. Beisker, A. Hartmann, M. Schloter, Detection methods for the expression of the dissimilatory copper-containing nitrite reductase gene (*DnirK*) in environmental samples, J. Microbiol. Meth. 55 (2003) 41–50.

Chapter 21

Denitrification and Agriculture

Jean Charles Munch and Gerard L. Velthof

21.1 Nitrogen in agricultural systems

Agricultural soils are managed systems, in view of human purposes. First attempt is providing the populations with food. Second is the production of plants of high quality, as well as animal feed with sufficient protein content. Furthermore, from the economic point of view, agriculture needs yields providing sufficient income. The essential plant nutrient N has to be used appropriately for these several goals. In a longer time appreciation, soil fertility is also to be considered, soil organic matter being the main reserve of plant nutrients. Providing growing plants with enough and especially with adequate amounts of N needs (i) the knowledge about the demand of the special crop, (ii) considerations on the organic N pool in soil and (iii) on its transformation in plant-available inorganic forms (NH_4^+ and NO_3^-) as well as (iv) taking in balance the potential biological N_2 fixation and the N deposition. The mineralization of N from the organic pool is provided by soil microorganisms, using soil organic matter as source of energy and nutrient. Weather conditions strongly affect the N transformations in soil. Because weather conditions

Biology of the Nitrogen Cycle
Edited by H. Bothe, S.J. Ferguson and W.E. Newton

are not predictable with enough precision, the calculation of N flow from the immobile into the plant-available form is generally a site-specific determination relying on the average weather data from the last few years. As an almost general rule, NO_3^- accumulates periodically in soils. However, the main accumulation of NO_3^- occurs (i) after application of certain fertilizers or manure, (ii) after harvest, when plant residues are degraded by soil microorganisms and (iii) in late winter before plant growth. When no growing plants take up the NO_3^-, it may be denitrified or water transferred to deeper soil parts and to groundwater or surface water.

21.2 Factors controlling denitrification in agricultural soils

The most important factors controlling denitrification are (i) the presence of an electron donor or energy source for the denitrifying bacteria, mostly available organic carbon, (ii) anoxic conditions and (iii) the NO_3^- content in the soil. If any of these conditions is not fulfilled, denitrification is unlikely. Factors affecting biological processes, such as the temperature, pH or the presence of toxic substances (e.g. heavy metals, organic compounds), may also affect denitrification. Considerable differences in the factors controlling denitrification exist between the topsoil (about 0–50 cm soil depth) and the subsoil (deeper than 50 cm).

21.2.1 Energy source

Organic C sources are available in most agricultural topsoils in the form of soil organic matter, decaying plant roots, crop residues and manure. The denitrification potential, i.e. its rate measured under anoxic conditions at 20°C at an excess of NO_3^-, is an indicator for the content of available C in the soil. Denitrification potentials in agricultural soils strongly decrease with increasing depth (Table 21-1). The denitrification potential of the topsoil is higher in grassland than in maize fields. In permanent grasslands, organic matter accumulates during ageing. The denitrification potential decreases in the order: peat soils > clay soils > loamy soils > sandy soil. In arable soils, the rates of denitrification and of N_2O emissions are higher when crop residues with easily available C are incorporated in the soil, such as leaves from sugar beet and vegetables, than when crop residues with stable organic C are incorporated, such as straw from wheat. Manures contain easily degradable C, such as volatile fatty acids, and application of manures to soils increases denitrification activity and N_2O emissions [1, 2]. Adjustment of animal feed and the time and method of manure storage affect the composition of manure and this may also affect denitrification and N_2O emissions after application to soil.

The situation is different for subsoils than for topsoils. Most subsoils contain little organic C. Transport of dissolved organic carbon from upper

Table 21-1 Potential denitrification rates (mg N per kg dry soil per day) in grassland and maize fields on different soils in the Netherlands.

Soil layers (cm)	Grassland Peat (n = 3)	Clay (n = 3)	Loam (n = 2)	Sand (n = 3)	Maize land Loam (n = 2)	Sand (n = 3)
0–20	267	151	65	26	20	11
20–40	317	125	30	4	9	4
40–60	116	5	1	0.1	1	0.1
60–80	61	0.9	0.3	0.5	0.3	0
80–100	39	0.6	0.2	0.2	0.1	0

Note: Denitrification is performed under anoxic conditions at 20°C with excess of NO_3^-.

soil to subsoils may occur and may increase denitrification there. However, the amount of C leached to the subsoil is much smaller than the amount of C added to the topsoil via crops residues and manures. In some subsoils and sediments, FeS_2 in pyrite may be used as an electron donor for denitrifying bacteria. Significant denitrification can occur even with no C available.

21.2.2 Anoxic conditions

The O_2 content of a soil is largely influenced by rainfall, irrigation, groundwater table, soil texture and plant root plus microbial respiration. In most agricultural topsoils aerobic conditions will prevail, because farmers will try to keep the soils aerated in order to stimulate crop growth. In the topsoil, O_2 abundantly occurs in general, thus not allowing denitrification to proceed. Generally, denitrification rates increase after rainfalls or irrigation and decrease again when the soil dries out. The chance for anoxic conditions is higher in soils with low porosity (clayey and loamy soil) than in soils with a coarse structure as in sandy soils [3]. The water-filled pore space, i.e. the percentage of the soil pores filled with water, is often used as an indicator for anoxic conditions. Supplementary knowledge about soil structure leads to the quantification of not aerated microsites in the soils. Denitrification rates in soils exponentially increase when water-filled pore space increases from about 90–100%.

Anoxic conditions or O_2 deficiency may also occur if the rate of O_2 consumption in the soil exceeds that of supply of O_2. High O_2 consumptions are found when the respiration activity in the soil is high, e.g. after application of an easily degradable source of organic matter (manures or crop residues). Since O_2-transport rates are much slower in wet than in dry soils, anoxic conditions and anoxic microsites are most probable in wet soils with a high respiration rate. However, local high rates of O_2

consumption (e.g. near a source of easily degradable organic matter) may cause enhanced denitrificaton in microsites in soil, even in dry soils. The O_2 content is also affected by soil compaction, e.g. by tractors or grazing cattle. Soil compaction may enhance denitrification, especially in wet soils [3]. The O_2 status of the subsoils may vary widely, depending on the groundwater table and the soil organic matter content. In most soils, O_2 concentrations decrease with increasing depth, where O_2 demand is however a limited one.

21.2.3 Nitrate content

There are many possible sources of NO_3^- in agricultural soils, including artificial fertilizers, manures, organic products, crop residues, atmospheric deposition and biological N_2 fixation. NO_3^- can be directly applied to soils via artificial fertilizers. Application of NH_4^+-containing fertilizers and manures may also increase the NO_3^- content of the soil by nitrification. Organic N can be mineralized to NH_4^+ and, by nitrification, to NO_3^- in the soil. Thus, application of organic N via manures or crop residues may also increase the NO_3^- content in the soil. NO_3^- can be removed from the topsoil by crop uptake, leached to deeper soil layers, groundwater and surface water, incorporated in soil organic matter and converted to gases by denitrification.

The NO_3^- content in agricultural soils strongly varies in time, because of the different sources and sinks of NO_3^-. Generally, highest NO_3^- contents in topsoils are found just after fertilizer or manure application. In this period, risk on denitrification and N_2O emission is highest. The NO_3^- contents decrease after the time following N application to soil, because of plant uptake. The importance of winter-time emissions has not been elucidated definitively but has to be taken into account [4].

In subsoils, the major source of NO_3^- is the NO_3^- leached from the topsoil. The major sink is denitrification and leaching to deeper layers, especially when there is no or limited crop N uptake from the subsoil. Variations in NO_3^- contents of the subsoil are less strong than those in the topsoil.

21.3 Agricultural consequences

Denitrification is an important source of N loss from agricultural soils. This loss of plant-available N can lead to yield depression and decline of quality (e.g. the protein content) of the harvested products. Therefore, farmers fertilize more N than the minimal level to plants, taking into account these losses. The mineral fertilizer consists of inorganic N synthesised by the Haber–Bosch procedure, needing high energy consumptions. However, fertilization is applicable with a high efficiency in periods of rapid plant development, when disposed in adequate formulations.

Manure consists of both of inorganic and organic N. The organic N becomes available for plants with a time lag after microbial attack and degradation that runs also in periods with not sufficient root development to absorb all bio-available N. Good knowledge about the contents of plant-available N in manure (i.e. inorganic N and easily mineralizable N) and site-dependent mineralization behaviour is required to tune the amount of applied manure N to the N uptake of the crop and to decrease N losses.

21.4 Environmental consequences of denitrification

From an environmental point of view, denitrification has both positive and negative effects. The positive effect of denitrification is that it decreases the leaching of NO_3^- to ground and surface waters. The negative effect is that denitrification is a major source of the greenhouse gas N_2O and is a loss of N otherwise available for the growth of plants.

21.4.1 Leaching of NO_3^-

Leaching of NO_3^- to ground and surface waters may result in high NO_3^- concentrations in these waters. Water with a high NO_3^- concentration is not suitable for drinking purposes, because it may pose a risk to human and health (it may cause methaemoglobinaemia). The standard for NO_3^- concentration in drinking water set by WHO is 50 mg L^{-1}. This value is adapted by the E.U. as a standard for groundwater in the Nitrate-Directive, which aims at decreasing the pollution of ground and surface waters by NO_3^-. In surface waters, NO_3^- and other nutrients may cause eutrophication, i.e. excessive growth of plants, including algae.

Denitrification largely influences the NO_3^- concentrations in ground and surface waters. In soils with high denitrification, such as clayey and peat soils, concentrations of NO_3^- in groundwater are generally low. In sandy soils, NO_3^- concentrations are higher, but large differences between soils may occur. In the Netherlands, sandy soils with shallow groundwater levels have lower NO_3^- concentrations than those with deep groundwater levels (Figure 21-1), indicating higher denitrification rates in the wet sandy soils.

During transport of NO_3^- to surface water, denitrification may remove NO_3^-. In riparian zones, denitrification is often enhanced because of the wet conditions and the presence of available organic matter. Denitrification also occurs when the NO_3^- diffuses into the anoxic sediments of the surface waters.

21.4.2 Nitrous oxide emission

N_2O is a stable and strong greenhouse gas. It has a lifetime of over 100 years and is 300 times more effective than CO_2. Soils are the dominant

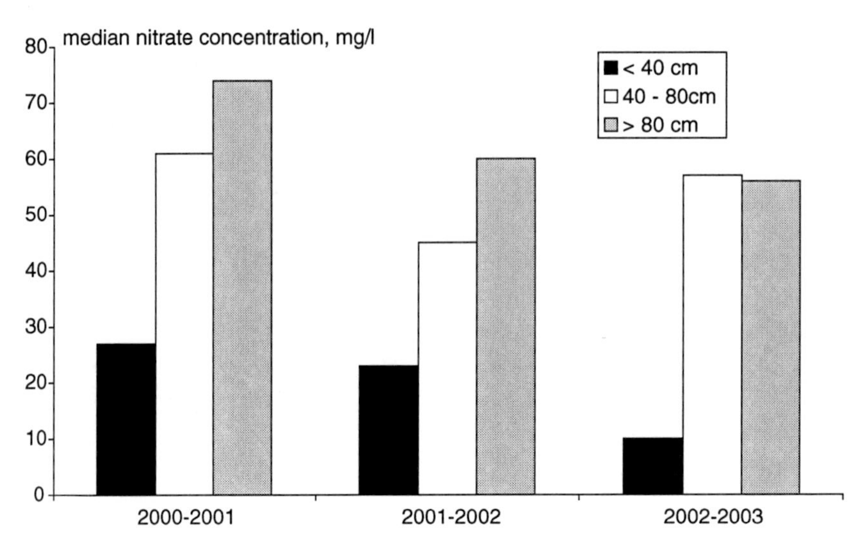

Figure 21-1 NO_3^- concentrations in the upper groundwater of sandy agricultural soils in the Netherlands in the period 2000–2003 at three classes of groundwater tables, i.e. most shallow groundwater level <40 cm (n = 70–72), 40–80 cm (n = 183–185) and >80 cm (n = 212–217).

source of atmospheric N_2O. In them, N_2O is produced mostly by denitrification and, to some extent, by nitrification. Highest N_2O emission rates are generally found during wet conditions in fertilized or freshly harvested soils and in thawing soils.

The factors controlling denitrification also affect N_2O emissions. However, there is not always a positive linear relationship between denitrification and N_2O emissions, because the ratio between the end products of denitrification, i.e. N_2O and N_2, is also affected by the controlling factors (Table 21-2).

Countries must report their N_2O-emissions to United Nations Framework Convention on Climate Change (UNFCCC), like all other greenhouse gas productions. The Intergovernmental Panel on Climate Change (IPCC) developed a simple method of calculating the N_2O emissions from agricultural sources. It is based on activity data and emission factors. The activity data include the amounts of fertilizer and manure applied to soil, soil type and number of animals. The emission factors express the amount of N_2O produced per unit of N input. For example, the N_2O emission factor for fertilizers is the percentage of the N applied as fertilizer that escapes as N_2O to the atmosphere. Emission factors are derived from experiments in which N_2O emission is measured. The variation in emission factors is high, because of effects of soil type, climate, fertilizer type, crop etc. The

Table 21-2 Effect of changes in factors on denitrification and on the N_2O/N_2 ratio of denitrification.

	Effect on denitrification	Effect on N_2O/N_2 ratio
Increasing NO_3^- content	+	+
Increasing O_2 content	−	+
Increasing available organic carbon content	+	−
Increasing temperature	+	−
Decreasing pH	−	+

IPPC defined default (i.e. average) emission factors, but countries can use country-specific emission factors. The default emission factor for nitrous oxide fertilizers and manures is 1.25% of the N applied. The IPCC method also gives estimates for the N_2O that is emitted during denitrification in ground and surface waters; 2.5% of the N leached to ground and surface water ultimately escapes as N_2O to the atmosphere.

21.5 Quantification of denitrification losses

21.5.1 Measurements

There are three approaches to quantify denitrification losses from agricultural soils, namely (i) measurement, (ii) N budget calculations and (iii) modelling.

Quantifying denitrification is difficult and no standard, absolute methods exist. Measurements have to encounter the difficulty that the major end product N_2 cannot easily be detected because of its high background concentration of almost 80% in ambient air. The most frequently used measurement methods are the acetylene-inhibition method and ^{15}N-labelling techniques.

C_2H_2 inhibits N_2O reductase, yielding N_2O as the end product of denitrification in a C_2H_2-treated soil [5]. Production of N_2O can readily be measured due to the low background concentration of N_2O approximately 315 ppb in ambient air. The C_2H_2-inhibition technique is the most widely used method to measure denitrification activity. However, there are several disadvantages of using this technique:

- C_2H_2 also blocks nitrification, at higher concentrations, by which denitrification may be made to slow down when nitrification is the major source of NO_3^- in soils;
- C_2H_2 diffusion into the soil and N_2O out of the soil may be a problem, especially in wet or heavy-textured soils;

- To facilitate a homogeneous distribution of C_2H_2 throughout the soil, mostly intact soil cores are taken from the field and incubated in closed bottles containing C_2H_2. Taking soil cores may disturb the soil structure and O_2 may diffuse into soil possibly also affecting denitrification;
- Microorganisms may adapt to C_2H_2, or even use it as energy source;
- The temporal and spatial variability of denitrification in the field is very high, because of the many controlling factors. Many samples are required to determine an accurate estimate of denitrification losses [6, 7];
- Quantification of denitrification in subsoil is difficult because samples have to be taken from deep soil layers and so under disturbed conditions.

A review of measured denitrification rates using C_2H_2-inhibition technique showed a large range of annual denitrification losses from agricultural soils from 0 to 239 kg N ha^{-1} and year [8]. Highest denitrification losses were measured from irrigated fertilized soils. The mean denitrification rate in agricultural soils was 13 kg N ha^{-1} and year. Most of the studies only measured denitrification in the topsoil, so that overall denitrification rates including those for the subsoil are probably higher.

The fate of ^{15}N-labelled substrate (fertilizer, manure, crop residue) can be measured, including ^{15}N-labelled N_2 and/or N_2O fluxes. The advantage of this method is that it can be part of 'normal' fertilization (no disturbance), used for long-term measurements, that N_2 production can be determined and that an N budget can be made. However, this method also has several disadvantages:

- The analysis of ^{15}N is only possible at specialized laboratories;
- Only denitrification from ^{15}N is measured. Denitrification from other N sources is not included (e.g. N mineralized from organic matter);
- It is assumed that ^{15}N is homogeneously distributed throughout the soil. A heterogeneous distribution of ^{15}N, i.e. due to adsorption to soil particles or aggregation of organic residues, may lead to errors in the estimation of denitrification;
- ^{15}N-labelled material can be expensive and can only be used on small plots.

N_2O emission is mostly measured with closed flux chambers. Boxes are placed on the soil and N_2O emission is calculated from the increase of the N_2O concentration in the headspace of the box in time [9, 10]. There is a large diversity in flux chamber techniques with different sizes (generally less than 1 m^2), materials, operation systems, gas analysis techniques and measurement time. Flux chambers are often used in field experiments with different treatments in order to obtain emission factors. Emissions of

N_2O can also be measured using micro-meteorological techniques. The N_2O emission is then calculated from the vertical N_2O gradient, using data of wind speed and turbulence. Sophisticated equipment is required for this method. Micrometeorological methods are used for quantification of N_2O emission at a larger spatial scale than with flux chamber methods, but are less suitable to get N_2O-emission factors.

21.5.2 N budgets

Another approach to estimate denitrification from soils is to estimate or measure the inputs of N (fertilizers, manures, etc.) and the outputs of N via the harvested crop and grazed grassland (Table 21-3). Leaching, NH_3 volatilization and storage in the soil can generally be estimated or measured more accurately than denitrification, so that denitrification can be roughly determined from the difference between the N inputs via fertilizer and manure and the outputs via harvest and the estimates of the other N losses. Budgets of N can be calculated from plot scale to global scale. The general drawback of estimation of denitrification rates from N budgets is that all errors in the estimates of the different inputs and outputs contribute to the final estimate of denitrification.

21.5.3 Modelling

Several computer models with a wide range of complexity have been developed to calculate denitrification rates. In simple models, denitrification is assumed to be determined by parameters that can be easily measured, such as NO_3^- content, water content and temperature. Some models convert a measured potential denitrification (Dpot) rate into an actual denitrification rate (Dact) by using reduction functions for NO_3^- content (fN), anoxic conditions (fS), temperature (fT) and soil acidity (fpH): Dact = Dpot × fN × fS × fT × fpH [11]. The reduction must be derived from experiments. The shapes of the reduction functions differ amongst the different studies, indicating that they may be site specific.

Table 21-3 N budget at the field scale.

Input of N	Output of N
Fertilizers	Harvested crop
Manures	Grazing by cattle and sheep
Excretion during grazing	
Organic products	Leaching to ground and surface water
Atmospheric deposition	Ammonia volatilization
Biological N_2 fixation	Storage in soil
Mineralization of soil organic N	Denitrification
Total N input	Total N output

Complex process models simulate the dynamics and metabolism of denitrifying bacteria as well as the effects of different environmental conditions on denitrification. These models can be used on a different scale (microbial culture studies to field studies) and mostly require detailed input data. These data can often not easily be determined.

Meta-models are based on large sets of data, either obtained experimentally or by detailed computer models. Denitrification is then related to soil, crop, management and environmental characteristics. The general drawback of meta-models is that they are only valid within the range of conditions they were determined for. Extrapolation outside these conditions or in time should not be done. The goodness of the models is not always large.

21.6 Mitigation of denitrification activity in the field

Mitigation of denitrification and N_2O emissions from agricultural and horticultural soils is possible within limits; its total elimination is, however, impossible. The is due to (i) the dependence of processes of weather conditions and (ii) need of management also in periods without crops, without uptake of NO_3^- by plants. Mitigation is possible by reducing generally N input in soils, which may also lead to yield depression and loss of product quality (as given in organic farming). A reduction of N losses by conservation of high yield and high plant quality is possible by splitting of N fertilization [12].

As recently found, N_2O emissions and NO_3^- leaching are also due to heterogeneity of soils in single fields, being managed homogeneously. Precision agriculture is a form of soil management using yield maps of fields and considering heterogeneity of soils in the single fields (often given by remote-sensing techniques – plane or satellites or by soil contacting radars). Plant density and N fertilization are realized according to this heterogeneity by GPS positioning of seeders and fertilizer engines.

Mitigation may also further be possible by adaptation of soil management. Each kind of soil tillage intensifies microbial soil activity by making available more soil organic matter from aggregates. Thus O_2 consumption is intensified and, upon O_2 deficiency, denitrification begins. From the economical point of view, no till is attractive for farmers, but it leads to accumulation of most soil organic matter in a finer soil zone. Microbial activity is intensified in this zone, with the hazard of O_2 deficiency and enhanced denitrification activity. O_2 deficiency occurs often in wet soils. Drainage of these sites reduces denitrification. Denitrification losses and N_2O emission may also be decreased by avoiding application of fertilizers and manure during wet conditions, especially NO_3-containing fertilizers [13].

References

[1] H. Flessa, R. Ruser, R. Schilling, N. Loftfield, J.C. Munch, E.A. Kaiser, F. Beese, N_2O and CH_4 fluxes in potato fields: automated measurement, management effects and temporal variation, Geoderma 105 (2002) 307–325.
[2] G.L. Velthof, J.A. Nelemans, O. Oenema, P.J. Kuikman, Gaseous N and carbon losses from pig manure derived from different diets, J. Environ. Qual. 34 (2005) 698–706.
[3] R. Ruser, A. Flessa, R. Russow, G. Schmidt, F. Buegger, J.C. Munch, Emissions of N_2O, N_2 and CO_2 from soil fertilized with nitrate: effect of compaction, soil moisture and rewetting, Soil Biol. Biochem. 38 (2006) 263–274.
[4] U. Sehy, J. Dyckmans, R. Ruser, J.C. Munch, Adding dissolved organic carbon to stimulate freeze-thaw related N_2O-emissions from soil, J. Plant. Nutr. Soil Sci. 167 (2004) 471–478.
[5] T. Yoshinari, R. Hynes, R. Knowles, Acetylene inhibition of nitrous oxide reduction and measurement of denitrification and nitrification in soil, Soil Biol. Biochem. 9 (1977) 177–183.
[6] E.A. Kaiser, R. Ruser, Nitrous oxide emissions from arable soils in Germany – An evaluation of six long term field experiments, J. Plant Nutr. Soil Sci. 163 (2000) 249–260.
[7] E.A. Kaiser, K. Kohrs, M. Kuecke, E. Schnug, O. Heinemeyer, J.C. Munch, Nitrous oxide release from arable soils: importance of N-fertilization, crops and temporal variation, Soil Biol. Biochem. 30 (1998) 1553–1563.
[8] L. Barton, C.D.A. McLay, L.A. Schipper, C.T. Smith, Annual denitrification rates in agricultural and forest soils: a review, Aust. J. Soil Res. 37 (1999) 1073–1093.
[9] G.L. Velthof, O. Oenema, Nitrous oxide fluxes from grassland in the Netherlands: I. Statistical analysis of flux chamber measurements, Eur. J. Soil Sci. 46 (1995) 533–540.
[10] E.A. Kaiser, J.C. Munch, O. Heinemeyer, Importance of soil cover box area for the determination of N_2O emissions from arable soils, Plant Soil 181 (1996) 185–192.
[11] M. Heinen, Simplified denitrification models: overview and properties, Geoderma (2005) (in press).
[12] U. Sehy, R. Ruser, J.C. Munch, Nitrous oxide fluxes from maize fields: relationship to yield, site-specific fertilization, and soil conditions, Agr. Ecosyst. Environ. (2003) 97–111.
[13] G.L. Velthof, O. Oenema, R. Postma, M.L. van Beusichem, Effects of type and amount of applied nitrogen fertilizer on nitrous oxide fluxes from intensively managed grassland, Nutr. Cycl. Agroecosyst. 46 (1997) 257–267.

Chapter 22

Denitrification and N-Cycling in Forest Ecosystems

Per Ambus and Sophie Zechmeister-Boltenstern

22.1 Introduction

Denitrification is the unique pathway whereby N in the terrestrial biosphere is transformed into atmospheric N_2. Although the total loss is small compared to the biosphere N-burden, many investigations have attempted to quantify denitrification in terrestrial systems. Denitrification in agricultural systems has received particular interest due to the discovery that up to 30% of applied fertilizer N could be lost via the process. More recently, the denitrification process also in other ecosystems has received increasing attention. One reason is the potential of the denitrifying process to alleviate environmental impacts of NO_3^- in ground- and surface waters via the enzymatic reduction of NO_3^- in stream sediments, constructed wetlands, riparian buffer strips and biological sewage treatment plants. Increased interest in denitrification in the biosphere has also developed in parallel with the increasing focus on atmospheric change and global warming as the denitrification process is a source of atmospheric NO and N_2O.

Biology of the Nitrogen Cycle
Edited by H. Bothe, S.J. Ferguson and W.E. Newton

The occurrence of denitrification in forest ecosystems has consequently been subject to much research during the past three decades. Globally, forest ecosystems cover 29.6% of the total land area and thus contribute a potential source of NO and N_2O effluxes into the atmosphere. The greater part of research in forest N-cycling and denitrification has been directed towards forests in the boreal and temperate regions, and knowledge from these studies will provide basis for the presentation in this chapter.

22.2 Characteristics of the N-cycle in forest ecosystems

N-cycling in many forest ecosystems can be characterized by a relatively closed nature where the rates of input from precipitation and biological N_2-fixation balances outputs by denitrification, volatilization and leaching losses (Figure 22-1). This, for example, was experimentally verified in a series of coniferous forest systems that exhibited a strong capacity to assimilate almost all of the NO_3^- produced [1]. Furthermore, the rates of input and output of forest systems are very small in comparison with the total storage of N in the forest and the quantity that is cycled annually in the vegetation. The soil pool of total organic N is the biggest N storage pool amounting to 9–15 tons of N ha^{-1}. It is followed by N storage in trees and understory, which can have 0.6–1.4 tons, contained in needles/leaves, wood, twigs, branches and bark, coarse and fine roots. The partitioning between these plant components varies, although N-concentrations are larger in young plant parts. In 12 European forests, plant N pools constituted 2–19% of the total ecosystem N pool [2]. The microbial N pool contains 30–300 kg N ha^{-1}; it is easily mineralized and can be partly released after extreme weather events, which may result in sudden flushes of N leaching and denitrification. The extractable soil inorganic N pool is transient and may contain 10–40 kg N ha^{-1}, some of which is absorbed to clay minerals and humus surfaces. N-fluxes are variable in different forest ecosystems and additionally they show interannual variation. Typically input rates are less than 10 kg N ha^{-1} yr^{-1}, which is counterbalanced by similar outputs, whereas the internal N-mineralization in a mature forest system may approach 100–500 kg N ha^{-1} yr^{-1}. About 25–30% of this N is taken up by plants, the rest may be re-immobilized by the microflora. Non-symbiotic N_2-fixation is of marginal relevance in most forests. However, forests containing N_2-fixing plant species, such as alder trees, may have additional biological N inputs of up to 100 kg N ha^{-1} yr^{-1}. Denitrification rates in such forests can reach values of up to 40 kg N ha^{-1} yr^{-1}. Drained peatlands make also an exception to the closed model N-cycle, as N mineralization is speeded up with consequent high denitrification rates. Finally, this ideal closed forest N-cycle is obviously interrupted in forests subject to harvest or logging and fertilization.

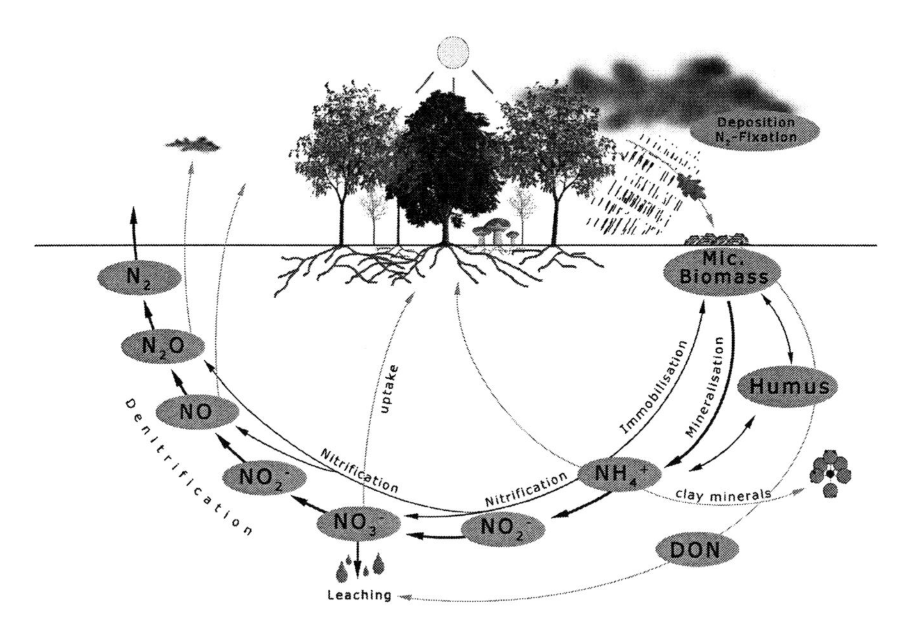

Figure 22-1 Outline of the nitrogen cycle in a forest ecosystem.

22.3 Impact of increased nitrogen inputs on denitrification

Forest exposed to enhanced N depositions, as is typical in urban areas or regions with intensive animal farming, may suffer from an interrupted "leaking" N-cycle with increased N-outputs. As a matter of fact, growing evidence over the past two decades has shown that forests exposed to increased atmospheric N depositions have become N saturated, indicated by the availability of mineral N in excess of combined plant and microbial nutritional demand [3]. Consequently, N saturation can be determined simply by the accumulation of mineral N, usually NH_4^+, in the forest soils or by increased losses of N via leaching and gaseous emissions to the environment compared with background levels in unaffected forests [3]. The celerity of the saturation process depends on soil conditions, forest type and forest history. Four general stages along this continuum have been discussed. The transit from stage 0 to stage 1 is characterized by a relief of N limitations on biological functions. At stage 2, net nitrification is induced. Both NO_3^--leaching and N_2O-effluxes are low but detectable during this period. The critical threshold occurs when N becomes a non-limiting element for plant growth at stage 3. At this point biological retention processes become less effective and NO_3^--leaching losses increase substantially. Effluxes of NO, N_2O and ultimately N_2 may increase because of nitrification and denitrification processes [4]. In agreement with this concept, studies in forest ecosystems exposed to increased N-deposition have demonstrated significant NO_3^--production in the organic layer and mineral soil underneath. It has also been shown that in N-saturated forests NH_4^+- and NO_3^--immobilization rates can be reduced, which was attributed to decreases in microbial biomass. This leaves more mineral N for nitrification and denitrification. A study with simulated N depositions (30 kg N ha^{-1} yr^{-1}) in a Norway spruce stand demonstrated accordingly that the denitrification loss increased from 1.7 kg N ha^{-1} in unaffected forest to 2.9 kg N ha^{-1} in areas receiving an additional N [5]. In addition, several studies have demonstrated increased losses of NO and N_2O in response to increased N-inputs. Within the EU-project NOFRETETE (http:// 195.127.136.75/ nofretete/index.html) a comparison of 13 European forests revealed that the two sites with highest N-loading situated in Germany and the Netherlands (32 and 40 kg N ha^{-1} yr^{-1}) also showed highest N-gas emissions with rates up to 16 kg N ha^{-1} yr^{-1} (NO–N + NO_2–N + N_2O–N + N_2–N). N-deposition rates range typically from 2 kg N ha^{-1} yr^{-1} in remote areas up to 50 kg N ha^{-1} yr^{-1} in polluted regions. This spatial scatter in N-input rates is very likely an essential reason for the high variability of leaching (4 and 21 kg N ha^{-1} yr^{-1}) and denitrification (<1 to 16 kg N ha^{-1} yr^{-1}) that has been observed across European forest systems.

22.4 Nitrogen inputs affect forest microbial communities

Global change can affect soil processes by either altering the functioning of existing organisms or by restructuring the community, modifying the fundamental physiologies that drive biogeochemical processes [6]. Although process control is fairly consistent among denitrifiers, differences in denitrifier communities produce different N_2O production rates under experimental conditions, where resources are either non-limiting or tightly controlled [6].

The soil biota in many terrestrial ecosystems evolved under low-N conditions and thus one anticipated consequence of elevated N inputs is a change in microbial community structure and function. Increases in N availability due to atmospheric deposition have been linked to reduced spore production, root colonization and species richness of mycorrhizal fungi. In temperate hardwood and pine forests fungal biomass was lower in N-fertilized compared to control plots resulting in significantly lower fungal:bacterial biomass ratios. The shift in microbial community composition is accompanied by a significant reduction in the activity of phenol oxidase, a lignin-degrading enzyme produced by white-rot fungi. N-enrichment also alters the pattern of microbial substrate use: the relative response to the addition of carboxylic acids and carbohydrates was significantly lower in the N-treated plots. These patterns are consistent with lower decomposition rates and altered N cycling. The conclusion is that saprotrophic fungi may be as susceptible to N-deposition as are mycorrhizal fungi [7].

A comparison of 13 European forests within the NOFRETETE project (mentioned above) also indicated that microbial stress symptoms, as indicated by a high ratio of cyclic fatty acids to precursor monounsaturated fatty acids, was highly correlated with N-deposition rates. The marker fatty acids for fungal biomass, 18:2(9,12) and 18:1(9), and the marker fatty acids for arbuscular mycorrhiza, 16:1(11), were depleted at high N-input sites. In contrast, the bacterial biomass was not affected. Consequently, the bacteria to fungi ratios were significantly enhanced and correlated with N-deposition as well as with NO_x emissions across sites.

22.5 Environmental regulation of denitrification

The presence of denitrifying enzymes are almost omnipresent and for the process to occur basically three requirements must congregate, viz. restricted availability of O_2, presence of NO_3^- (or other NO_x) and availability of degradable C-sources. The relative importance of each of these denitrification controllers varies among habitats; but for ecosystems exposed to atmospheric air, such as forest soils, O_2-availability is often the most critical. Each of these three fundamental components are regulated by a complex hierarchy of environmental controllers of varying importance including climate, soil type and texture, vegetation type and structure, precipitation, T, litter fall and anthropogenic impacts (N-deposition, soil tillage, drainage). As an almost inevitable consequence of the complexity and dynamics in this suite of denitrification regulators, it has been acknowledged by many investigators that denitrification is highly variable with respect both to seasonality and spatial distribution. In terrestrial upland soils, particularly high denitrification activity is often associated with the so-called "hotspots", e.g. the proximal vicinity of degrading organic matter (litter, roots, and animals) where O_2-free zones have developed due to high respiratory activity simultaneously with relatively high abundance of NO_3^- and C. The vertical distribution of organic matter in forest soil profiles, most often characterized by an undisturbed pattern with organic-rich litter and humus layer on top of the mineral subsoil, has often been found to mirror the distribution denitrifying capacity in the soil. The denitrification activity in soil "hotspots" may approach the maximum denitrifying capability of the soil limited by the level of expressed enzymes and be responsible for almost all denitrification activity in a particular site. However, usually "hotspots" are spatially scattered and short lived (few days), which makes the assessment of total denitrification overwhelming due to the requirement for at comprehensive spatial sampling scheme as no integrative method exists to measure total denitrification. In addition to this spatial variability, denitrification also exhibits marked temporal variability. It is well recognized that precipitation events are important triggers of denitrification activity being stimulated by the restricted O_2-diffusion into wetted soils. Such rainfall-driven events usually last only few days depending on soil drainage status and amount of precipitation. Variability in denitrification at larger temporal scales can be observed for example associated with the seasonality in forest ecosystem N-cycling, nutrient dynamics, litter fall and soil structure. In a study comparing nine different deciduous forest sites [8], denitrification activity was highest in spring and autumn, and lowest during summer, a pattern that could be closely tied to the forest tree activity. The end of the spring pulse of denitrification activity coincided with the bud break and thus increased competition for NO_3^- between trees and microorganisms. The onset of the autumn pulse was coincident with leaf senescence and litter fall giving rise to an increased input of C to the soil environment in combination with reduced nutrient uptake by the trees. A low denitrification activity in summer was possibly caused by NO_3^--limitation induced by increasing soil aggregation from root growth. In contrast, another study [9] found that denitrification activity peaked during the summer in a North American sugar maple-basswood forest site, which was explained by increased NO_3^--supply from nitrification during this period. However, in an edaphically and floristically similar sugar maple-oak site there was no peak in denitrification activity during summer. This was ascribed to the different overstory vegetation where in particular the presence of oak trees influenced litter chemistry with an impeding effect on nitrification activity.

22.6 Species involved in denitrification

Denitrifying bacteria represent 10–15% of the bacterial population in soil, water and sediment [10]. Denitrifiers use a broad range of inorganic and organic compounds as sources of C and energy with an efficiency of phosphorylation which is 67–71% [10]. Denitrification is a functional trait found within more than 50 genera. Most bacteria with this functional trait belong to a wide range of various subclasses of α-, β-, γ-, ε-proteobacteria and high and low GC Gram-positive bacteria [11]. Denitrifiers in forests include *Alcaligenes*, *Pseudomonas* and *Bacillus* species as well as Actinobacteria (especially S*treptomycetes*). They are aggregated in the upper (5 cm) soil layer and around roots and show highest numbers during the cold season [12]. Attention has to be given to denitrifiers found among fungi. Most of the distinct denitrifiers found belong to *Fusarium* and the related teleomorphic genera (*Gibberella* and *Nectria*) and anamorphic genera (*Cylindrocarpon* and Trichoderma). *Talaromyces flavus*, however, belongs to a different class (Plectomycetes) [13]. Among 67 strains of yeasts, 16 were able to evolve distinct amounts of N_2O, among them *Candida*, *Saccharomycopsis* and *Trichosporon*. Yeasts are primary colonizers of leaf litter, where they use simple sugars for growth. As many of the denitrifying systems of fungi and actinomycetes cannot reduce N_2O to N_2 this might be of consequence for forests, where these organisms are abundant.

22.7 Importance of denitrification for N_2O-emissions

In addition to its function in losses of N from forest ecosystems, denitrification is also a significant source of N-oxide trace gases. The atmospheric concentration of N_2O increased 17% over the past 250 years, which is subject to a good deal of concern. Researchers have devoted much effort to understand sources and mechanisms for N_2O emissions from forests and other ecosystems. Denitrification is also a source of NO, which is important for chemical processes in the troposphere and stratosphere. However, nitrification most often is the dominant process for NO formation in upland soils, rather than denitrification.

On a global scale, forest ecosystems emit ca. 4 Tg N_2O-N yr^{-1}, which is composed of 3 Tg from tropical forests and 1 Tg from forests in the temperate/boreal region. The total global N_2O emissions as implied by IPCC amounts to 16.4 Tg N pointing at a total forest contribution of ca. 24% [14].

The contribution of nitrification and denitrification to total N_2O production has been the subject of several studies in forest ecosystems. A German study [15] on the seasonality in N_2O emission strengths and source relationships in a beech forest identified denitrification as the primary process for N_2O production in the summer months. Other processes, e.g. nitrification, also contributed in the spring and autumn. Previous studies in North American forests [16] also showed that denitrification may be a primary source for N_2O production. A comparison across 11 different European forests (Table 22-1) also demonstrated that denitrification was the primary process for N_2O production [17]. However, the study also observed that N_2O production in these soils was strongly predicted by the nitrification activity, which suggests an essential role of nitrification for regulating denitrification through NO_3^- supply. This observation confirms the hypothesis that accelerated nitrification in response to increased N-deposition may lead to enhanced emissions of N_2O, and possibly also total denitrification losses.

Table 22-1 Occurrence of N_2O from soil NH_4^+ and NO_3^- in 11 different European forest soils. Most numbers are average of at least two observations, one in the spring and one in the autumn, respectively.

Forest type (country)	Contribution to N_2O formation (%)	
	Ammonium-N	Nitrate-N
Scots pine (SF)	25	54
Sitka spruce (GB)	32[a]	19–53[b]
Douglas fir (NL)	20	61
Norway spruce (D)	92[a]	100
Spruce-beech (A)	100	90
Norway spruce (HU)	n.a.	~100[a]
Pine (I)	0	26
Beech (DK)	9	50
Beech (A)	7	74
Mixed oak, poplar, ash (I)	37	57
Poplar (I)	22[a]	53[a]

Note: Data from Ref. [17]. [a]Only one observation. [b]Range for two observations. n.a., Data not available.

22.8 Effect of forest type on denitrification

N mineralization–immobilization processes in forest soils is influenced by litter quality parameters such as C:N ratio, contents of N, lignin and phenolic compounds. Deciduous tree species might generally increase nutrient cycling and microbial activities compared to coniferous tree species [18]. In accordance with this general perception, incubation of samples from European coniferous and deciduous forest soils indicated that soils under deciduous forest types emit more N_2O than soils under coniferous forest types, however, total denitrification was not examined in this study [17]. Other soil incubation studies have also confirmed that soils under deciduous forests may emit more N_2O compared to soils under coniferous forests [18]. A similar difference between these two main forest types has also been observed in German field studies finding that a beech forest emitted approximately twice as much N_2O as a spruce forest [19]. Few studies examined the effects of different forest types on total denitrification. As mentioned previously, changes in overstory composition may have an influence on total denitrification activity, and earlier studies in North America [16] demonstrated that coniferous forest communities exhibited less denitrification activity compared to hardwood forests in the same region. In addition to impacts on litter chemical properties, different shapes of tree litters may also have an influence on the gas diffusivity through the litter layer. Litters from broad-leaved trees may, in particular when they are wet, restrict O_2-diffusion into the soil to a greater extent than litters from spruce or pine needles with variable impacts on the denitrification activity.

22.9 Competition between plants and microorganisms

Microbial mineralization and nitrification are generally thought to be the rate-limiting steps in the N cycle. Microorganisms are assumed to acquire inorganic N before plants. While forest trees are often N-limited, heterotrophic soil bacteria depend on the availability of C in the soil. The C:N ratio of the litter is an important determining factor in whether N is mineralized or immobilized. The rate of mineralization of organic N depends first on the C:N ratio of the substrate being decomposed and second on the decomposer community's need for N relative to its need for C. Fungi generally assimilate substrate more efficiently than bacteria, which have a smaller C:N ratio and consequently a larger N demand per unit C. At C:N ratios greater than about 30:1, both fungi and bacteria will require additional N and net immobilization will occur. Litter quality is not only a feature of C- and N-concentrations, but also of the chemical resilience of the litter material. Because of the higher C:N ratio of fungi and, as mycorrhizal fungi get C from their host plants there is competition for N in the mycorrhizosphere where a specially adapted bacterial community lives.

Measurements of gross N-transformation have indicated that microbial heterotrophs and autotrophic nitrifiers compete for NH_4^+ in the soil, and that nitrifiers are successful competitors under conditions of stationary or declining populations of heterotrophs. Microbial immobilization of NO_3^- can be substantial in forest soils and the lack of a net increase in soil NO_3^- pool sizes during incubation is not an unequivocal indication that the nitrification process is insignificant or absent [1]. Within a given soil pool, the mean residence time of N may be a better indicator of N-dynamics than the changes in pool size or fluxes into or out of the pool.

The mean residence time of small amounts of NO_3^- in undisturbed coniferous forests was 15 h, indicating intense microbial activity. On a longer timescale pulses of NH_4^+ in an acid woodland soil were short lived, persisting for a few weeks [20]. On a short timescale soil microorganisms do compete better than plants for the added N, particularly for NH_4^+: its uptake by microorganisms was fivefold faster than that by plants. The NO_3^--uptake rate by microorganisms doubled that of plants. However, microbial cells turn over more rapidly than plant roots do, therefore releasing N back into the soil, whereas plants are able to retain captured N over a longer period. The relative turnover times of roots and microorganisms are likely to be a key determinant of competitive success: it is the rapid turnover of the microbial biomass that gives the roots an opportunity to capture released N through microbial cell lysis. However, there can be large differences between microbial populations in turnover time. Mycelial structures as found in fungi can recycle N internally and their N turnover therefore can be much slower than for bacteria. The principal weapon that plants possess is their ability to sequester N for much longer periods than most microorganisms. Grazing of the microbial community by higher trophic levels (e.g. protozoa, nematodes and invertebrates) releases NH_4^+ back into the root soil environment and stimulates mineralization rates.

In nutrient-impoverished conditions that prevail in forests, ectomycorrhizal fungi colonize at least 90% of the feeding roots of the tree [21]. Plants that have mycorrhizal symbioses are at a distinctive advantage compared with non-mycorrhizal roots. Most of the N in forests will enter the soil as protein or other complex forms such as chitin. Due to the wider enzymatic capabilities with the mycorrhizal fungi compared to plant roots and because of their size, the fungal hyphae are better able to penetrate to the sites of organic matter decomposition and compete directly with other microorganisms for the decomposition products. The mycorrhizal symbiosis helps the tree to take up organic compounds intact and successfully acquire N from organic patches in soil and thereby strengthens the competitiveness for N.

22.10 The N_2O:N_2 ratio and in situ quantification

Quantitative loss of N-gases from denitrification activity is usually associated with N_2O and N_2. The proportion of produced N_2O vs. total gas production from denitrification, i.e. $N_2O/[N_2O+N_2]$, varies substantially regulated by environmental conditions combined with enzymatic characteristics of the dominant microbial community. Most important for the N_2O to N_2 emission ratio is water content and availability [22], as shown in the conceptual model pictured in Figure 22-2. In addition, NO_3^--concentrations, pH and substrate are important. It can be generalized that the more the microorganisms are depending on NO_3^- as electron acceptor, the more complete they will denitrify and release N_2. In addition, a viable microflora not exposed to stress (such as acidification) is hypothesized to fully denitrify N_2. In field measurements, N_2O emissions tend to be low under complete waterlogging, compared to changing water tables. However, under laboratory conditions many soils react positively to waterlogging and only few switch to 100% N_2O emission. Production of N_2O solely by denitrification is rare and quantifying the total gaseous N-loss requires measuring of both N_2O and N_2. The most common method applied to measure gas exchange between soil and atmosphere is the chamber method. The principle of this method is to restrict the exchange of air with the atmosphere

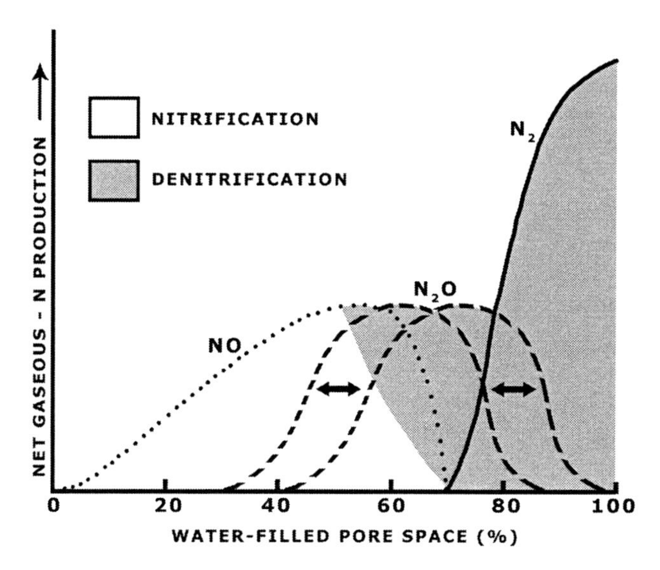

Figure 22-2 A conceptual model describing losses of NO, N_2O and N_2 from nitrification and denitrification as a function of soil moisture, after Ref. [22].

so that the emitted (or absorbed) gas can be detected readily in a chamber headspace. However, due to the fact that atmospheric air contains 79% N_2, the direct quantification of denitrification N-losses from forest floors is associated with a methodological challenge. The two most commonly used methods for quantification of denitrification in forests are substrate ^{15}N-labelling and C_2H_2-blockage of the N_2O reductase. However, both methods are invasive to the system and associated with vital drawbacks affecting the denitrification substrate availability. Either by introduction of additional substrate as in the case with ^{15}N-labelling of the NO_3^--pool or by the elimination of NO_3^--supply with C_2H_2-blockage due to the fact that nitrifying organisms are irreversibly extinguished by low concentrations of C_2H_2. More recently the barometric process separation (BaPS) method was introduced to assess denitrification in soil [23]. However, this method may be hampered, if other gases than the measured O_2, CO_2 and N_xO_x are metabolized or produced in the investigated soil.

22.11 Modelling

There are different approaches to achieve landscape or countrywide inventories of N_2O and denitrification emission from forests.

The process-based model EXPERT N was used to predict N_2O emissions for German forests [24]. The forest soils were stratified into seasonal emission types (broad-leaved forests with soil pH < 3.3) and stratified emission types, which are considered to emit only minimal amounts of N trace gases.

The process-oriented model PnET-N-DNDC (Photosynthesis and Evapotranspiration-Nitrification-Denitrification and Decomposition) was specially developed to simulate the emissions of NO and N_2O from forest soils on a daily scale [25]. The model can be used to calculate regional inventories of N trace gas emission from forest soils and was applied to make predictions for various future scenarios. The model is based on a forest physiology model used for predicting photosynthesis, respiration, organic C-production and allocation and litter production (Figure 22-3). Another input comes from a soil biogeochemistry model, namely DNDC, used for prediction of soil organic matter decomposition and N-turnover in agricultural soils. Six submodels predict forest growth, soil climate, decomposition, nitrification and denitrification, as well as methane decomposition. The decomposition sub-model tracks turnover of the litter

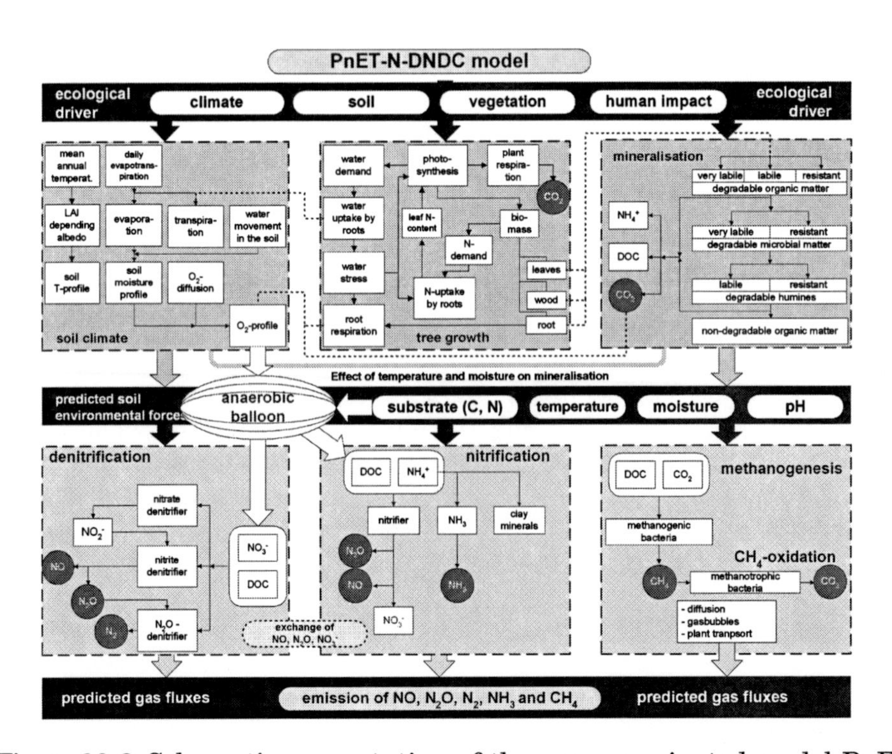

Figure 22-3 Schematic presentation of the process-oriented model PnET-N-DNDC for predictions of losses of N-gases and CH_4 in forest ecosystems. For detailed discussion, see Ref. [25].

and other organic matter in the soil and passes NH_4^+, NO_3^- and dissolved organic C (DOC) to the nitrification or denitrification submodels. The nitrification submodel predicts growth and death of nitrifiers, with the nitrification rate as well as NO and N_2O productions from nitrification depending on soil temperature, moisture, NH_4^+ and DOC concentration. The denitrification submodel simulates a function of soil temperature, moisture and substrate concentrations. The denitrification-induced NO-, N_2O- and N_2-fluxes are calculated on the basis of dynamics of soil aeration status, substrate limitation and gas diffusion. As a further source of NO emissions from forest soils, chemodenitrification occurs only when soil pH is <5. Rates of nitrification in forest soils are usually higher than rates of denitrification. Therefore it is assumed that the main source of nitrite in soils is nitrification. Since nitrification and denitrification can simultaneously occur in aerobic and anaerobic microsites, the concept of an "anaerobic balloon" is used to calculate the anaerobic fraction of soil in a given soil layer in dependency of O_2-diffusion and magnitude of soil and root respiration [25]. To simulate N trace gas emissions for a specific site the model requires the following input parameters: daily climate date (precipitation, minimum and maximum air temperature; optional: radiation), soil properties (texture, clay content, pH, soil organic C-content, stone content, humus type) and forest data (forest type and age, aboveground and belowground biomass, plant physiology parameters).

Inventories of N_2O and NO emissions from European forest soils show that the model performs very well especially for NO emissions, which amount to approximately 94 kt N yr^{-1}, less <1% of pyrogenic NO_x emissions. However, during summer time and in rural areas, NO-emissions from soils can contribute 22% to NO_x-pollution of the air. Highest emissions of NO from forest soils (>3 kg N ha^{-1} yr^{-1}) were simulated for highly N-affected forest areas in the Benelux states and northern Germany [26].

NO emitted due to microbial activity in forest soils must not necessarily reach the atmosphere above the forest canopy. Complex reactions between the trace gases NO, NO_2 and O_3 take place in the air, and with tree and soil surfaces. This is especially important during the night and for forests emitting large amounts of NO. N_2O emissions from forest soils are somewhat smaller than NO emissions with 82 kt N yr^{-1}, which is approximately 14.5% of the source strength coming from agricultural soils. The average N_2O emissions from forest soils were around 0.6 kg N ha^{-1} yr^{-1}, with elevated emissions in Central Europe and western Spain, but also in parts of Scandinavia where soils with high organic C content in the forest floor are found [26]. These values may seem small compared to those in agricultural soils; however, it has to be considered that in many countries forests cover more than half of the total land area.

N-trace gas fluxes may vary substantially from year to year. Future predictions say that increases of temperature and precipitation as well as N-deposition would all stimulate NO_x emissions from European forest soils.

22.12 Austrian case – the missing nitrogen

The comparison of 13 European forest sites within the NOFRETETE project revealed that some forests responded asymptotic to high N loads. In particular, one soil, situated in the Tyrolean Limestone Alps, was affected by a relatively high load of N of about 17 kg N ha^{-1} yr^{-1}, but this additional N could not be tracked within the system. Although all possible inputs and outputs were measured and a complete N-budget for this site was attempted, the disappearance of 13 kg N ha^{-1} yr^{-1} could not be explained. In order to get further insight into magnitudes of the N-cycling rates, the PnET-N-DNDC model was applied to simulate N-turnover in the site. The model showed some unexpected results, which afterwards was confirmed experimentally. It turned out, that most of the "missing" N was denitrified to N_2 and so rendered harmless and reemitted to the atmosphere. This fact was attributed to the neutral pH in the limestone soil and to a comparably high microbial biomass concentration, indicating optimum growth conditions for microbes, which may have favoured complete denitrification.

References

[1] J. M. Stark, S.C. Hart, High rates of nitrification and nitrate turnover in undisturbed coniferous forests, Nature 385 (1997) 61–64.
[2] E.-D. Schulze, Carbon and Nitrogen Cycling in European Forest Ecosystems (Ecological Studies 142), Springer, Berlin, 2000.
[3] J.D. Aber, K.J. Nadelhoffer, P. Steudler, J.M. Melillo, Nitrogen saturation in northern forest ecosystems, BioScience 39 (1989) 378–386.
[4] J.N. Galloway, J.D. Aber, J.W. Erisman, S.P. Seitzinger, R.W. Howarth, E.B. Cowling, B.J. Cosby, The nitrogen cascade, BioScience 53 (2003) 341–356.
[5] J. Mohn, A. Schürmann, F. Hagedorn, P. Schleppi, R. Bachofen, Increased rates of denitrification in nitrogen-treated forest soils, Forest Ecol. Manag. 137 (2000) 113–119.
[6] J.P. Schimel, J. Gulledge, Microbial community structure and global trace gases, Glob. Change Biol. 4 (1998) 745–758.
[7] S.D. Frey, M. Knorr, J.L. Parrent, R.T. Simpson, Chronic nitrogen enrichment affects the structure and function of soil microbial community in temperate hardwood and pine forests, Forest Ecol. Manag. 196 (2004) 159–171.

[8] P.M. Groffman, J.M. Tiedje, Denitrification in north temperate forest soils: spatial and temporal patterns at the landscape and seasonal scales, Soil Biol. Biochem. 21 (1989) 613–621.
[9] A.G. Merrill, D.R. Zak, Factors controlling denitrification rates in upland and swamp forests, Can. J. Forest Res. 22 (1992) 1597–1604.
[10] S. Casella, W.J. Payne, Potential of denitrifiers for soil environment protection, FEMS Microbiol. Lett. 140 (1996) 1–8.
[11] L. Philippot, S. Hallin, Finding the missing link between diversity and activity using denitrifying bacteria as a model functional community, Curr. Opin. Microbiol. 8 (2005) 234–239.
[12] A. Mergel, K. Kloos, H. Bothe, Seasonal fluctuations in the population of denitrifying and N_2-fixing bacteria in an acid soil of a Norway spruce forest, Plant Soil 230 (2001) 145–160.
[13] H. Shoun, D. Kim, H. Uchiyama, J. Sugiyama, Denitrification by fungi, FEMS Microbiol. Lett. 94 (1992) 277–282.
[14] IPCC: Climate change 2001: The scientific basis. Contribution of working group I to the third assessment report of the Intergovernmental Panel on Climate Change: J.T. Houghton, Y. Ding, D.J. Griggs, M. Noguer, P.J. van der Linden, X. Dai, K. Maskell, C.A. Johnson (Eds.) Cambridge University Press, United Kingdom, New York, NY, 2001.
[15] I. Wolf, R. Brumme, Contribution of nitrification and denitrification sources for seasonal N_2O emissions in an acid German forest soil, Soil Biol. Biochem. 34 (2002) 741–774.
[16] G.P. Robertson, J.M. Tiedje, Denitrification and nitrous oxide production in successional and old-growth Michigan forests, Soil Sci. Soc. Am. J. 48 (1984) 383–389.
[17] P. Ambus, S. Zechmeister-Boltenstern, K. Butterbach-Bahl, K, Sources of nitrous oxide emitted from European forest soils, Biogeosciences 3 (2006) 135–145.
[18] O.V. Menyailo, B.A. Hungate, W. Zech, The effect of single tree species on soil microbial activities related to C and N cycling in the Siberian artificial afforestation experiment, Plant Soil 242 (2002) 183–196.
[19] K. Butterbach-Bahl, R. Gasche, I. Breuer, H. Papen, Fluxes of NO and N_2O from temperate forest soils: impact of forest type, N deposition and of liming on the NO and N_2O emissions, Nutr. Cycl. Agroecosys. 48 (1997) 79–90.
[20] A. Hodge, D, Robinson, A. Fitter, Are microorganisms more effective than plants at competing for nitrogen, Trends Plant Sci. 5 (2000) 304–308.
[21] S.W. Simard, D.A. Perry, M.D. Jones, D.D. Myrold, D.M. Durall, R. Molina, Net transfer of carbon between ectomycorrhizal tree species in the field, Nature 388 (1997) 579–582.

[22] E.A. Davidson, M. Keller, H.E. Erickson, L.V. Verchot, E. Veldkamp, Testing a conceptual model of soil emissions of nitrous and nitric oxides, BioScience 50 (2000) 667–680.
[23] J. Ingwersen, K. Butterbach-Bahl, R. Gasche, O. Richter, H. Papen, Barometric process separation: new method for quantifying nitrification, denitrification, and nitrous oxide sources in soils, Soil Sci. Soc. Am. J. 63 (1999) 117–128.
[24] H. Schulte-Bisping, R. Brumme, E. Priesack, Nitrous oxide emission inventory of German forest soils, J. Geophys. Res. 108 (2003) 4132.
[25] K. Butterbach-Bahl, F. Stange, H. Papen, C. Li, Regional inventory of nitric oxide and nitrous oxide emissions for forest soils of Southeastern Germany using biogeochemical model PnET-N-DNDC, J. Geophys. Res. 106 (2001) 34155–34166.
[26] M. Kesik, P. Ambus, R. Baritz, N. Brüggemann, K. Butterbach-Bahl, M. Damm, J. Duyzer, L. Horvath, R. Kiese, B. Kitzler, A. Leip, C. Li, M. Pihlatie, K. Pilegaard, G. Seufert, D. Simpson, U. Skiba, G. Smiatek, T. Vesala, S. Zechmeister-Boltenstern, Inventories of N_2O and NO emissions from European forest soils, Biogeosciences 2 (2005) 353–375.

Chapter 23

Denitrification in Wetlands

Oswald Van Cleemput, Pascal Boeckx, Per-Eric Lindgren and Karin Tonderski

23.1 Introduction

Wetlands are situated between upland systems and true aquatic systems. The global wetland area may be estimated at 700–1000 Mha [1, 2] of which almost half is situated in the tropics. Almost one-fifth is occupied by lowland rice fields. They are sources, sinks and transformers of nutrients and carbon.

Nitrogen in wetlands occurs in many forms covering a range of valence states from −3 (in NH_4^+) to +5 (in NO_3^-). Its fixation from the atmosphere (N_2 fixation) and return (denitrification) to it depends on transformations between these states. The importance of N_2 fixation and denitrification in wetlands on a global scale is shown in Table 23-1. Wetlands are important sinks for NO_3^-, but, when used in agriculture, e.g. paddy rice, they are also important sources of NH_3–N, which is emitted by volatilization from the floodwater.

Definite prerequisites for denitrification are anoxic conditions and the presence of NO_3^- as electron acceptor. In wetland soils, anoxic conditions

Biology of the Nitrogen Cycle
Edited by H. Bothe, S.J. Ferguson and W.E. Newton

Table 23-1 Nitrogen fixation and denitrification in wetlands on a global scale [3, 4].

	N_2 fixation Mean rate ($g\ N\ m^{-2}\ yr^{-1}$)	Total ($Tg\ N\ yr^{-1}$)	Denitrification Mean rate ($g\ N\ m^{-2}\ yr^{-1}$)	Total ($Tg\ N\ yr^{-1}$)
Temperate				
Bogs/fens	1.0	3.0	0.4	1.2
Floodplains	2.0	6.0	1.0	3.0
Tropical				
Bogs	1.0	0.5	0.4	0.2
Swamps	3.5	7.8	1.0	2.2
Floodplains	3.5	5.2	1.0	1.5
Rice fields	3.5	5.0	7.5	10.8
Total		27.5		18.9

predominate because the chemical and microbial demands for O_2 greatly exceeds its supply, and the diffusion of O_2 in water is about 10^4 slower than in air.

Dissimilatory reduction of NO_3^- to NH_4^+ (DNRA) is of less importance in wetland soils because concentrations of NH_4^+ and organic N are in general high in anaerobic environments. However, under strongly reducing conditions ($Eh < -250$ mV) reduction of N_2 to NH_4^+ (N_2 fixation) is thermodynamically possible.

23.2 Wetlands as an environment for denitrification

In natural wetlands as well as in paddy soils, NO_3^- for denitrification can either be supplied or formed. Application is rather unusual but formation depends on the intensity of nitrification. Aerobic nitrification and anaerobic denitrification can occur in close proximity. Atmospheric O_2 diffuses through the floodwater creating an aerobic soil layer with formation of NO_3^- as a consequence. This NO_3^- diffuses to the underlying anaerobic layer where it can be denitrified. In paddy soils or soils grown with aquatic plants, an oxidized zone is also formed around their roots. Indeed, O_2 is transported through the plants and excreted along the roots, creating conditions for nitrification. The thickness of the oxidized zone depends on the oxygen consumption rate and can reach a few millimeters to a few centimeters. Identification of the oxidized and reduced zone can be done through measurement of a redox potential profile [5]. The co-existence of oxidized and reduced zones or layers is illustrated for flooded soils and around a macrophyte root in Figure 23-1. Both zones can occur over large

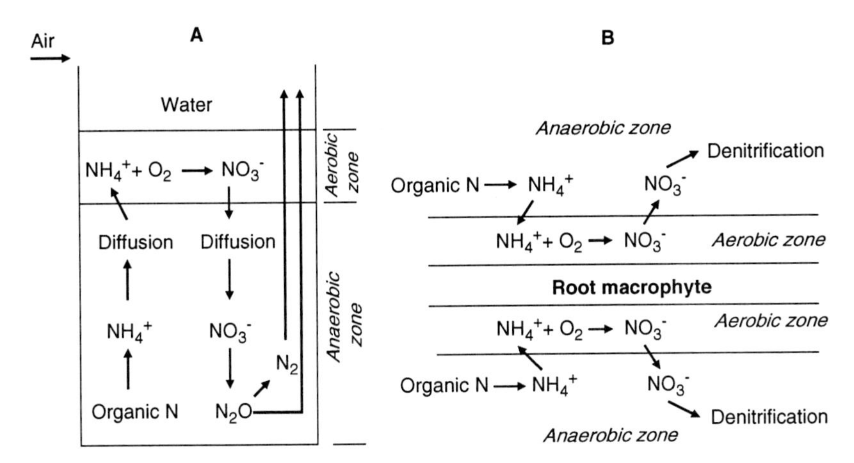

Figure 23-1 Scheme representing the interaction between aerobic and anaerobic zones in a flooded soil (A) and around a macrophyte root (B) (after [9])

soil volumes, but on a microscale they can be near each other. The stimulating effect of plant roots on microbial activity was demonstrated in constructed wetlands (CWs) [6]. The whole root zone was postulated to be influenced by bacteria and hence the process of nitrogen transformation [6]. In another study, from 60 to 75% of NO_3^- applied on the surface of flooded rice fields was lost by denitrification over 2–3 weeks [7].

Plants may have also other functions in supporting denitrification in wetlands. The epiphytic community of mature and senescent shoots of submerged macrophytes has been shown to be sites for significant denitrifying activity [8]. The activity was 3- to 10-fold higher on senescent shoots suggesting a succession in the dominance from algal to heterotrophic organisms in the epiphytic biofilm as the shoots age. Furthermore, the denitrifying capacity of epiphyton adapted to high nutrient loadings (i.e. in a wastewater reservoir) was about a hundred times higher than that of epiphyton adapted to a lower nutrient loading (an eutrophic lake) [8]. Other studies also showed that nutrient enrichment promotes the growth of bacteria in biofilms. However, the high activity per community biomass suggested that also the activity of the denitrifying microorganisms might be stimulated by high N loading [8].

23.3 Molecular diversity of denitrifying bacteria

During the last decade culture independent methods for investigating denitrifying bacteria have been developed. These methods are mainly based on PCR-amplification of key enzyme genes in the denitrification process. Two structurally different NO_2^- reductases, Cu-Nir (encoded by *nir*K gene) and cd_1-Nir (encoded by *nir*S gene), are found among denitrifying bacteria. Together with N_2O-reductase gene, *nos*Z, the NO_2^--reductase genes are the main targets for investigating the distribution and diversity of the denitrifying bacteria. An even distribution and diversity of *nir*K and *nir*S sequences was found in a study including 35 different CWs with a broad variety of physico-chemical properties, in agricultural areas in Southern Sweden as investigated by denaturing gradient gel electrophoresis (DGGE) analysis [10]. Also, a high diversity of *nos*Z sequences in most of the wetlands investigated was noted in that study. In forested areas only *nir*S fragments could be amplified from marsh soils, while both *nir*K and *nir*S sequences could be retrieved from an upland soil in the US [11]. So far, it has not been conclusively shown whether wetlands support a denitrifying bacterial community consisting of a distribution of *nir*K/*nir*S differing from other environments.

23.4 Denitrification in riparian wetlands

A riparian zone is the interface between a terrestrial and aquatic ecosystem. Next to a number of functions (hydrology, biodiversity, erosion), it also serves as a zone for NO_3^- retention from groundwater through denitrification, plant uptake and (microbial and abiotic) immobilization, and dissimilatory NO_3^- reduction to NH_4^+. A technique was successfully developed to quantify groundwater NO_3^- retention processes in a riparian zone using the variation of natural abundance of ^{15}N in NO_3^- [12]. Hereby, site-specific ^{15}N enrichment factors for denitrification and plant uptake and a seasonal enrichment factor for groundwater NO_3^- retention were used. It was shown that, although NO_3^- uptake by plants fractionate minimally against ^{15}N, increasing $\delta^{15}N-NO_3^-$ values in groundwater NO_3^- retention studies should not be interpreted as a contribution of denitrification solely. Especially in ecosystems with high N loads, ^{15}N isotope fractionation of NO_3^- by plants can be of significant importance. The technique should only be applied during the growing season, as in this period denitrification and plant uptake are the major NO_3^--retention processes. During winter, decomposition of riparian plant material may influence the $\delta^{15}N-NO_3^-$ signals through liberation of large amounts of labile C stimulating other NO_3^--retention processes such as DNRA and N immobilization. The relative importance of denitrification and plant uptake to groundwater NO_3^- retention in a riparian buffer zone was 49 and 51% during spring, 53 and 47% during summer and 75 and 25% during autumn, respectively [12].

Recent studies indicate that climate has little impact on the overall N removal in riparian wetlands. Across a range of climatic conditions, climate had little impact on the annual N-removal efficiency [13]. This led to the conclusion that biological N-removal mechanisms were more closely related to NO_3^- load and hydraulic conditions than to climatic parameters. Water table elevations are among the prime determining factors for N dynamics and formation of the end product upon reduction [14]. In riparian areas with water tables of 30 cm below the soil surface, denitrification only occurred in fine textured soils showing the additional importance of geological characteristics. The relevance of hydro-geo-morphology is also emphasized by studies pointing out the importance of buried C for NO_3^- removal in riparian zones [15].

23.5 Denitrification in constructed wetlands

Constructed wetlands are used for N removal in both wastewater and runoff water carrying high loads of NO_3^-. CWs are basically divided in those with predominantly subsurface flow in a sand or gravel bed, and those with surface flow, i.e. a water depth of usually 0.2–2 m. Only the latter ones are discussed here. Depending on the dominant chemical compounds of N in the water, different processes are rate limiting for N removal. Most studies of N removal in CWs are based on measurements of mass inflow and outflow, and a large variation in N removal has been observed (Table 23-2). Denitrification is assumed to account for most of the N removal in highly loaded CW. If this is a correct assumption, denitrification rates in CWs are one to two orders of magnitude higher than in natural wetlands (Table 23-1). Indeed, extremely high N removals in CWs have been observed in experiments. In North Carolina, a CW removed 64% of the wastewater load of 3.6 g N m^{-2} d^{-1} [16] with NO_3^-–N concentrations of >100 mg L^{-1}. The removal was more than double of the removal observed when the wetland was loaded with wastewater containing little NO_3^-–N. This shows the importance of nitrification as a rate-limiting factor for denitrification in CWs. Generally, observed N removal and denitrification are dependent on the N load to the wetland ecosystem [17, 18]. However, other factors also limit denitrification in CWs, some of which are discussed below.

Table 23-2 Removal of nitrogen in free water surface wetlands (after [19]).

Parameter	Mean concentration (mg L^{-1})		Mean of mass flow (g m^{-2} yr^{-1})		
	In	Out	In	Out	Removal
Tot N	19	11	432	268	164
Min – max	9–116	4–93	71–2300	13–1800	40–760

Constructed wetlands, receiving high loads of NO_3^-–N, can be used for elucidating some of the rate-limiting factors for denitrification. A distinct T-effect on N removal in wetlands has been observed, e.g. removal rates were 10 g m^{-2} d^{-1} N during summer (17°C) and 4 g $m^{-2}d^{-1}$ during winter (3°C) in a wetland loaded with wastewater [20]. Later studies on denitrification in the same wetland showed no temperature dependency above 12°C, whereas the T- coefficient for lower T was estimated at 1.01 [21]. By adding exogenous C, however, the N removal could be increased by an order of magnitude emphasizing the importance of energy availability for the denitrifying bacteria.

In (more) natural wetlands, the macrophyte organic matter serves as a major source of energy for denitrifying bacteria. This has been demonstrated in studies where N removal has been higher in wetlands with vegetation than in controls without vegetation [22, 23]. In a recent study of denitrification with intact sediment cores using acetylene-inhibition technique, the potential denitrification varied significantly between detritus layers from different plant species [24]. Hypothetically, this also implies that there is a seasonal difference in denitrification capacity between different plant communities as the organic matter decomposes at different rates. The potential for denitrification was, however, still limited by C availability as demonstrated by the significant increase in the denitrification rate after addition of acetate to the cores. Another critical factor for denitrification in wetlands is the diffusion of NO_3^- and O_2, depending on the water flow rate in the wetland [25]. High flows can oxygenate the bottom water, hence decreasing denitrification, but low flows on the other hand may limit the supply of NO_3^- to the O_2-free zones.

23.6 Emission of nitrogen gases from wetlands

23.6.1 Influencing factors

Especially, nitrification–denitrification and chemodenitrification [26] are considered to be the main processes producing N_2O and NO. Through diffusion and mass transport, these gases can easily move from aerobic to anaerobic zones and vice versa, being emitted to the atmosphere. However, in paddy rice the emission is not important if the fields are kept continuously submerged during the growing season [27]. Through flooding, conditions are sufficiently reducing and the availability of organic substances is sufficiently large so that denitrification proceeds to N_2. However, these conditions also promote formation and emission of CH_4. Choosing the appropriate fluctuating water regime in combination with the presence of organic substances might create redox potential conditions with minimal N_2O as well as CH_4 emissions [28, 29].

During the rice-growing season, the N_2O fluxes were barely detectable although small emissions (<3.5 mg N m^{-2} day^{-1}) occurred after N application. During fall periods, significant N_2O emission can occur [30, 31].

23.6.2 Flux measurement techniques

Several methods exist for measuring emissions of denitrification gases (NO, N_2O). The static box method is widely used for determining of N_2O emissions. Nitrous oxide, emitted from a wetland, can be captured in a closed box covering a certain area, whereby the increase in concentration can be converted to flux. The detection of N_2O is mainly done by gas chromatography (electron capture detector) or by photo acoustic infrared detection (FAID). In systems with growing macrophytes, care should be taken to include the plants in the box as an amount of N_2O is conveyed through the stem of the plants. A survey of this technique has been published by IAEA [32] and in Ref. [33]. Nitric oxide fluxes, although of minor importance in wetlands, can be determined by a dynamic box method coupled to a chemiluminescence detector. Flux gradient techniques using tunable diode lasers, as analytical technique, are less suitable in wetlands due to their heterogenic nature.

References

[1] I. Aselman, P.J. Crutzen, Global distribution of natural fresh-water wetlands and rice paddies: their net primary productivity, seasonality and possible methane emissions, J. Atm. Chem. 8 (1989) 307–358.

[2] W.J. Mitsch, J.Q. Gosselink, Wetlands, J. Wiley & Sons, New York, 2004.

[3] T.V. Armentano, J.T.A. Verhoeven, Biogeochemical cycles: global, in: B.C. Patten et al. (Eds.) Wetlands and Shallow Continental Water Bodies, vol. 1: Human and Natural Relationships, SPB Academic Publishers, The Hague, 1990, pp. 281–311.

[4] G. Kirk, The Biogeochemistry of Submerged Soils, J. Wiley & Sons, England, 2004, 291p.

[5] W.H. Jr. Patrick, R.D. DeLaune, Characterization of the oxidized and reduced zones in flooded soils, Soil Sci. Soc. Amer. Proc. 36 (1972) 573–576.

[6] C. Münch, P. Kuschk, I. Röske, Root stimulated nitrogen removal – only a local effect or important for the water treatment? Proceedings of 9th International Conference on Wetland Systems for Water Pollution Control, Avignon, Sept. 2004, pp. 395–402.

[7] R.J. Buresh, M.I. Samson, S.K. De Datta, Quantification of denitrification in flooded soils as affected by rice establishment methods, Soil Biol. Biochem. 25 (1993) 843–848.

[8] P.G. Eriksson, S.E.B. Weisner, Functional differences in epiphytic microbial communities in nutrient rich freshwater ecosystems: an assay of denitrifying capacity, Freshwater Biol. 36 (1996) 555–562.

[9] G. Hofman, O. Van Cleemput, Soil and Plant Nitrogen. International Fertilizer Industry Association (IFA), Paris, France, 2004, 48p.

[10] S. Milenkovski, G. Thiere, K. Samuelsson, J. Svensson, S. Weisner, O. Berglund, P.E. Lindgren, Determination of microbial community composition using DGGE in thirty-five constructed wetlands in Southern Sweden, Manuscript to be submitted.

[11] A. Priemé, G. Braker, J. Tiedje, Diversity of nitrite reductase (*nir*K and *nir*S) gene fragments in forested upland and wetland soils, Appl. Environ. Microb. 68 (2002) 1893–2000.

[12] K. Dhondt, P. Boeckx, O. Van Cleemput, G. Hofman, Quantifying nitrate retention processes in a riparean buffer zone using the natural abundance of ^{15}N in NO_3^-, Rapid Commun. Mass Spectrom. 17 (2003) 2597–2604.

[13] S. Sabater, A. Butturini, J.-C. Clement, T.P. Burt, D. Dowrick, M. Hefting, V. Maître, G. Pinay, C. Postolache, M. Rzepecki, F. Sabater, Nitrogen removal by riparian buffers along a European climatic gradient: patterns and factors of variation, Ecosystems 6 (2003) 20–30.

[14] M. Hefting, J.C. Clément, D. Dowrick, A.C. Cosandey, S. Bernal, C. Cimpian, A. Tatur, T.P. Burt, G. Pinay, Water table elevation controls on soil nitrogen cycling in riparian wetlands along a European climatic gradient, Biogeochemistry 67 (2004) 113–134.

[15] K. Dhondt, P. Boeckx, G. Hofman, O. Van Cleemput, Temporal and spatial patters of denitrification enzyme activity and nitrous oxide fluxes in three adjacent vegetated riparian buffer zones, Biol. Fert. Soils 40 (2004) 243–251.

[16] M.E. Poach, P.G. Hunt, M.B. Vanotti, K.C. Stone, T.A. Matheny, M.H. Johnson, E.J. Sadler, Improved nitrogen treatment by constructed wetlands receiving partially nitrified liquid swine manure, Ecol. Eng. 20 (2003) 183–197.

[17] R.H. Kadlec, R.L. Knight, Treatment Wetlands, CRC Press, Boca Raton, FL, 1996.

[18] R.H. Kadlec, Nitrogen farming: the use of wetlands for nitrate removal, J. Environ. Safety Health (in press).

[19] J. Vymazal, Types of constructed wetlands for wastewater treatment: their potential for nutrient removal, in: J. Vymazal (Ed.) Transformations of Nutrients in Constructed Natural and Constructed Wetlands, Backhuys Publishers, Leiden, 2001, p. 496.

[20] P.S. Burgoon, R.H. Kadlec, M. Henderson, Treatment of potato processing wastewater with engineered natural systems, Water Sci. Techn. 40 (1999) 211–215.

[21] P.S. Burgoon, Denitrification in free water surface wetlands receiving carbon supplements, 7th International Conference on Wetland Systems for Water Pollution Control, Lake Buena Vista, University of Florida, 2000.

[22] T. Zhu, F. J. Sikora, Ammonium and nitrate removal in vegetated and unvegetated wetlands, Water Sci. Techn. 32 (1995) 219–228.

[23] P.A.M. Bachand, A.J. Horne, Denitrification in constructed free-water surface wetlands: I. Very high nitrate removal rates in a macrocosm study, Ecol. Eng. 14 (2000) 9–15.

[24] S. Kallner Bastviken, P.E. Eriksson, A. Premrov, K.S. Tonderski, Potential denitrification of wetland sediments with different plant species detritus, Ecol. Eng. 25 (2005) 183–190.

[25] J.F. Martin, K.R. Reddy, Interaction and spatial distribution of wetland nitrogen processes, Ecol. Mod. 105 (1997) 1–21.

[26] O. Van Cleemput, Subsoils: Chemo- and biological denitrification, N_2O and N_2 emissions, Nutr. Cycl. Agroecosys. 52 (1998) 187–194.

[27] A.X. Hou, G.X. Chen, Z.P. Wang, O. Van Cleemput, W.H. Jr. Patrick, Methane and nitrous oxide emissions from a rice field in relation to soil redox and microbiological processes, Soil Sci. Soc. Amer. J. 64 (2000) 2180–2186.

[28] K.W. Yu, Z.P. Wang, A. Vermoesen, W.H. Jr. Patrick, O. Van Cleemput, Nitrous oxide and methane emissions from different soil suspensions: effect of soil redox status, Biol. Fert. Soils 34 (2001) 25–30.

[29] K.W. Yu, W.H. Jr. Patrick, Redox window with minimum global warming potential contribution from rice soils, Soil Sci. Soc. Amer. J. 68 (2004) 2086–2091.

[30] K.F. Bronson, H.U. Neue, U. Singh, E.B.H. Abao, Automated chamber measurement of methane and nitrous oxide flux in a flooded rice soil: I. Residue, nitrogen, and water management, Soil Sci. Soc. Amer. J. 61 (1997) 981–987.

[31] K.F. Bronson, H.U. Neue, U. Singh, E.B.H. Abao, Automated chamber measurement of methane and nitrous oxide flux in a flooded rice soil: II. Fallow period emissions, Soil Sci. Soc. Amer. J. 61 (1997) 988–993.

[32] IAEA, Manual on measurement of methane and nitrous oxide emissions from agriculture, IAEA-TECDOC-674, International Atomic Energy Agency, Vienna, Austria, 1992.

[33] O. Van Cleemput, P. Boeckx, Measurement of greenhouse gas fluxes, in: R. Lal (Ed.) Encyclopedia of Soil Science, Marcel Dekker, New York, 2002, pp. 626–629.

Chapter 24

Organisms of the Nitrogen Cycle Under Extreme Conditions: Low Temperature, Salinity, pH Value and Water Stress

Blaž Stres, Maria José Bonete,
Rosa Maria Martínez-Espinosa, Ivan Mahne
and Hermann Bothe

24.1 Introduction

More than one-third of the terrestrial surface is exposed to low temperatures (T) and one-third is represented by dry habitats including saline, alkaline environments and dry seabeds. The main limitations for denitrification in agricultural soils, availabilities of O_2, H_2O, C and NO_3^-, are coupled to interplays of T, pH, salinity and severe desiccation in more unusual habitats. Denitrifying microorganisms can adapt to low or high pH value, take advantage of sporadic rain events in deserts, maintain their activity below 0°C, survive multiple freeze–thaw cycles and take advantage of environmental conditions in hypersaline environments. The aim of this chapter is to provide a general understanding of the importance of denitrification

Biology of the Nitrogen Cycle
Edited by H. Bothe, S.J. Ferguson and W.E. Newton

and other reactions of the N-cycle in these environments to global ecology, the principles governing their activities and ongoing adaptations that enable the microbes to thrive in such environments.

24.2 Low temperature

24.2.1 Interactions at low temperatures

Almost one-third of Earth's terrestrial surface is exposed to T below 0°C, a condition that prevails during 3–9 months a year. Generally, the cold seasons can be divided into a period of gradual decrease in T towards 0°C, a prolonged period of completely frozen state and a period of gradual melting of ice with more frequent freeze–thaw cycles. The extent and duration of each period, as well as the N-transformation rates, vary drastically both locally and annually.

To survive at low T, microbes can reduce their cell size and their capsular polysaccharide coat thickness; they can also change their fatty acid and phospholipid composition. Moreover, they can decrease the fractional volume of cellular H_2O, increase the fraction of ordered cellular H_2O and extract energy by catalysing redox reactions of ions in aqueous veins in ice or in thin aqueous films covering solid particle surface [1]. The lowest T capable of sustaining microbial life in aqueous environments or in soils is supposed to be about −20°C [2], but no evidence exists for a threshold or cutoff in metabolic rate at T down to −40°C [1]. Metabolic activity is, however, influenced by the availability of liquid water. The amount of unfrozen water in soil is dependent on pore size distribution, void ratio, particle size, surface area and concentration of water-soluble ions [3]. At −5°C for several days, 8−20% of the soil water remained unfrozen [4], and at −10°C and −20°C, 2−3% and 1−2% of the water did not freeze respectively [5]. In frozen soils, bacteria are covered by thin film of water. This means that nutrients reach the cells and waste products are eliminated by diffusion through narrow channels of unfrozen H_2O. The thickness of the H_2O-films decreases from about 15 nm at −1.5°C to about 5 nm at −10°C. The buildup of diffusion gradients progressively slows and eventually stops the movement of nutrients and waste products to and from cells.

The unfrozen microsites, surrounded by ice, have limited gas exchange, which leads to the development of O_2-deficiency thereby favouring denitrification. In these water films, the availability of labile C is high due not only to microorganisms, being killed by freezing or hygroscopic effects, but also to organic matter from broken aggregates [6]. Furthermore, nutrient concentrations in liquid water films also increase by ion exclusion from the growing ice grid [4]. Such ion exclusions significantly contribute to the establishment of C- and N-enriched microsites covered with liquid water below 0°C.

However, the depth of soil frost penetration is dependent upon the amount of organic litter, vegetation type, water status of soil and snow cover [7, 8]. Low soil T limits soil N-mineralisation, but earlier and deeper snow conditions with the associated warmer winter soil T dramatically increase over-winter N-mineralisation [9]. This process and nitrification were found to proceed below zero. N-mineralisation, nitrification and denitrification were measured simultaneously in frozen soils down to −4°C. The minimum T at which mineralisation could occur in Swedish cold adapted soils was estimated to be −9.3°C [10].

24.2.2 The conceptual model of denitrification at low temperature

In denitrification, low T most severely affects the N_2O-reduction process, likely due to the suppression of N_2O-reductase activity [11]. In laboratory studies with soil cores, the N_2O/N_2 ratio increases in parallel with the T decrease [12]. N_2O-reductase activity is markedly affected also by the availability of easily decomposable organic matter [13]. The enhanced level of NO_3^- in the remaining liquid H_2O film may also increase the N_2O/N_2 ratio because the affinity for electrons from the respiratory chain might be higher for NO_3^--reductase than for N_2O-reducing enzyme [14]. This increase in N_2O/N_2 ratio with lower T is not due to differences in activation energies between N_2O-reduction and N_2O-production, but due to the differences in T response of N_2O diffusion [15].

From results of soil incubation experiments in which T and moisture conditions [7, 15] were varied, Öquist et al. [10] proposed a conceptual model of N_2O formation from denitrification in relation to T, integrating also effects of soil water, T and O_2 (Figure 24-1). During freezing, the diffusion barriers increase with simultaneous raise in the number of anoxic microsites leading to higher rates of N_2O-formation than expected if T were the only determining factor between 0°C and −4°C. Below −4°C, N_2O formation likely drops owing to decreasing amounts of free H_2O at lower T [10]. Higher T enhances soil O_2-consumption and induces anoxic conditions that increase the production of N_2O. In this respect, the conceptual model and the established mechanism accounting for anomalous N_2O-formation rates at higher T [16] are similar.

Until recently, most models and global budget estimates assumed that trace gas exchange stopped when soil was covered by snow or soil T dropped close or below 0°C, although studies showed that frozen soils can contribute up to 70% of annual atmospheric load [17–19]. Most of the studies exploring gas emissions from soil at low T have focused on periods of soil thaw, when peaking emission rates can be observed [10, and references therein]. In addition, during T decrease, emissions close to 0°C may represent up to 14% of emissions encountered at thawing [20]. Chemodenitrification in frozen soils was reported to be another source of N_2O [21]. However, experiments with soil treated by X-ray [17], $CHCl_3$ [22] or autoclaving [10]

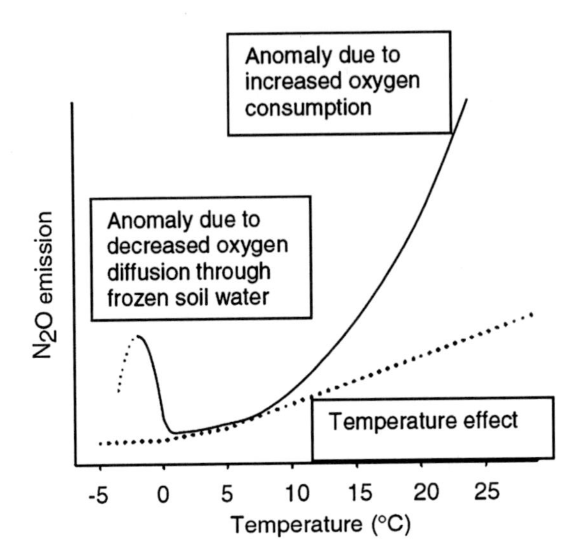

Figure 24-1 Conceptual illustration of the T effect on the production of N_2O in soils. The dotted line includes only the direct T effects, while the sold line combines both the direct T response and increased anoxic conditions from increased oxygen consumption due to increased heterotrophic activity at increasing T (adopted from [10]).

showed that chemodenitrification, in these cases, was negligible. Altogether, our knowledge about chemical and biological reactions, as well as about the organisms involved, in soils at frozen–unfrozen boundaries with transiently high solute concentrations is still fragmentary.

Finally, the T responses of N_2O-production and reduction rates might significantly differ among different soil types depending upon composition of the microbial communities. Owing to an estimated half-life of about 120 years and a global warming potential about 310 times that of CO_2, the understanding of the influence of low T on N_2O-production and consumption is of key importance, since even a small accumulation may cause destructive effects for centuries [23].

24.3 Denitrification in halophilic environments

24.3.1 Description of halophilic environments and their strategies to cope with salts

Hypersaline habitats can be grouped into two categories: thalassohaline and athalassohaline. Thalassohaline habitats are created by evaporation of seawater, have Na^+ and Cl^- as dominant anions and pH values close to neutrality. When evaporation proceeds, the ionic composition may change due to the precipitation of gypsum ($CaSO_4 \cdot 2\,H_2O$) and other minerals beyond their solubility threshold. NaCl-saturated thalassohaline brines, such as found in saltern crystalliser ponds, often display a bright red colouration due to the large number of pigmented microorganisms (*Halorubrum* and others). The ionic compositions of athalassohaline environments differ from that of seawater, often contain high concentrations of CO_3^{2-} and HCO_3^- and their pH can exceed a value of 10 [24]. The term sodic soil is often used for such inland salt habitats dominated by Na_2CO_3 or Na_2SO_4 [25]. The vegetation around such habitats comprises typical plants, "the halophytes". Their composition is remarkably rather similar worldwide, as distinct belt formation can be observed as a function of the salt load [25].

The term "halophile" includes organisms that require NaCl or another salt for growth ("salt-resistant organisms") and those which thrive in both non-saline and saline habitats. The latter ("salt tolerant organisms"), owing to their capacity to endure higher salt loads, are more effective competitors in salt marshes than non-adapted organisms. An example for salt tolerant bacteria is the denitrifying *Azospirillum halopraeferens*, which grows best when 0.25% NaCl is present in the medium [26]. Diversity of halophilic environments is reflected by the great variety within the microbial communities adapted to life under harsh conditions and is expressed both at the phylogenetic (halophiles are found in all domains of life) and physiological levels (most modes of energy generation described in non-halophilic organisms are also present in halophilic counterparts) [27].

The strategies used by halophilic organisms to cope with the adverse effects of high salt loads have been discussed elsewhere in detail [27]. One strategy involves accumulation of K^+ and Cl^- ions to maintain osmotic balance. This is used by haloarchaea of the family Halobacteriaceae such as the denitrifier *Haloferax mediterranei* or by members of the Halanaerobiales in the domain Bacteria [28]. The presence of high intracellular salt concentrations requires far-reaching adaptations of the enzymatic machinery so as to be active at high molar salt concentration [29]. In the second, widely used, strategy of osmotic adaptation, organisms exclude salts from the cytoplasm and accumulate organic solutes (sugar alcohols, amino acids and derivates such as glycine, betaine and ectoine) to provide the osmotic balance [30]. The excretion of sodium ions from the interior of the cells is achieved by Na^+/H^+ antiporters [28]. The major stress factor for organisms thriving in NaCl, sodic or potash soils does not appear to be the salt itself, but the extremely negative water potential of such soils, as noted by earlier botanists for the halophytes [31]. The dissociated ionised salts strongly bind water, which then can hardly be mobilised by the organisms for their growth, particularly under longer periods of drought. This "physiological drought" may be the major cause for the inability of most organisms to thrive in saline habitats.

24.3.2 Typical halophilic microorganisms

The aerobic halophilic Archaea of the family Halobacteriaceae are the halophiles par excellence. They are the main component of the microbial biomass in environments such as the Dead Sea, hypersaline soda lakes, saltern crystallizer ponds and potash mines. Typical genera of the Halobacteriaceae in such sites include *Halorubrum*, *Halogeometricum*, *Haloferax, Halosimplex* and also the alkoliphilic *Natronococcus* and *Natronomonas* [32]. The methanogenic branch of the Euryarchaeota also contains halophilic representatives; nonetheless, no halophiles have yet been identified within the Crenarchaeota kingdom [27]. Within the domain Eukarya, halophiles are scarce, the green alga *Dunaliella* is the only eukaryote described at the moment in the Dead Sea and other locations. The Bacteria domain contains many types of haloresistant and halotolerant microorganisms into the α-, β-or γ-Proteobacteria and Gram-positive Bacteria groups [33]. Most halophiles within the Bacteria domain are moderate rather than extreme halophiles, with the exception of the more recently described *Salinibacter* sp. that has been retrieved from several extreme saline habitats [34]. Many novel haloalkaliphilic sulfur-oxidising bacteria of the genera *Thioalkalimicrobium* and *Thioalkalivibrio* have been described from soda lakes [35] and grow in extremely alkaline ($pH > 7.5$) and saline environments. Only one of those isolated, described as *Thialkalivibrio denitrificans* ALJD^T, is able to grow under anaerobic conditions [36].

PCR technology now reveals that halophilic bacteria are seemingly cosmopolites. Almost identical sequences to *Halobacillus locisalis* of a saltern at the Yellow Sea in Korea, or to *Loktanella* recently found in salt brines in the Vestfold Hill region of Antarctica have now been detected in soils surrounding a potash mine of Northern Germany [32]. Endemism among bacteria is at best rare and seemingly does not occur within halophilic organisms. However, sequencing of PCR products obtained from DNA of any habitat indicates that a whole realm of not-yet-revealed bacteria thrives on Earth. Recent evidence indicates that the salt load in soils does not dictate the distribution of bacteria in salt marshes, in contrast to the situation with plants. Extreme halophilic and non-salt-tolerant bacteria are found site-by-site at all salt concentrations in the soil [32].

24.3.3 Microorganisms from halophilic environments with activities in the N-cycle

The most representative prokaryotic denitrifiers described from halophilic environments are listed in Table 24-1. In the Douro River estuary (Portugal), e.g. salinity apparently does not regulate denitrification, suggesting that halotolerant microorganisms dominate the denitrifying communities [37]. In contrast, denitrification activity decreased with

Table 24-1 Selected examples of bacteria occurring in saline habitats.

Taxon	Genus and species	Habitat
Archaea domain		
	Haloarcula hispanica	Marine saltern
	Haloarcula vallismostis	Salt pool
	Haloarcula marismotui	Dead Sea
	Haloferax mediterranei	Solar saltern
	Haloferax denitrificans	Solar saltern
Bacteria domain		
Proteobacteria; α-subclass	*Erythrobacter* sp. OCh114	Marine
	Paracoccus halodenitrificans	Brines
	Azospirillum halopraeferens	Salt marshes (moderate haline)
Proteobacteria; β-subclass	*Spirillum* sp.	Marine
	Thiobacillus denitrificans	Water (fresh and marine)
Proteobacteria; γ-subclass	*Alteromonas denitrificans*	Marine
	Halomonas subglaciescola	Hypersaline lake,
Antarctica		
	Pseudomonas nautica	Marine
	Pseudomonas putrefaciens	Marine
	Thiomicrospira denitrificans	Marine tidal flat
Gram-positive bacteria	*Bacillus halodenitrificans*	Solar saltern

increasing salinities in Danish estuarine sediments where denitrifying communities are represented by non-halophilic microorganism [38].

Among the Archaea domain, only the family Halobacteriaceae has been shown to have denitrifying members. Haloarchaea show important differences in the structure and regulation of enzymes and transport systems involved in the denitrification pathway, with interesting evolutionary implications [33, 39]. Halophilic bacterial and archaeal *nar* genes differ in their organization [39]. Some denitrifying halophilic Archaea such as *Haloarcula marismortui* do not reduce N_2O to N_2, although other haloarchaea such as *Haloferax denitrificans* are able to reduce N_2O in a complete denitrification pathway [39]. Among the bacteria, *Thialkalivibrio denitrificans* ALJD^T also does not express dissimilatory NO_3^--reductase but grows with NO_2^- or N_2O as electron acceptor and thiosulfate or polysulfide as electron donor [36].

Nitrification apparently operates in saline areas [35]. Sequences closely related to those from *Nitrospira marina* have recently been retrieved from DNA of a soil surrounding a potash mine [32]. The same site contains sequences of apparently nitrifying bacteria that are closely related to those obtained from marine sediments of the Chesapeake Bay on the Eastern coast of the the United States [40] Sequences of planctomycetes were detected in both potash mine soil [32] and hypersaline environment of the stromatolites of the Hamelin pool in Australia [41]. It is, however, not known whether such bacteria adapted to such high salinity perform anammox.

24.3.4 Effect of pH and water stress on denitrification

The relationship between soil pH and denitrification seems to be particularly complicated as the rate of the denitrification process itself, and also the ratio of its gaseous products depends on pH value. Generally, at low pH value, overall denitrification activity is low and the fraction of N_2O produced is high. Extensive information on the relationship between soil pH and denitrification is compiled in [42]. These authors stated that the expression "optimum pH for denitrification" has little or no meaning without proper specification of the particular attributes of denitrification to which it is applied.

Dry and desert ecosystems appear to be unsuitable for microbial denitrification, since desert soils are hot, dry and nutrient-poor and receive high solar irradiation. Surface soil T can exceed 60°C in the summer and drops below freezing in winter. However, rates of denitrification in wet desert soil were found to be comparable to those found in more mesic ecosystems [43]. A significant fraction was attributed to a persistent pool of desiccation-tolerant enzymes within the bacteria thriving there, as was also shown for other soils [44]. Rapid response by denitrifying and also other bacteria may be of particular significance in desert ecosystems since moisture availability can be highly sporadic. Many studies report a rapid increase in nutrient availability and microbial activity after wetting dry soils [45, and references herein]. Denitrification also resumes shortly after wetting of desert soils indicating a quick establishment of anaerobic microsites, which are apparently coupled to soil moisture, carbon availability and T decrease. Detailed information on the bacterial tolerance to desiccation is currently not available.

References

[1] P. Buford Price, T. Sowers, Temperature dependence of metabolic rates for microbial growth, maintenance, and survival, Proc. Natl. Acad. Sci. USA 101 (2004) 4631–4636.

[2] L. Finegold, Molecular and biophysical aspects of adaptation of life to T below freezing point, Adv. Space Res. 18 (1996) 87–95.

[3] T.P. Burt, P.J. Williams, Hydraulic conductivity in frozen soils, Earth Surf. Proc. 1 (1974) 349–360.

[4] M. Stähli, D. Stadler, Measurement of water and solute dynamics in freezing soil columns with time domain reflectometry, J. Hydrol. 195 (1997) 352–369.

[5] E.M. Rivkina, E.I. Friedman, M.C.P. McKay, D.A. Gilichinsky, Metabolic activity of permafrost bacteria below the freezing point, Appl. Environ. Microbiol. 66 (2000) 3230–3233.
[6] S. Christiensen, J.M. Tiedje, Brief and vigorous N_2O production by soil at spring thawing, J. Soil Sci. 41 (1990) 1–4.
[7] R. Teepe, R. Brumme, F. Beese, Nitrous oxide emissions from frozen soils under agricultural, fallow and forest land, Soil Biol. Biochem. 32 (2000) 1807–1810.
[8] J.P. Schimmel, C. Bilbrough, J.M. Welker, Increased snow depth affects microbial activity and nitrogen mineralization in two Arctic tundra communities, Soil Biol. Biochem. 36 (2004) 217–227.
[9] S.K. Schmidt, D.A. Lipson, Microbial growth under the snow: implications for nutrient and allelochemical availability in temperate soils, Plant Soil 259 (2004) 1–7.
[10] M. G. Öquist, M. Nilsson, F. Sörensson, Å. Kasimir-Klemedtsson, T. Persson, P.W. L. Klemedtsson, Nitrous oxide production in a forest soil at low temperature processes and environmental controls, FEMS Microb. Ecol. 49 (2004) 371–378.
[11] J. Melin, H. Nommik, Denitrification measurements in intact soil cores, Acta. Agric. Scand. 44 (1983) 1797–1806.
[12] L.D. Bailey, E.G. Beauchamp, Effects of temperature on NO_3^- and NO_2^- reduction, nitrogenous gas production, and redox potential in a saturated soil, Can. J. Soil Sci. 53 (1973) 213–218.
[13] L. Dendoven, J.M. Anderson, Maintenance of denitrification potential in pasture soil following anaerobic events, Soil Biol. Biochem. 27 (1995) 1251–1260.
[14] L. Dendooven, P. Splatt, J.M. Anderson, D. Scholefield, Kinetic of the denitrification process in a soil under permanent pasture, Soil Biol. Biochem. 26 (1994) 361–370
[15] L. Holtan-Hartwig, P. Dörsch, L.R. Bakken, Low temperature control of soil denitrifying communities: kinetics of N_2O production and reduction, Soil Biol. Biochem. 34 (2002) 1797–1806.
[16] K.A. Smith, The potential for feedback effects induced by global warming on emissions of nitrous oxide by soils, Global Change Biol. 3 (1997) 327–338.
[17] M. Röver, O. Heinemeyer, E.-A. Kasier, Microbial induced nitrous oxide emissions from arable soil during winter, Soil Biol. Biochem. 30 (1998) 1859–1865.
[18] M.C. Ryan, R.G. Kachanoski, R.W. Gillham, Overwinter soil nitrogen dynamics in seasonally frozen soils, Can. J. Soil Sci. 80 (2000) 541–550.
[19] E.-A. Kaiser, K. Kohrs, M. Kücke, E. Schnug, O. Heinemeyer, J.C. Munch, Nitrous oxide release from arable soil: importance of N-fertilization, crops and temporal variation, Soil Biol. Biochem. 30 (1998) 1553–1563.

[20] H. Koponen, L. Flöjt, P.T. Martikainen, Nitrous oxide emissions from agricultural soils at low T: a laboratory microcosm study, Soil Biol. Biochem. 36 (2000) 757–766.
[21] C.B. Christianson, C.M. Cho, Chemical denitrification of nitrite in frozen soils, Soil Sci. Soc. Amer. J. 47 (1983) 38–42.
[22] L.L. Goodrad, D.R. Keeney, Nitrous oxide emissions from soils during thawing, Can. J. Soil Sci. 64 (1984) 187–194.
[23] W.C. Trogler, Physical properties and mechanisms of formation of nitrous oxide, Coord. Chem. Rev. 187 (1999) 303–327.
[24] R.L. Mancinelli, Microbial life in brines, evaporites and saline sediments: the search for life on Mars, in: T. Tokano (Ed.) Water on Mars and Life, Adv. Astrobiol. Biogeophys., (Vol.4) Springer, Berlin, 2005, pp. 277–297.
[25] M. Landwehr, U. Hildebrandt, P. Wilde, K. Nawrath, T. Tóth, B. Biró, H. Bothe, The arbuscular mycorrhizal fungus *Glomus geosporum* in European saline, sodic and gypsum soils, Mycorrhiza 12 (2002) 199–211.
[26] B. Reinhold, T. Hurek, I. Fendrik, B. Pot, M. Gillis, K. Kertsers, D. Tielemans, D.J. Ley, *Azospirillum halopraeferens* sp. nov., a nitrogen-fixing organism associate with roots of Kallar grass [Leptochloa fusca (L. Kunth)], Int. J. Syst. Bacetriol. 37 (1987) 43–51.
[27] A. Oren, Diversity of halophilic microorganism: environments, phylogeny, physiology and applications, J. Ind. Microbiol. Biotechnol. 28 (2002) 56–63.
[28] A. Oren, Bioenergetic aspects of halophilism, Microbiol. Mol. Biol. Rev. 63 (1999) 334–348.
[29] K.L. Britton, P.J. Baker, M.F.S. Ruzheinikov, D.J. Gilmour, M.J. Bonete, J. Ferrer, C. Pire, J. Esclapez, D.W. Rice, Analysis of protein solvent interactions in glucose dehydrogenase from the extreme halophile *Haloferax mediterranei,* Proc. Natl. Acad. Sci. USA, 13 (2005) 4846–4851.
[30] A. Ventosa, J.J. Nieto, A. Oren, Biology of aerobic moderately halophilic bacteria, Microbiol. Mol. Biol. Rev. 62 (1998) 504–544.
[31] O. Stocker, Das Halophytenproblem, in: K. v. Frisch, R. Goldschmidt, W. Ruhland, H. Winterstein (Eds.) Ergebnisse der Biologie, Vol. 3, Springer, Berlin, 1928, pp. 266–353.
[32] S. Eilmus, C. Rösch, H. Bothe, Community analysis of total, N_2-fixing, denitrifying and nitrifying bacteria in a potash marsh, Environ. Pollut. (2006) in press.
[33] W. Zumft, The denitrifying prokaryotes. in: A. Balows, H.G. Trüper, M. Dworkin, W. Harder, K.H. Schleifer (Eds.) The Prokaryotes. A Handbook on the Biology of Bacteria: Ecophysiology, Isolation, Identification, Applications, Vol. 1, Springer, New York, 1992, pp. 554–582.

[34] J. Antón, R. Roselló-Mora, F. Rodríguez-Valera, R.I. Amann, Extremely halophilic bacteria in cystrallizer ponds from solar salterns, Appl. Environ. Microbiol. 66 (2000) 3052–3057.
[35] D.Yu. Sorokin, J.G. Kuenen, Haloalkaliphilic sulfur-oxidizing bacteria in soda lakes, FEMS Microbiol. Rev., 29 (2005) 685–702.
[36] D. Yu. Sorokin, J.G. Kuenen, M.S. Jetten, Denitrification at extremely high pH values by the alkaliophilic, obligately chemolithoautotrophic, sulfur-oxidizing bacterium *Thioalkalivibrio denitrificans* strain ALJD, Arch. Microbiol. 175 (2001) 94–101.
[37] C.M. Magalhaes, S.B. Joye, R.M. Moreira, W.J. Wiebe, A.A. Bordalo, Effect of salinity and inorganic nitrogen concentrations on nitrification and denitrification rates in intertidal sediments and rocky biofilms of the Douro River estuary, Portugal, Water Res., 39 (2005) 1783–1794.
[38] S. Rysgaards, P. Thastum, T. Dalsgaard, P.B. Christensen, N.P. Sloth, Effects of salinity on NH_4^+ adsorption capacity, nitrification, and denitrification in Danish estuarine sediments, Estuaries 22 (1999) 21–31.
[39] B. Lledó, R.M. Martínez-Espinosa, F.C. Marhuenda-Egea, M.J. Bonete, Respiratory nitrate reductase from haloarchaeon *Haloferax* mediterranei: biochemical and genetic analysis, Biochim. Biophys Acta 1674 (2004) 50–59.
[40] L.A. Francis, G.D. O'Mullan, B.B. Ward, Diversity of *amoA* genes across environmental gradients in Chesapeake Bay sediments, Geobiol. 1 (2003) 129–140.
[41] B.P. Burns, F. Goh, M. Allen, B.A. Neilan, Microbial diversity of extant stromatolithes in the hypersaline marine environment of Shark Bay, Australia, Environ. Microbiol. 6 (2004) 1096–2011.
[42] M. Šimek, J. E. Cooper, the influence of soil pH on denitrification: progress towards the understanding of this interaction over last 50 years, Eur. J. Soil Sci. 53 (2002) 345–354.
[43] R.A. Virginia, W.M. Jarrell, E. Franco-Vizcaino, Direct measurement of denitrification in a *Prosopis* (mesquite) dominated Sonoran Desert ecosystem, Oecologia 53 (1982) 120–122.
[44] P.M. Groffman, J.M. Tiedje, Denitrification in the north temperate forest soils: relationship between denitrification and environmental factors at the landscape scale, Soil Biol. Biochem. 21 (1989) 621–626.
[45] W.T. Peterjohn, Denitrification: enzyme content and activity in desert soils, Soil Biol. Biochem. 23 (1991) 845–855.

Chapter 25

Nitrous Oxide Emission and Global Changes: Modeling Approaches

Lars Bakken and Peter Dörsch

25.1 Introduction

Anthropogenic disturbance of the biogeochemical cycles is perhaps today's greatest environmental challenge, and C- and N-cycling are probably most profoundly affected. The annual input of biologically reactive N to the biosphere by human activities (through synthetic fertilizers, biological N-fixation in agriculture, and NO_x from combustion) is roughly equal to the prehistoric (pre-industrial) annual input by biological fixation in natural ecosystems [1], and will increase further as the world population grows in number and prosperity. This situation results in a net accumulation of reactive N (be it in biomass, humic substances, or mineral forms of N), until counterbalanced by an equal rate of bacterial reduction of NO_3^- to N_2. The slow reaction rates of the various N pools imply that it will take long time to reach equilibrium [2, 3], and that the detrimental side effects of anthropogenic N loading will persist long after N loading has ceased.

Biology of the Nitrogen Cycle
Edited by H. Bothe, S.J. Ferguson and W.E. Newton

Denitrification is, therefore, a key process in the global changes driven by N enrichment. In addition to being the only significant process, which removes reactive N from the biosphere, denitrification produces N_2O as an inevitable side product. The anthropogenic impact on the N-cycle is thus clearly detectable as a net accumulation of atmospheric N_2O, which coincides with the increasing inputs of reactive-N in modern times. Stable isotope signatures of atmospheric N_2O strongly suggest that soil emissions make a substantial contribution [4, 5]. The ongoing accumulation of N_2O in the atmosphere is of great concern because it contributes to global warming [6] and destruction of stratospheric ozone [7] (see also Figure 0-2 in the Preface).

Historical records of N_2O in the atmosphere reflect climate changes rather closely [8–10] suggesting that the ongoing global warming in itself will enhance N_2O emission. The relationship between temperature and emissions, although apparently clear when comparing global changes over thousands of years (as done in ice core analyses), is not so obvious at higher resolution in time and space. And we need to know the regulation of fluxes at such high resolution today, not only to achieve a better understanding and quantification of the ongoing N_2O loading, but also to assess the expected changes in emission patterns in response to the ongoing N enrichment and global warming. These considerations underscore the need for dynamic models of denitrification and N_2O fluxes in soils and sediments as driven by hydrology and temperature [11].

Herein, a brief survey of the critical aspects of denitrification modeling starts with a discussion of the product stoichiometry as controlled by the biology of denitrifiers. It is followed by a discussion of attempts to model O_2 distribution within the soil matrix and then by examples of various approaches to simulate denitrification and N_2O emission as a part of complex soil–plant biogeochemical models.

25.2 Product stoichiometry of denitrification

Denitrification results in emission of three gases, NO, N_2O, and N_2, and the stoichiometry of the emitted gas mixture depends on the relative activities of the three enzymes, NO_2^--, NO-, and N_2O-reductases, which are encoded by the *nir*, *nor*, and *nos* genes, respectively. For survival, organisms need to keep the activities of both NO_2^-- and NO-reductases strictly synchronized to ensure that the NO concentration remains in the nanomolar range [12]. This is probably one reason why NO emissions normally represent only a moderate fraction of the products from denitrification in intact soils [13–15]. Short-term anaerobic incubations of soil slurries, however, have demonstrated that NO may represent 3–30% of the denitrification product, depending on soil type [15], suggesting that denitrifying communities may become severely dysfunctional in response to perturbations, such as dispersion or a sudden switch to anaerobic conditions.

N_2O, on the other hand, has no known toxicity to bacteria, which means that a low relative N_2O-reductase activity is not harmful, unless the availability of electron acceptors other than N_2O is severely limited. Several denitrifying bacteria persist under natural conditions without intact *nos* genes, and the organisms that do carry an intact *nos* gene do not always express it [16]. Further, when bacteria are transferred from aerobic to anaerobic conditions, *nos* expression appears to lag behind expression of the genes for the other reductases, resulting in transient accumulation of N_2O in cultures. This observation suggests that most natural habitats exert a rather weak selection pressure for both the preservation of N_2O-reductase and the synchronization of its formation with that of the other reductases. Thus, the absence of N_2O-reductase activity would be of only minor importance for the bacteria, although it could represent a disaster for the global environment!

The transient accumulation of all intermediates (NO_2^-, NO, and N_2O) is a recurring observation in denitrification. It has been ascribed to enzyme kinetics either alone [17, 18] or together with sequential gene expression [19]. Thus, product stoichiometry is strongly affected by both enzyme kinetics and the relative amounts of NO_3^--, NO_2^--, NO-, and N_2O-reductases present. Relative N_2O-reductase activity (compared to that of the other reductases) is severely decreased by low pH value [20, 21] and O_2 [22] either because N_2O-reductase is unstable or only partially functional under such conditions, or because its formation is repressed. In contrast, several denitrifying bacteria are known to express *nar* (encoding NO_3^--reductase) and *nir* [hence also *nor*] in the presence of O_2 [23]. Experiments with denitrifying communities are consistent with these observations. When a mixed bacterial community is confronted with anaerobic conditions, the relative N_2O-reductase activity is often low initially, but increases in response to prolonged (20–40 h) anaerobic incubation [24]. Thus, a common regulatory pattern appears to exist in denitrifying bacteria, but with variations, which have implications for the propensity for N_2O emissions. Comparisons of denitrifying communities from different soils show persistent differences in their N_2O production/reduction kinetics [25, 26].

In summary, the biology of denitrifying bacteria suggests a variety of response patterns regarding the synchronization of the activities of NO_3^--, NO_2^--, NO-, and N_2O-reductases, with implications for the stoichiometry of products. The role of denitrifier community composition and functioning can easily be overlooked, however, when measuring N_2O emissions from soils to the atmosphere because of the daunting spatial and temporal variability of denitrification and N_2O emission rates as caused by variations in soil moisture, temperature, respiration, and NO_3^- concentration as well as gas-diffusion rates within the soil matrix. Gas-flux patterns are thus an extremely blurred picture of the biological reality within the soil matrix and this causes potential heuristic conflicts between the microbiologist/biochemist and the ecologist. The detailed phenomena studied by microbiologists/biochemists are rarely of direct relevance for interpreting observations at the field scale, and few of the observations from field experiments are meaningful (even as hypothesis-generating observations) for microbiologists/biochemists. Mathematical modeling, however, could potentially bridge this gap between disciplines. Dynamic biogeochemical models of heat and water transport, plant growth, and microbial mineralization of C and N are potential platforms for simulating the biology of denitrifying bacteria, as they affect the performance of the whole system in terms of N_2 and N_2O emissions. These models would represent a "Systems Biology" approach to denitrification at the ecosystem level. Owing to the complexity of interactions and spatial heterogeneity of soils and their microbial communities, all attempts to model denitrification have, so far, been based on gross simplifications. New and more refined models can be constructed to start bridging the gap between gas-flux data and the biology of denitrifying bacteria. This task will be Herculean in nature, however, and the following section goes some way in explaining why.

25.3 Models of soil anaerobiosis as a regulator for denitrification

Oxygen concentration is the master variable for denitrification; hence a proper calculation of the distribution of O_2 and anaerobic sites within the soil matrix is a prerequisite for modeling denitrification. Simple empirical models commonly lack an explicit representation of pO_2, but use soil moisture content and respiration as surrogate variables. The simplest versions use soil moisture as a switch, which – when passing a critical value – turns denitrification on or off. This highly simplistic approach to control denitrification rates was used in an early version of the coupled DeNitrification–DeComposition (DNDC) model [27], where denitrification is triggered by precipitation and continues as long as soil moisture is above 40%.

A more refined approach uses a power function of soil-moisture content as a dimensionless reduction factor (0–1), which is multiplied by an estimated "maximum" or "potential" denitrification rate. The shape of such soil-moisture functions defines the response of denitrification to changing soil moisture as a surrogate for anoxic volumes [28]. A range of empirical models have been developed based on this approach, and some of them incorporate microbial respiration as a second regulating variable [29]. One of the more elaborated versions is the NGAS model in DAYCENT [30], in which a sigmoidal arctang function of the water-filled pore space (WFPS) is used to scale denitrification (equation [1]). The potential denitrification rate is then calculated as in equation (2).

$$Fd(WFPS) = 0.5 + (arctang(0.6 * \pi(0.1 * [WFPS] - \alpha)))/\pi \quad (1)$$

where WFPS is the water-filled pore space and α is a function of respiration and the gas-diffusion coefficient at field capacity.

$$Rate = Fd(WFPS) * min(0.1 * R^{1.3}, Fd[NO_3^-]) \quad (2)$$

where R is the soil-respiration rate [in μg CO_2–C g^{-1} soil day^{-1}], and $Fd[NO_3^-]$ is a function of the nitrate concentration.

The strength of this model compared to numerous other empirical models is its use of more elaborate soil physical calculations to derive the gas-diffusion coefficient, based on bulk density, actual soil-moisture content, and the soil-moisture content at field capacity.

A different approach is used in the revised "anaerobic balloon" version of the DNDC model [31]. Here, the O_2-diffusion coefficient in soil is calculated by a simple power function of air-filled porosity divided by total porosity (or by the air-filled porosity at field capacity to account for soil structure). The model calculates pO_2 numerically as a steady state of O_2 diffusion and consumption for each soil layer, and the anaerobic fraction as a linear function of pO_2 in a soil layer relative to pO_2 in air.

In contrast to empirical models, mechanistic models of soil pO_2 explicitly account for O_2 distribution within the soil matrix as a function of respiration and soil structure. Several attempts have been made to find simplified representations of soil structure, including those based on spherical aggregates [32, 33]. The aggregates are assumed to be distributed lognormally [33] or packed hexagonally [32], and both models apply radial diffusion by Fick's law to soil aggregates, thereby accounting for the macro (inter-aggregate) and micro (intra-aggregate) structure of soil ("dual porosity models"). With a given microbial O_2 consumption (often assumed to be zero-order and homogeneously distributed throughout the aggregate), aggregate cores may turn anoxic and support denitrification. Consequently, both the occurrence and extent of anaerobic zones in a structured soil can be modeled as a function of aggregate diameter. Moreover, by modeling NO_3^- diffusion from the aerobic aggregate surfaces to the anoxic cores, the "optimal" aggregate diameter for denitrification can be determined [34]. However, to derive the anaerobic volume of a soil layer, assumptions have to be made as to the distribution of aggregate sizes. Although conceptually interesting, aggregate models are difficult to parameterize because aggregate distributions are not commonly reported. Moreover, they are not applicable to nonaggregated soils, such as peat or single-grained sandy soils.

An alternative way to account for soil structure is to consider the air-filled pore structure of a soil. Pore models conceive soil porosity as a system of parallel, cylindrical pores [35, 36]. Similar to aggregate models, a size distribution (here the diameters of air-filled pores) is needed. This distribution can be derived from the water retention curve, which is a parameter set commonly used for soil characterization and not restricted to structured soils. For the calculation of the anoxic volume, it is assumed that O_2 diffuses radially from cylindrical air-filled pores into the saturated soil matrix, which consists of solids and water-filled pores, thereby creating an oxygenated zone or "aerobic cylinder" around the pore. At a certain distance from the air-filled pore O_2 diffusion and O_2 consumption become equal and pO_2 is zero. Assuming steady state, this distance can be calculated numerically, yielding the radius r_a of the aerobic cylinder (Figure 25-1). In contrast to other diffusion-based models, which have to rely on estimated O_2 diffusivity in soil, pore models can use the well-defined diffusion coefficient for O_2 in water ($1.5 \times 10^{-9} m^2 s^{-1}$) with minor corrections for tortuosity and impediment by soil particles [35]. Finally, the anaerobic fraction is calculated from the surface area of aerobic cylinders, the number of air-filled pores (calculated from the total volume fraction of air-filled pores), and the spatial distribution of aerobic cylinders in a cross-sectional unit. An apparently unresolved problem is the potential error due to overlap of estimated oxygenated zones [37].

From Figure 25-1, it is obvious that the volume (or cross section) of the aerobic zone around an air-filled pore depends on the pore diameter. As a simplification, the use of a "typical air-filled pore diameter" has been devised, calculated as the geometrical mean of the minimum (equal to a function of the actual water potential) and the maximum (equal to a pore

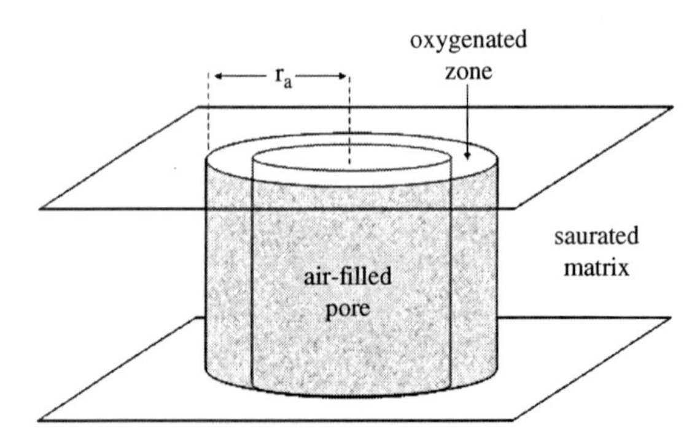

Figure 25-1 Pore models. O_2 diffuses radially from cylindrical air-filled pores into a saturated matrix, creating an "aerobic cylinder" around the pore; the radius r_a at steady state is defined by the point where O_2 consumption equals O_2 diffusion into the soil matrix and pO_2 is zero.

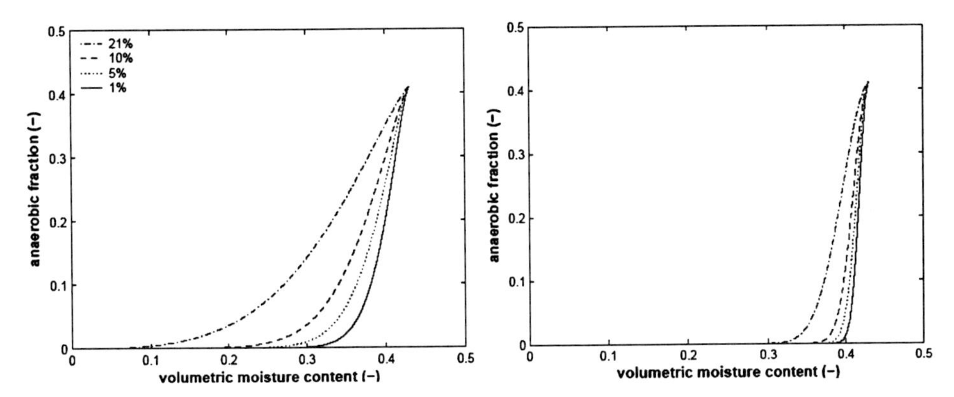

Figure 25-2 Predicted anaerobic fraction as a function of soil moisture content and O_2 concentrations in soil air of a sandy soil based on the "average pore" model (left) and the "pore class" model (right). From Ref. [38].

corresponding to a pressure head of 0.1 m) size [35, 36]. However, in theory, the relation between r_a and pore radius cannot be constant because small air-filled pores contribute disproportionably more to the aerobic fraction than larger pores [38]. As a consequence, anaerobic volumes tend to be overestimated at low soil moisture contents (many air-filled small pores). So instead, the size of the aerobic cylinder might be better calculated based on a range of "air-filled pore size" classes rather than for a "typical" average pore [38]. This modified "pore-class model" appears to give more adequate estimates of anaerobic volumes over a range of soil moistures in different soil textures. Compared to the "average pore" model [35], the "pore class" model shows a sharper response to changes in soil moisture within a critical intermediate range of soil-moisture levels (Figure 25-2), similar to that predicted by soil-aggregate models [34].

The sharp response of anaerobiosis to soil moisture predicted by aggregate models and the "pore class" model suggests that soil moisture acts almost as an on/off switch with respect to denitrification. In real soils, however, such sharp thresholds would be softened by the heterogeneous spatial distribution of respiration (roots and microbes). The response of the "average pore" model may, therefore, be closer to reality than the "pore class" model, i.e., it might be right, but for the wrong reason. This is, in fact, an excellent illustration of a scaling problem.

25.4 Denitrification and N_2O flux in soil biogeochemical models

Several of the biogeochemical models that simulate both the C- and N-transformations in soil–plant ecosystems are available and they work at different scales. A common feature of most of these models is that they rely on an explicit simulation of heat and water transport in the soil, thus predicting temporal fluctuations in moisture and temperature through the soil profile. Further, they provide daily estimates of primary production and the microbial mineralization of organic C (plant litter and soil organic matter) and N. Prominent examples are DAISY [39], DAYCENT[40, 41], DNDC [27, 31], ExpertN [42], ANIMO [36], and SOILN [28]. In theory, the models provide all the necessary variables for estimating O_2 distribution, denitrification, and N_2O emission. Thus, the models are useful platforms for implementing denitrification models. The task is far from trivial, however, considering the complexity of interactions between regulating factors [43].

Ecological modeling necessarily implies simplification; only a fraction of the phenomena (genes, enzymes, populations, communities, modes of regulation and interactions, etc.) can be explicitly treated [44]. There are many reasons for this. One major reason is that we lack direct observations at organism level; the measured denitrification rates and N_2O emissions from a soil (be it a soil sample or a soil profile) represent the average activity of many organisms, which experience a range of different conditions depending on their position in the soil matrix. A second reason is that these systems are very complex with regard to their physical structure, chemistry, and individual populations. Any attempt to simulate activities at individual organism/population level is futile because it cannot be verified by observations, becomes computationally demanding, and is unlikely to be of any value as a predictive tool due to the extreme number of parameters needed. This is particularly true when scaling up to larger areas, where simple empirical relationships/models are generally preferred [45].

Simplified models of denitrification are based on empirically established relationships between N_2 and N_2O production by denitrification and a set of measurable soil variables, such as soil porosity, soil moisture content, temperature, respiration, NO_3^- content, pH value, etc. (see equation [3]). A very simplistic approach is the "hole in the pipe" model [46], where NO and N_2O fluxes from nitrification and denitrification are estimated as a regression function of N-transformation rates (either directly measured rates, mineral-N pool sizes, or an index for each soil) and volumetric soil-moisture content. Denitrification models that are incorporated in dynamic C- and N-cycling models are normally more refined, taking advantage of the models' estimation of a number of relevant regulating factors. Linear or nonlinear response functions are then established, based on empirical studies in laboratory or field trials (see equation [4]).

$$F_{denit} = f(T, \Theta, pH, [NO_3^-], C_{min} \ldots) \qquad (3)$$

where F_{denit} is any denitrification-related flux rate, T is soil or aboveground temperature, Θ is the volumetric soil moisture, pH is the soil's acidity, $[NO_3^-]$ is the concentration of nitrate and nitrite, and C_{min} is either the concentration of decomposable C or the rate of mineralization.

$$F_{denit} = \alpha * f(T) * f(\Theta) * f([NO_3^-]) ... * f(x) \quad (4)$$

where α is the potential denitrification rate under specified (optimal) conditions and sometimes named maximum potential denitrification, and $f(\Theta)*f([NO_3^-]) ... *f(x)$ are the response functions of variables listed under equation (3).

For example, the response to temperature may be described by a Q_{10}-, Arrhenius-, or square-root function [47]. The response to NO_3^- may be represented by a first-order or Michaelis–Menten function and the response to pH by an optimum curve. Alternatives are numerous, such as power functions, sigmoidal functions (Boltzmann functions for pH response [31] or arctang functions [30]), or simple threshold values (on/off switches). More than 50 empirical denitrification models have been published, mostly within biogeochemical models of C- and N-cycling (for an overview, see Ref. [29]). These models are similar in their basic structure, but vary by the shape of their response functions and the way in which they are generated. Their response functions are largely based on fitting equations to measured data, suggesting that the overall performance is to some degree confined to the soil or site they were developed for. A weakness of most empirical models is that they implicitly assume a unique and static denitrification potential. Variations in denitrification rates are thought to be exclusively due to changes in external factors. Neither the physiology nor population dynamics of the microorganisms that mediate the process are considered.

NGAS-DAYCENT [30, 48] is one of the more sophisticated empirical models, which predicts denitrification and its product stoichiometry (N_2O/N_2) as a function of NO_3^- content, soil respiration, and diffusion constraint. The diffusion constraint is a function of soil moisture, total porosity, and the pore-size distribution (soil moisture at field capacity) as described in the previous section. The product stoichiometry (N_2/N_2O ratio) is controlled by a combination of soil moisture and respiration rates, with increasing N_2/N_2O ratio, increasing respiration rate, and increasing soil moisture, which agrees well with observations.

25.5 Microbial kinetics of denitrification in biogeochemical models

Explicit modeling of denitrifier population dynamics in biochemical models was pioneered by Leffelaar and Wessel [49]. In the model, the relative reduction rates for NO_3^-, NO_2^-, and N_2O are simulated by assuming three different bacterial populations carrying NO_3^-, NO_2^--, or N_2O-reductase

(NO formation and reduction is ignored). Growth and decline of each population is based on classical bacterial-growth equations [50], where metabolic rates are double Michaelis–Menten functions of substrates (carbon and electron acceptors), and population growth is a function of metabolic rate, growth yield per mole of substrate, and maintenance energy demand. The maintenance energy demand then ensures population decline either if substrates are lacking or if conditions are aerobic. This approach successfully simulates transient accumulation of N_2O during anaerobic incubations of small soil samples because the sequential induction of the enzymes is emulated by differences in initial growth rates for the different "reductase populations".

This approach has also been used for large-scale predictive modeling of N_2O emission within the ecosystem models, DNDC [31] and NLOSS [51]. In the DNDC model, a further refinement was introduced by assuming that the different "reductase populations" had different sigmoidal responses to pH; the population carrying N_2O-reductase being most sensitive to acid conditions. This modification ensures that the N_2O/N_2 product ratio increases with decreasing soil pH, in agreement with numerous empirical observations [20]. The growth parameters, originally adopted by Leffelaar and Wessel [49] based on literature data on pure cultures, have been used in all subsequent mechanistic denitrification models. It seems worthwhile to challenge the validity of growth parameters in a wider parameter space as given in the soil environment.

Another example of explicit biogeochemical modeling of microbial metabolism is that of grant and Pattey [52], who use a very complex model for both nitrification and denitrification with numerous parameters, which are derived from experiments with cultured bacteria. In contrast to the DNDC and NLOSS models, transient accumulation of N_2O is ensured by concentration-dependent competition between the electron acceptors (as formulated in the general denitrification model of Almeida et al. [18]) and not by growth of any single population. Another feature of this model is its sophisticated treatment of gas release (by volatilization and ebullition) from the soil matrix. The model performed reasonably well in predicting N_2O emission (measured with Eddy covariance) from a 1.5 ha area, with R^2 values of 0.28 and 0.37 for the spatial and temporal variability, respectively. These simulations of the spatial and temporal variability have implications for future flux measurements, especially with high time resolution at representative spots.

The above two examples serve to illustrate that implementation of explicit microbial and biochemical kinetics in biogeochemical models can be used for predictive purposes. It is striking, however, that their predictive power appears not to depend on a true representation of the microbial reality within the soil matrix. For example, the denitrifier growth models are based on the assumption of individual populations of denitrifiers, each carrying only one of the relevant reductases. This is obviously incorrect, although such cases exist (except for *nir*). Further, the model takes no account of regulation despite the fact that the relative expression of the reductases in question is dependent on low O_2 availability. It emulates regulation, however, by the independent growth and decline of the individual "populations." Finally, the model assumes that denitrification only takes place in the absence of O_2, which is again not true, at least for a large fraction of denitrifying bacteria.

When four biogeochemical models [53] with process-oriented nitrification/denitrification algorithms of varying complexity (CENTURY-NGAS [54]; DNDC [27]; ExpertN [42]; NASA-CASA [55]) were tested against N_2O flux measurements from agricultural soils, it was found that total annual N_2O emissions could be modeled fairly accurately, but emission dynamics were not captured correctly. So far, no full-fledged sensitivity analysis has been undertaken to test whether the models' inability to predict temporal dynamics is due to either the denitrification submodels or an inadequate description of C- and N-cycling or O_2 distribution. New models with more elaborate and legitimate representations of the biology of denitrifying bacteria may hypothetically improve predictions, but the epistemological gain is probably a more important reason for such exercises.

REFERENCES

[1] J.N. Galloway, W.H. Schlesinger, H. Levy, A. Michaels, J.L. Schnoor, Nitrogen fixation: anthropogenic enhancement-environmental response, Global Biogeochem. Cycles 9 (1995) 235–252.

[2] J.N. Galloway, J.D. Aber, J.W. Erisman, S.P. Seitzinger, R.W. Howarth, E.B. Cowling, B.J. Cosby, The nitrogen cascade, Bioscience 53 (2003) 341–356.

[3] L.R. Bakken, M.A. Bleken, Temporal aspects of N-enrichment and emission of N_2O to the atmosphere, Nutr. Cycl. Agroecosys. 52 (1998) 107–121.

[4] T. Sowers, A. Rodebaugh, N. Yoshida, S. Toyoda, Extending records of the isotopic composition of atmospheric N_2O back to 1800 AD from air trapped in snow at the South Pole and the Greenland Ice Sheet Project II ice core, Global Biogeochem. Cycles 16 (4) (2002) 1–10.

[5] T. Röckmann, I. Levin, High-precision determination of the changing isotopic composition of atmospheric N_2O from 1990 to 2002, J. Geophys. Res. Atmospheres 110 (D21304) (2005) 1–8.

[6] J. Hansen, M. Sato, Greenhouse gas growth rates, Proc. Natl. Acad. Sci. USA 101 (2004) 16109–16114.

[7] I.G. Dyominov, A.M. Zadorozhny, Greenhouse gases and recovery of the Earth's ozone layer, Adv. Space Res. 35 (2005) 1369–1374.

[8] J. Fluckiger, A. Dallenbach, T. Blunier, B. Stauffer, T.F. Stocker, D. Raynaud, J.M. Barnola, Variations in atmospheric N_2O concentration during abrupt climatic changes, Science 285 (1999) 227–230.
[9] T. Sowers, R.B. Alley, J. Jubenville, Ice core records of atmospheric N_2O covering the last 106 000 years, Science 301 (2003) 945–948.
[10] K.J. Meissner, E.D. Galbraith, C. Volker, Denitrification under glacial and interglacial conditions: a physical approach, Paleoceanography 20 (PA3001) (2005) 1–13.
[11] D. Schimel, N.S. Panikov, Simulation models of terrestrial trace gas fluxes at soil microsites to global scales, in: A.F. Bouwman (Ed.) Approaches to Scaling of Trace Gas Fluxes in Ecosystems, Elsevier, Amsterdam, 1999), pp. 185–203.
[12] M. Kunak, I. Kucera, R.J.M. van Spanning, Nitric oxide oscillations in *Paracoccus denitrificans*: the effects of environmental factors and of segregating nitrite reductase and nitric oxide reductase into separate cells, Arch. Biochem. Biophys. 429 (2004) 237–243.
[13] N.E. Ostrom, L.O. Hedin, J.C. von Fischer, G.P. Robertson, Nitrogen transformations and NO_3^- removal at a soil-stream interface: a stable isotope approach, Ecol. Appl. 12 (2002) 1027–1043.
[14] A. Bollmann, R. Conrad, Acetylene blockage technique leads to underestimation of denitrification rates in oxic soils due to scavenging of intermediate nitric oxide, Soil Biol. Biochem. 29 (1997) 1067–1077.
[15] R.E. Murray, R. Knowles, Trace amounts of O_2 affect NO and N_2O production during denitrifying enzyme activity (DEA) assays, Soil Biol. Biochem. 36 (2004) 513–517.
[16] D. Cheneby, S. Perrez, C. Devroe, S. Hallet, Y. Couton, F. Bizouard, G. Iuretig, J.C. Germon, L. Philippot, Denitrifying bacteria in bulk and maize-rhizospheric soil: diversity and N_2O-reducing abilities, Can. J. Microbiol. 50 (2004) 469–474.
[17] M.R. Betlach, J.M. Tiedje, Kinetic explanation for accumulation of nitrite, nitric oxide, and nitrous oxide during bacterial denitrification, Appl. Environ. Microbiol. 42 (1981) 1074–1084.
[18] J.S. Almeida, M.A.M. Reis, M.J.T. Carrondo, A unifying kinetic model of denitrification, J. Theor. Biol. 186 (1997) 241–249.
[19] L. Philippot, P. Mirleau, S. Mazurier, S. Siblot, A. Hartmann, P. Lemanceau, J.C. Germon, Characterization and transcriptional analysis of *Pseudomonas fluorescens* denitrifying clusters containing the *nar*, *nir*, *nor* and *nos* genes, Biochim. Biophys. Acta 1517 (2001) 436–440.
[20] M. Simek, J.E. Cooper, The influence of soil pH on denitrification: progress towards the understanding of this interaction over the last 50 years, Eur. J. Soil Sci. 53 (2002) 345–354.
[21] J.R. Firth, C. Edwards, Effects of cultural conditions on denitrification by *Pseudomonas stutzeri* measured by membrane inlet mass spectrometry, J. Appl. Microbiol. 87 (1999) 353–358.

[22] M.K. Firestone, M.S. Smith, R.B. Firestone, J.M. Tiedje, Influence of nitrate, nitrite, and oxygen on the composition of the gaseous products of denitrification in soil, Soil Sci. Soc. Am. J. 43 (1979) 1140–1144.
[23] J.O. Ka, J. Urbance, R.W. Ye, T.Y. Ahn, J.M. Tiedje, Diversity of oxygen and N-oxide regulation of nitrite reductases in denitrifying bacteria, FEMS Microbiol. Lett. 156 (1997) 55–60.
[24] L. Holtan-Hartwig, P. Dörsch, L.R. Bakken, Comparison of denitrifying communities in organic soils: kinetics of NO_3^- and N_2O reduction, Soil Biol. Biochem. 32 (2000) 833–843.
[25] L. Holtan-Hartwig, P. Dörsch, L.R. Bakken, Low temperature control of soil denitrifying communities: kinetics of N_2O production and reduction, Soil Biol. Biochem. 34 (2002) 1797–1806.
[26] P. Dörsch, L.R. Bakken, Low-temperature response of denitrification: comparison of soils, Eurasian Soil Sci. 37 (2004) 102–106.
[27] C.S. Li, S. Frolking, T.A. Frolking, A model of nitrous-oxide evolution from soil driven by rainfall events 1. Model structure and sensitivity, J. Geophys. Res. Atmospheres 97 (1992) 9759–9776.
[28] H. Johnsson, L. Bergstrom, P.E. Jansson, K. Paustian, Simulated nitrogen dynamics and losses in a layered agricultural soil, Agr. Ecosyst. Environ. 18 (1987) 333–356.
[29] M. Heinen, Simplified denitrification models: overview and properties, Geoderma (in press, doi: 10.1016/j.geoderma.2005.06.010, 2006) 1–20.
[30] S.J. Del Grosso, W.J. Parton, A.R. Mosier, D.S. Ojima, A.E. Kulmala, S. Phongpan, General model for N_2O and N_2 gas emissions from soils due to dentrification, Global Biogeochem. Cycles 14 (2000) 1045–1060.
[31] C.S. Li, J. Aber, F. Stange, K. Butterbach-Bahl, H. Papen, A process-oriented model of N_2O and NO emissions from forest soils: 1. Model development, J. Geophys. Res. Atmospheres 105 (2000) 4369–4384.
[32] P. Leffelaar, Simulation of partial anaerobiosis in a model soil in respect to denitrification, Soil Sci. 128 (1979) 110–349.
[33] K.A. Smith, A model of the extent of anaerobic zones in aggregated soils and its potential application to estimates of denitrification, J. Soil Sci. 31 (1980) 263–277.
[34] J.R.M. Arah, K.A. Smith, Steady-state denitrification in aggregated soils – a mathematical model, J. Soil Sci. 40 (1989) 139–149.
[35] J.R.M. Arah, A.J.A. Vinten, Simplified models of anoxia and denitrification in aggregated and simple-structured soils, Eur. J. Soil Sci. 46 (1995) 507–517.
[36] P. Groenendijk, J.G. Kroes, Modelling the nitrogen and phosphorus leaching to groundwater and surface water with ANIMO 3.5, Report 144, Winand Staring Centre, Wageningen, 1999.
[37] C.A. Langeveld, P.A. Leffelaar, Modelling below-ground processes to explain field-scale emissions of nitrous oxide, Ecol. Model. 149 (2002) 97–112.

[38] G. Schurgers, P. Dörsch, L.R. Bakken, P. Leffelaar, L.E. Haugen, Modelling soil anaerobiosis from water retention characteristics and soil respiration, Soil Biol. Biochem. (2006) (doi:10.1016/ j.soilbio. 2006.04.016), in press.
[39] S. Hansen, H. Jensen, N. Nielsen, H. Svendsen. DAISY - soil plant atmosphere system model, Danish simulation model for transformation and transport of energy and matter in the soil plant atmosphere system, A10, Copenhagen, Denmark, The National Agency for Environmental Protection, NPO report, 1990.
[40] S.J. Del Grosso, W.J. Parton, A.R. Mosier, D.S. Ojima, A.E. Kulmala, S. Phongpan, General model for N_2O and N_2 gas emissions from soils due to dentrification, Global Biogeochem. Cycles 14 (2000) 1045–1060.
[41] W.J. Parton, E.A. Holland, S.J. Del Grosso, M.D. Hartman, R.E. Martin, A.R. Mosier, D.S. Ojima, D.S. Schimel, Generalized model for NOx and N_2O emissions from soils, J. Geophys. Res. Atmospheres 106 (2001) 17403–17419.
[42] T. Engel, E. Priesack, Expert-N, a building block system of nitrogen models as a resource for advice, research, water management and policy, in: H. Eijsackers, T. Hamers (Eds.) Integrated Soil and Sediment Research: A Basis for Proper Protection, Kluwer Academic Publishers, Dordrecht, 1993, pp. 503–507.
[43] J. Tiedje, Ecology of dentrification and dissimilatory nitrate reduction to ammonium, in: A.J.B. Zehnder (Ed.) Biology of Anaerobic Microorganisms, Wiley & Sons, New York, 1988, pp. 179–244.
[44] V.I. Gertsev, V.V.Gertseva, Classification of mathematical models in ecology, Ecol. Modelling 178 (2004) 329–334.
[45] D.S. Schimel, C.S. Potter, Process modelling and spatial extrapolation, in: P. Matson, R.C. Harris (Eds.) Biogenic Trace Gases: Measuring Emissions from Soil and Water, Blackwell Scientific Publishing, Cambridge, 1995, pp. 358–383.
[46] E.A. Davidson, M. Keller, H.E. Erickson, L.V. Verchot, E. Veldkamp, Testing a conceptual model of soil emissions of nitrous and nitric oxides, Bioscience 50 (2000) 667–680.
[47] D.A. Ratkowsky, R.K. Lowry, T.A. McMeekin, A.N. Stokes, R.E. Chandler, Model for bacterial culture growth rate throughout the entire biokinetic temperature range, J. Bacteriol. 154 (1983) 1222–1226.
[48] W.J. Parton, A.R. Mosier, D.S. Ojima, D.W. Valentine, D.S. Schimel, K. Weier, A.E. Kulmala, Generalized model for N_2 and N_2O production from nitrification and denitrification, Global Biogeochem. Cycles 10 (1996) 401–412.
[49] P.A. Leffelaar, W.W. Wessel, Denitrification in a homogenous, closed-system experiment and simulation, Soil Sci. 146 (1988) 335–349.

[50] S.J. Pirt, Maintenance energy of bacteria in growing cultures, Proc. Roy. Soc. Lond. B. 163 (1965) 224–231.
[51] W.J. Riley, P.A. Matson, NLOSS: a mechanistic model of denitrified N_2O and N_2 evolution from soil, Soil Sci. 165 (2000) 237–249.
[52] R.F. Grant, E. Pattey, Modelling variability in N_2O emissions from fertilized agricultural fields, Soil Biol. Biochem. 35 (2003) 225–243.
[53] S.E. Frolking, A.R. Mosier, D.S. Ojima, C. Li, W.J. Parton, C.S. Potter, E. Priesack, R. Stenger, C. Haberbosch, P. Dörsch, H. Flessa, K.A. Smith, Comparison of N_2O emissions from soils at three temperate agricultural sites: simulations of year-round measurements by four models, Nutr. Cycl. Agroecosys. 52 (1998) 77–105.
[54] W.J. Parton, A.R. Mosier, D.S. Ojima, D.W. Valentine, D.S. Schimel, K. Weier, A.E. Kulmala, Generalized model for N_2 and N_2O production from nitrification and denitrification, Global Biogeochem. Cycles 10 (1996) 401–412.
[55] C.S. Potter, R.H. Riley, S.A. Klooster, Simulation modeling of nitrogen trace gas emissions along an age gradient of tropical forest soils, Ecol. Modelling 97 (1997) 179–176.

Chapter 26

Interactions among Organisms that Result in Enhanced Activities of N-Cycle Reactions

Hermann Bothe and Harold Drake

26.1 Introductory statement

There are many such interactions among organisms that might lead to increased N-cycle activity. It is not always clear whether or not these interactions are really beneficial and whether they are actually due to enhanced N-cycle activity or to some other effect. The most commonly encountered of these interactions are outlined below.

26.2 Associative bacteria are potentially beneficial to the health of plants

Microorganisms can interact with plants or animals both in the free-living and symbiotic state. Any effort to characterize such interactions is confronted with the enormous complexity of the microbial community.

Biology of the Nitrogen Cycle
Edited by H. Bothe, S.J. Ferguson and W.E. Newton

Soils, for example, harbor approximately 10^4 different ribotypes (~ bacteria plus archaea). It is often stated that only a small percentage of bacteria, perhaps about 1%, can be cultured as yet [1], although this percentage may be significantly increased when cultivation techniques, e.g., choice of media, inoculum size, and incubation time, are improved [2]. Methods to characterize the microbial community structure by molecular techniques have been critically evaluated elsewhere in this book (Chapter 20). A more recent approach comes from computer-based programs, which identify bacteria by their terminal restriction length polymorphisms *using multiple restriction enzymes* [3, 4]. Such algorithms now allow us to grossly characterize complex microbial structures, including the uncultured bacteria, with predicted accuracy somewhat akin to those of human parliament elections. In future, the characterization by such algorithms may be combined with quantitative real-time PCR determinations that will allow us to assess the concentration of a gene in an environmental sample for its comprehensive characterization. Nonetheless, cloning and sequencing of amplicons of 16S rRNA (or any other functional gene), obtained by PCR of DNA extracted from environmental samples, might indicate that a realm of not yet recognized microorganisms thrives in nature and likely harbors fascinating biological properties.

Owing to the excretion of assimilates into the plant apoplasm, the bacterial abundance is higher in roots than in the bulk, plant-free soil. Model experiments showed that bacteria, such as *Azospirillum*, when associated with plants, e.g., wheat, may perform either N_2-fixation (C_2H_2-reduction) or denitrification, depending on the culture conditions [5]. However, it remains uncertain whether such activities significantly contribute to N-gains or N-losses in the field. Even in the tropics, *Azospirillum* hardly exceeds 10^5 cells per g of roots, which amounts to less than 0.1% of the root weight, whereas in the legume root nodules, there are about 10^{11} to 10^{12} rhizobial cells per g nodule fresh weight. Thus, any significant or sustainable contribution of *Azospirillum* or any other associative bacterium to the N-budget of plants seems unlikely unless special conditions, like *Gluconacetobacter diazotrophicus* in sugarcane phloem [6], are met. Contrasting views on this matter exist [7].

If not by means of N_2-fixation, beneficial bacteria could promote plant growth by phytohormone production. Many bacteria excrete auxins (indolylacetic acid) and, in minor amounts, other phytohormones. However, classical studies show that the concentration of auxins and other phytohormones are already close to the optimum in roots. Any addition of phytohormones can result in, at best, a small increment in root growth but, in most instances, the addition of phytohormones is growth inhibitory. Maybe beneficial bacteria are simply more effective than pathogenic organisms in utilizing the limiting amounts of nutrient in roots and thereby suppress their growth. Most importantly, programs designed to exploit beneficial bacteria have to cope with a major problem; bacteria externally supplemented into a habitat must compete with the indigenous population and, in most cases, they cannot and are overgrown.

The complexity of N_2-fixation and associated regulatory cascades makes it difficult to transfer the full complement of genes coding for N_2-fixation to eukaryotic cells and all such attempts thus far have failed. Nitrogenase is generally soluble in bacterial cells with the remarkable exception of the associative bacterium *Azoarcus*, in which nitrogenase is particle bound [8]. N_2-fixing spheroid bodies of cyanobacterial origin, as in the diatom *Rhopalodia gibba* [9], may offer perspectives for transferring N_2-fixation to higher plants. In this context, the, as yet poorly characterized, symbiotic systems, such as the flagellate *Petalomonas sphagnophila* [10] and its endocytobiotic cyanobacteria, are possible alternatives.

26.3 Plants eating animals (carnivorous plants) to get access to an extra N-source

The biology of plants that eat animals (carnivorous plants) is fascinating. Carnivorous plants (*Drosera*, *Nepenthes*, *Dionaea*, *Sarracenia*, *Pinguicula*, *Utricularia*, and others) belong to diverse plant families of unrelated taxonomic affinities. They generally occur at N-limited sites, such as bogs, swamps, or moors. The main aim of carnivorous plants is to mobilize amino acids from the proteins of their prey. Classical botanical experiments show that carnivorous plants are not absolutely dependent on the acquisition of fixed-N from the animals, rather the additional animal-derived N significantly enhances both flower and seed formation. In the water-filled leaves of the northern pitcher plants, *Sarracenia purpurea* L., bacteria likely play a major role in the acquisition of nutrients by excreting digestive enzymes. The breakdown products are then absorbed by the plants either directly or later on when the bacteria decompose [11, 12]. This interesting microbial community structure awaits characterization by modern molecular techniques.

26.4 The role of mycorrhizal fungi in mobilizing soil nutrients, in particular nitrogen

More than 80% of all higher plants are colonized by arbuscular mycorrhizal fungi (AMF), particularly under diverse stress conditions, such as mineral deficiency, drought, pollution by salt or heavy metals, and others. The fine hyphae of the AMF more efficiently exploit soil particles and more effectively mobilize water and minerals, such as phosphorus, Zn or Cu and others, than do roots [13]. Although the phrase "plants do not have roots, they have mycorrhizas" [14] may be an exaggeration, but nutrient supply to higher plants indeed proceeds, to a large extent, via AMF. In the field, apparently mycorrhizal plants are always colonized by AMF, which could mean that plants always suffer from stress, even under best soil conditions, e.g., in chernozems. In addition, they also mobilize organic litter [15], which may indicate that AMF symbiosis developed from a saprophytic mode of life.

After some initial controversy [16], it is now clear that AMF assimilate inorganic nitrogen (NO_3^-, NH_4^+). AMF assimilate inorganic nitrogen by the glutamine synthetase (GS)/GOGAT pathway. The NO_3^- or NH_4^+ taken up by the extraradical mycelium of the fungi is metabolized via glutamine to arginine, which is then transferred to the intraradical hyphae [17]. By using arginine, cells can mobilize four N-atoms per molecule. In the intraradical hyphae, arginine is likely degraded via urea to NH_4^+, which is subsequently transported to the plant host cells. Organic carbon is, however, retained in the fungal cells and probably transferred back to the extraradical hyphae. It remains to be shown whether this is the only N-allocation pathway from the soil to the plant in the AMF symbiosis. In tomato, expression of a specific NO_3^- transporter is drastically enhanced on AMF colonization [18], suggesting that NO_3^- may also, to a certain degree, be directly translocated from the fungal cells to the plant cells where it can be incorporated by the GS/GOGAT pathway. The activities of the AMF enzymes are apparently not decreased within the plant cells after mycorrhizal colonization. The current view is depicted in Figure 26-1.

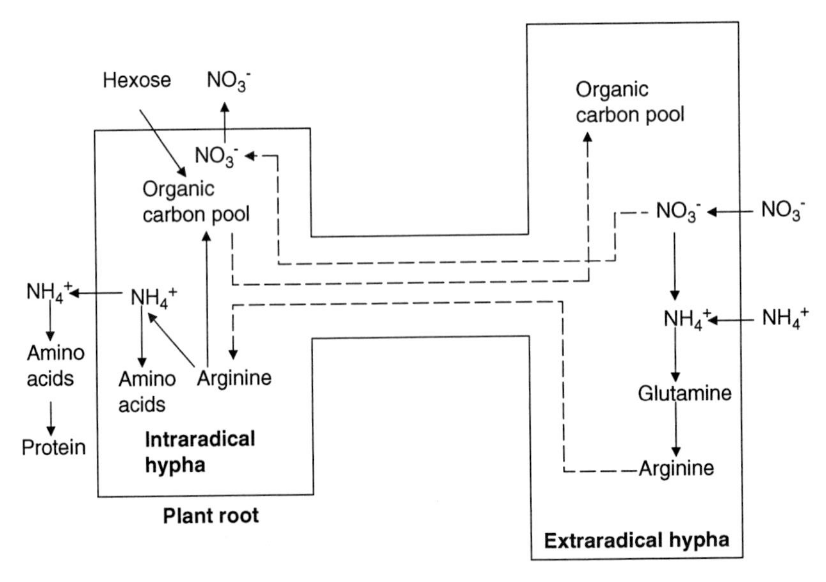

Figure 26-1 The exchange of metabolites in the interaction between arbuscular mycorrhizal fungi (AMF) and plants (modified from Ref. [17]).

In ectomycorrhizal fungi, which form symbioses mainly with trees, NO_3^- taken up by the fungi is reduced to NH_4^+. NH_4^+ can also be utilized directly from soils and is then used to form an amino acid, which is subsequently transferred to the plants [19].

Mycorrhizal fungi interact with N-metabolizing bacteria. A bacterium related to *Burkholderia* resides inside the spores of the AMF *Gigaspora margarita* and also expresses nitrogenase [20]. However, N_2-fixation has not been demonstrated for this association, and a recent bacterial isolate from spores of this fungus, termed "*Candidatus Glomeribacter gigasporarum*" [21] had apparently lost the *nif* genes. Bacteria support fungal growth in various ways and vice versa [13, 22]. For example [23], two isolates of *Paenibacillus validus* promoted fungal growth until spore fertility was reached without any dependence on a plant. Thus, AMF are not necessarily obligate biotrophs, but the factors that determine AMF growth are not entirely clear. Furthermore, interactions between mycorrhizal fungi and N_2-fxing bacteria are manifold and complex [24] and the impacts of denitrifying bacteria on mycorrhiza have not been assessed at all.

26.5 The earthworm gut as a transient habitat for terrestrial denitrifiers

Earthworms colonize many terrestrial ecosystems and, due to their feeding and burrowing habits, have profound impact on the structure and fertility of soil [25]. The gut of the earthworm is anoxic [26], and soil microbes capable of anaerobic metabolism should be stimulated by the in situ conditions of the gut. This likelihood is illustrated by the earthworm's emission of the greenhouse gas, N_2O, via soil denitrifying bacteria that are ingested and activated in the gut [27]. The in vivo emission of N_2O by earthworms occurs at a relatively linear rate; it can constitute a significant part of the total N_2O emitted from earthworm-colonized soils and it could yield 3×10^8 kg N_2O globally per year [27]. N_2, the end product of complete denitrification, is also emitted from earthworms at a rate that approximates or exceeds that of N_2O [28].

The gut of the earthworm constitutes a unique transient habitat for soil microbes, and the most important factors (Figure 26-2) in stimulating the in situ activity of denitrifiers during gut passage are anoxia, the NO_2^- concentration, and high concentrations of electron donors [26].

The carbohydrate-rich mucus that is synthesized in the pharyngeal glands is a unique constituent of the earthworm gut. This mucus aids passage of ingested material and, together with soil-derived organic matter, is subject to digestion by diverse exoenzymes from both the earthworm and ingested soil microbes [29, 30]. The high concentrations of sugars,

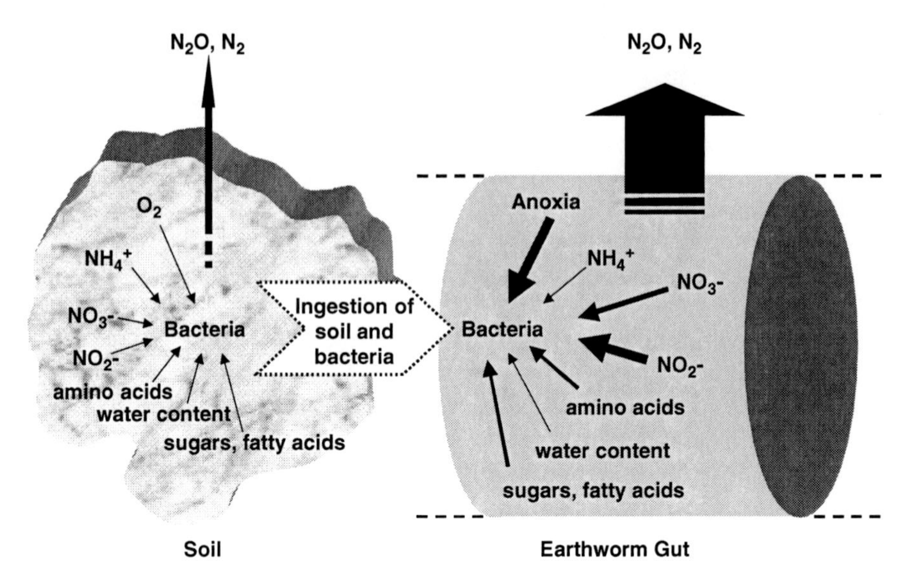

Figure 26-2 Hypothetical model illustrating as to which factors stimulate the production of N_2O and N_2 by bacteria ingested into the gut of the earthworm. The relative concentrations of compounds are reflected in the font sizes, and the relative effect of each compound on the production of N_2O and N_2 in the gut is indicated by the thickness of the arrows.

e.g., up to 80 mM glucose, and organic acids, e.g., 5 mM acetate and 2 mM lactate on average, in the aqueous phase of gut contents likely fuel denitrification and appear to be primarily derived from this mucus via its hydrolysis and fermentation in the gut [26, 27].

Denitrification is greatly accelerated in the anoxic core of the earthworm gut [26]. It is catalyzed by bacteria that are phylogenetically similar to soil denitrifiers, as indicated by the similar *nosZ* libraries produced from both soil and earthworm gut content [31]. However, well-known PCR and DNA extraction biases limit the conclusions of molecular-based studies in general, including those of Ref. [31]. The *nosZ* primers used in Ref. [31] may not be fully adequate for assessing many phylogenetically diverse denitrifiers (e.g., Gram-positive denitrifiers are not targeted by currently available *nosZ* primers), thus emphasizing the need for a more complete comparative analysis (e.g., with *narG* or *nirK/S*) of denitrifiers in soil and gut content, assuming that PCR amplification of the extracted DNA provides quantitative data on the bacterial composition of both communities. Denitrifying bacteria, e.g., *Dechloromonas denitrificans* [32], appear to be the primary source of the N_2O emitted by earthworms. However, other nitrate- and nitrite-reducing microbes, including those capable of fermentation, might also be involved because N_2O-producing nitrate-respiring and nitrite-detoxifying fermentative bacteria, e.g., *Paenibacillus anaericanus*, can be readily isolated from gut contents of earthworms [32]. The growth of denitrifiying bacteria appears to be significantly enhanced during gut passage, indicating that the metabolic status and detectable viability of this microbial group is affected by the in situ conditions of the earthworm gut. The diversity of denitrifiers of earthworm-colonized soils might be significantly affected if they were not periodically subjected to the in situ conditions of the earthworm gut.

References

[1] T. Torsvik, K. Salte, R. Sørheim, J. Goksøyr, Comparison of phenotypic diversity and DNA heterogeneity in a population of soil bacteria, Appl. Environ. Microbiol. 56 (1990) 776–781.
[2] K.E.R. Davis, S.J. Joseph, P.H. Janssen, Effect of growth medium, inoculum size, and incubation time on culturability and isolation of soil bacteria, Appl. Environ. Microbiol.71 (2005) 826–834.
[3] A.D. Kent, D.J. Smith, B.J. Benson, E.W. Triplett, Web-based phylogenetic assignment tool for analysis of terminal restriction fragment length polymorphism profiles of microbial communities, Appl. Environ. Microbiol. 69 (2003) 6768–6776.
[4] C. Rösch, H. Bothe, Improved assessment of denitrifying, N_2-fixing, and total-community bacteria by terminal restriction fragment length polymorphism analysis using multiple restriction enzymes, Appl. Environ. Microbiol. 71 (2005) 2026–2035.
[5] G. Neuer, A. Kronenberg, H. Bothe, Denitrification and nitrogen fixation by *Azospirillum*, III. Properties of a wheat-*Azospirillum* association, Arch. Microbiol. 141 (1985) 364–370.
[6] M. Sevilla, R.H. Burris, N. Gunapala, C. Kennedy, Comparison of benefit to sugarcane plant growth and ^{15}N incorporation following inoculation of sterile plants with *Acetobacter diazotrophicus* wild type and nif⁻ mutant strains, Mol. Plant Microb. Interact. 14 (2001) 358–366.
[7] T. Hurek, L.L. Handley, B. Reinhold-Hurek, Y. Piche, *Azoarcus* grass endophytes contribute fixed nitrogen to the plant in an uncultured state, Mol. Plant Microb. Interact. 15 (2002) 233–242.
[8] T. Hurek, M. Van Montagu, E. Kellenberger, B. Reinhold-Hurek, Induction of complex intracytoplasmatic membranes related to nitrogen fixation in *Azoarcus* sp. BH72, Mol. Microbiol. 18 (1995) 225–236.
[9] J. Prechtl, C. Kneip, P. Lockhart, K, Wenderoth, U.-G. Maier, Intracelluar spheroid bodies of *Rhopalodia gibba* have nitrogen- fixing apparatus of cyanobacterial origin, Mol. Biol. Evol. 21 (2004) 1477–1481.

[10] E. Schnepf, I. Schlegel, D. Hepperle, *Petolomonas sphagnophila* (Eugenophyta) and its endocytobiotic cyanobacteria: a unique form of symbiosis, Phycologia 41 (2002) 153–157.
[11] D.L. Cochran-Stafira, C.N. von Ende, Integrating bacteria into food webs:studies with *Sarracenia purpurea* inquilines, Ecology 79 (1998) 880–898.
[12] R.M. Goodman, J.-M. Weisz, Plant-microbe symbioses: an evolutionary survey, in: J.T. Stanley, A.-L. Reysenbach (Eds.) Biodiversity of Microbial Life, Wiley-Liss, New York, 2002, pp. 237–287.
[13] P. Jeffries, S. Gianinazzi, S. Perotto, K. Turnau, J.-M. Barea, The contribution of arbuscular mycorrhizal fungi in sustainable maintenance of plant health and soil fertility, Biol. Fertil. Soils 37 (2003) 1–16.
[14] C. Azcón-Aguilar, J.M. Barea, Arbuscular mycorrhizas and biological control of soil-borne plant pathogens – an overview of the mechanisms involved, Mycorrhiza 6 (1996) 457–464.
[15] A. Hodge, C.D. Campbell, A.H. Fitter, An arbuscular mycorrhizal fungus accelerates decomposition and acquires nitrogen directly from organic material, Nature 413 (2001) 297–299.
[16] E. George, K.-U. Häusler, D. Vetterlein, E. Gorgus, H. Marschner, Water and nutrient translocation by hyphae of *Glomus mosseae*, Can. J. Bot. 70 (1992) 2130–2137.
[17] M. Govindarajulu, P.E. Pfeffer, H. Jin, J. Abubaker, D.D. Douds, J.W. Allen, H. Bücking, P.J. Lammers, Y. Shachar Hill, Nitrogen transfer in the arbuscular mycorrhizal symbiosis, Nature 435 (2005) 819–823.
[18] U. Hildebrandt, E. Schmelzer, H. Bothe, Expression of nitrate transporter genes in tomato colonized by an arbuscular mycorrhizal fungus, Physiol. Plant. 115 (2002) 125–136.
[19] M. Chalot, A. Javelle, D. Bludez, R. Lambilliote, R. Cooke, H. Sentenac, D. Wipf, B. Botton, An update on nutrient transport processes in ectomycorrhizas, Plant Soil 244 (2002) 165–175.
[20] D. Minerdi, R. Fani, R. Gallo, A. Boarino, P. Bonfante, Nitrogen fixation genes in an endosymbiotic *Burkholderia* strain, Appl. Environ. Microbiol. 67 (2001) 725–732.
[21] P. Jargeat, C. Cosseau, B. Ola'h, B.A. Jauneau, P. Bonfante, J. Batut, G. Bécard, Isolation, free-living capacities, and genome structure of "*Candidatus Glomeribacter gigasporarum*", the endocellular bacterium of the mycorrhizal fungus *Gigaspora margarita*, J. Bacteriol. 186 (2004) 6876–6884.
[22] D. Roesti, K. Ineichen, O. Braissant, D. Redecker, A. Wiemken, M. Aragno, Bacteria associated with spores of the arbuscular mycorrhizal fungi *Glomus geosporum* and *Glomus constrictum*, Appl. Environ. Microbiol. 71 (2005) 6673–6679.
[23] U. Hildebrandt, F. Ouziad, F.-J. Marner, H. Bothe, The bacterium *Paenibacillus validus* stimulates growth of the arbuscular mycorrhizal

fungus *Glomus intradices* up to the formation of fertile spores, FEMS Lett. 254 (2006) 258–267.
[24] J.-M. Barea, M.J. Pozo, R. Azcón, C. Azcón-Aguilar, Microbial co-operation in the rhizosphere, J. Exper. Bot. 56 (2005) 1761–1778.
[25] P. Lavelle, D. Bignell, M. Lepage, Soil function in a changing world: the role of ecosystem engineers, Eur. J. Soil Biol. 33 (1997) 159–193.
[26] M.A. Horn, A. Schramm, H.L. Drake, The earthworm gut: an ideal habitat for ingested N_2O-producing microorganisms, Appl. Environ. Microbiol. 69 (2003) 1662–1669.
[27] H.L. Drake, A. Schramm, M. Horn, Earthworm gut microbial biomes: their importance to soil microorganisms, denitrification, and the terrestrial production of the greenhouse gas N_2O, in: H. König, A. Varma (Eds.) Intestinal Microorganisms of Termites and Other Invertebrates, Springer, Heidelberg, New York, 2006, 65–87.
[28] M.A. Horn, R. Mertel, M. Kästner, M. Gehre, H.L. Drake, In vivo emission of dinitrogen by earthworms via denitrifying bacteria in the gut, Appl. Environ. Microbiol. 72 (2006) 1013–1018.
[29] A. Merino-Trigo, L. Sampedro, F.J. Rodriguez-Berrocal, S. Mato, M.P. de la Cadena, Activity and partial characterization of xylanolytic enzymes in the earthworm *Eisenia andrei* fed on organic wastes, Soil Biol. Biochem. 31 (1999) 1735–1740.
[30] F. Urbasek, V. Pizl, Activity of digestive enzymes in the gut of five earthworm species (Oligochaeta, Lumbricidae), Rev. Ecol. Biol. Soil 28 (1991) 461–468.
[31] M.A. Horn, H.L. Drake, A. Schramm, Nitrous oxide reductase genes *(nosZ)* of denitrifying microbial populations in soil and the earthworm gut are phylogenetically similar, Appl. Environ. Microbiol. 72 (2006) 1019–1026.
[32] M.A. Horn, J. Ihssen, C. Matthies, A. Schramm, G. Acker, H.L. Drake, *Dechloromonas denitrificans* sp. nov., *Flavobacterium denitrificans* sp. nov., *Paenibacillus anaericanus* sp. nov., and *Paenibacillus terrae* strain MH72, N_2O-producing bacteria isolated from the gut of the earthworm *Aporrectodea caliginosa*, Int. J. Syst. Evol. Microbiol. 55 (2005) 1255–1265.

Index

- B -

- C -

- T -

Colour Plate Section

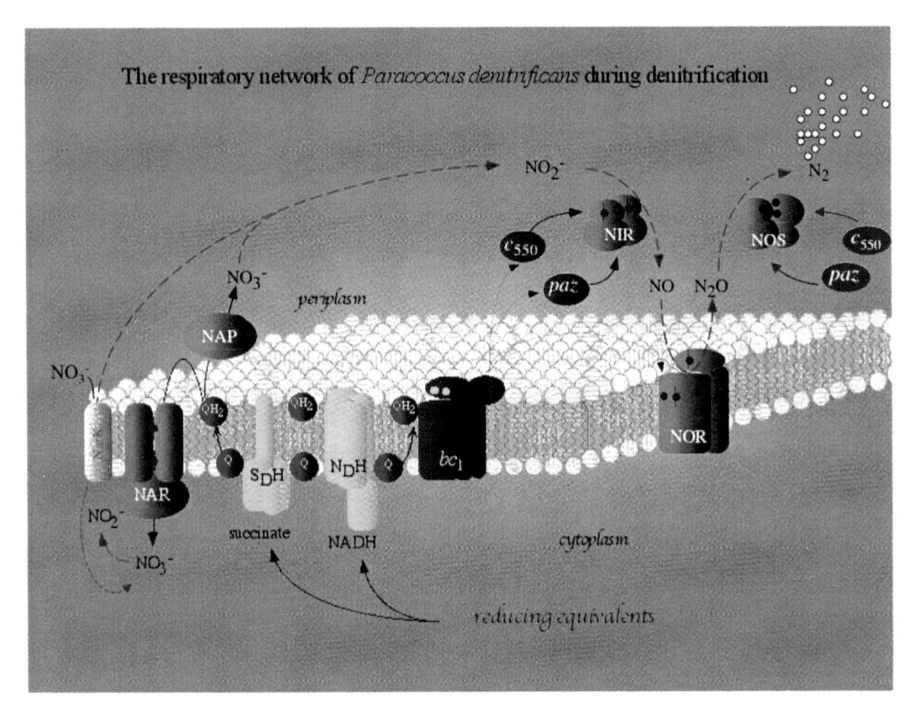

Plate 1 Scheme of a full denitrification process in *Paracoccus denitrificans*. Dashed arrows, N-oxide transport; straight arrows, e^--transport. SDH, succinate dehydrogenase; NDH, NADH dehydrogenase; Q, quinone; bc_1, cytochrome bc_1 complex; c_{550}, cytochrome *c*; paz, pseudoazurin; NAR, membrane-bound NO_3^--reductase; NAP, periplasmic NO_3^--reductase; NIR, cd_1-type NO_2^--reductase; NOR, bc-type NO-reductase; NOS, N_2O-reductase; NarK, NO_3^-/NO_2^- antiporter. (See page 5, this volume)

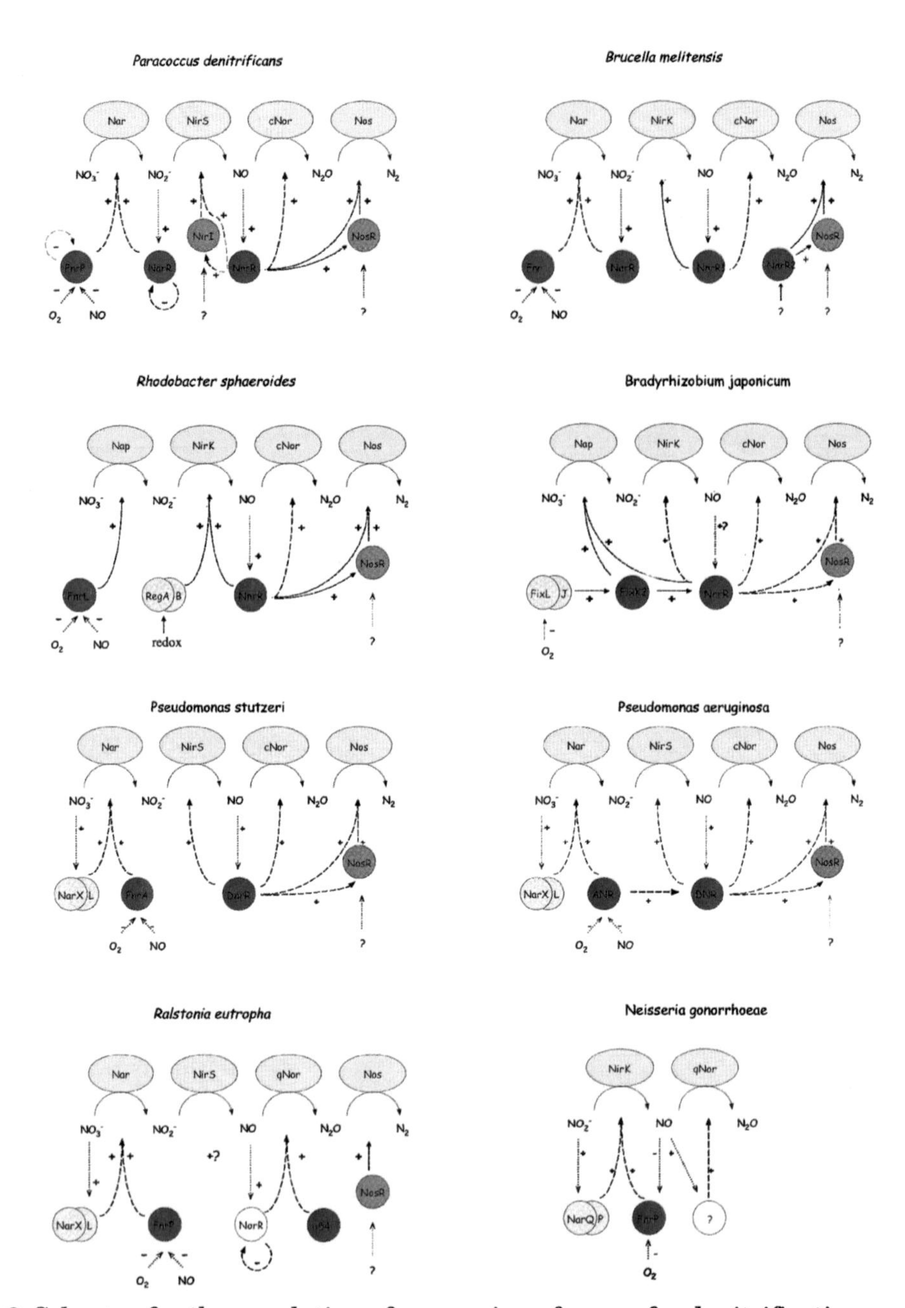

Plate 2 Schemes for the regulation of expression of genes for denitrification in some denitrifiers. Nar, NO_3^--reductase; NirS or NirK, cd_1-type or Cu-type NO_2^--reductases respectively; cNor or qNor, *bc*-type or quinol dependent NO-reductases respectively; Nos, N_2O-reductase. Dotted lines, activating (+) or inactivating (−) signals; dashed lines, positive (+) or negative (−) transcription regulation of the corresponding genes by the regulators. (See page 15, this volume)

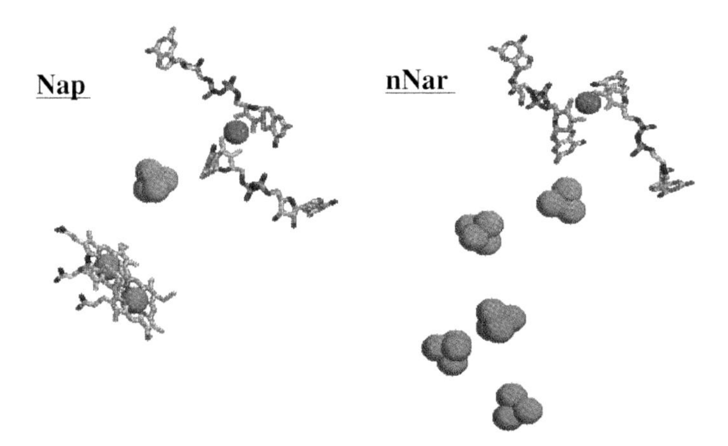

Plate 3 The organization of the nNar and Nap redox centres (figure kindly provided by Dr. C. Butler). (See page 25, this volume)

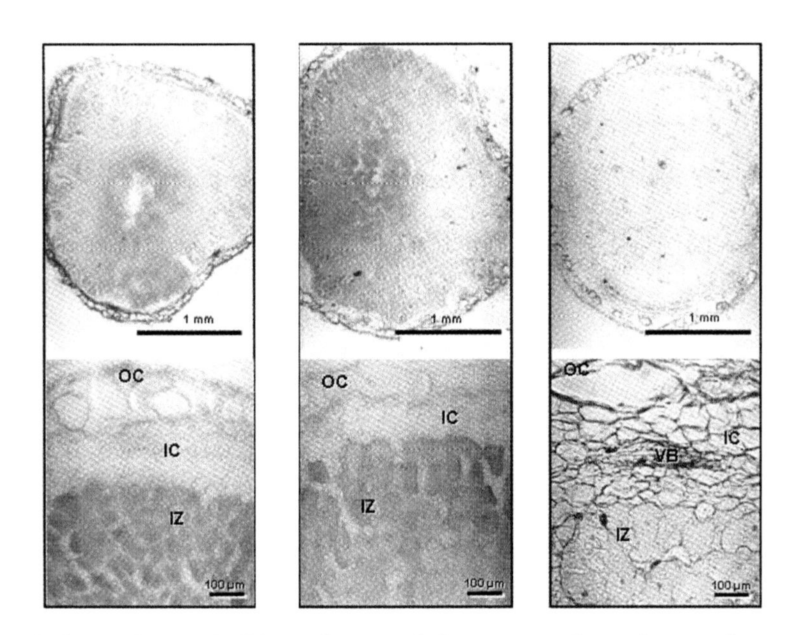

Plate 4 Colour Figure D Histochemical detection of β-galactosidase activity in soybean nodules isolated from plants inoculated with B. *japonicum* cells containing a *nirK-lacZ* fusion. Nodules were obtained from plants grown in the absence (panel A) or in the presence (panel B) of 4 mM KNO_3. Nodule sections from nitrate-treated soybeans inoculated with *B. japonicum* USDA110 containing a promoterless *lacZ* gene were used as a control (panel C). OC, outer cortex; IC, inner cortex; IZ, infected zone; VB, vascular bundles. (See page 87, this volume)

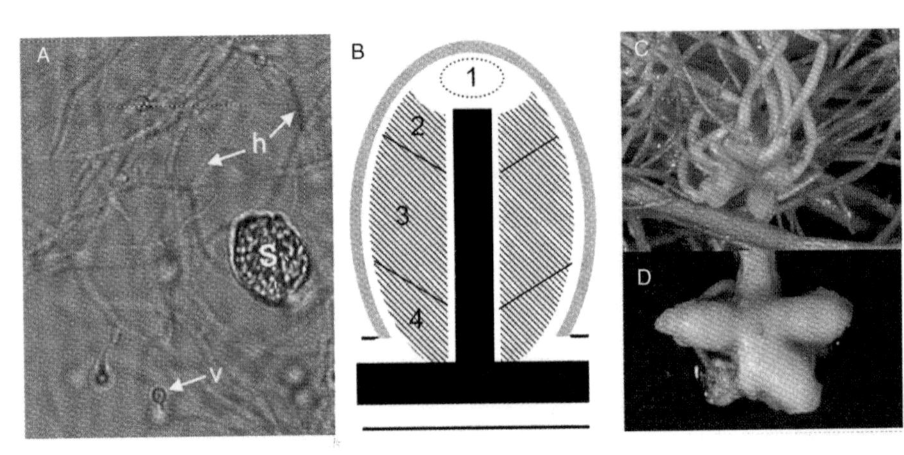

Plate 5 (A) *Frankia* culture grown in N-free medium under 21% O_2. h, hyphae; v, vesicle; s, sporangium. (B) Scheme of an actinorhizal nodule lobe. Due to the activity of the apical meristem (1) cortical cells are arranged in a developmental gradient in the infection zone, (2) cortical cells become gradually filled with *Frankia* hyphae, and afterward, vesicles differentiate in the fixation zone, (3) *Frankia* fixes N_2 in the infected cells and in the senescence zone, and (4) *Frankia* hyphae and vesicles are degraded. Nodule lobes are surrounded by a periderm, which is shown in dark gray. (C) Photograph of a ca. 4—week-old nodule from *Casuarina glauca*. Aeration is provided by agraviotropically growing nodule roots that emerge from the tips of nodule lobes. (D) Photograph of a ca. 8-week-old nodule from *Datisca glomerata*. (See page 168, this volume)